AF443518

86

Advances in Biochemical Engineering/Biotechnology

Series Editor: T. Scheper

Springer

Berlin
Heidelberg
New York
Hong Kong
London
Milan
Paris
Tokyo

New Trends and Developments in Biochemical Engineering

With contributions by
M. Beyer, M. M. Borah, P. S. J. Cheetham, A. Chenchik,
M. Dutta, N. N. Dutta, K. Friehs, M. N. Gupta,
M. Herrler, R. Kumar, T. Madhusudhan,
A. Mukhopadhyay, I. Roy, K. Schügerl, G. Seidel,
S. Sharma, P. D. Siebert, C. Tollnick, B. Zhumabayeva

Springer

Advances in Biochemical Engineering/Biotechnology reviews actual trends in modern biotechnology. Its aim is to cover all aspects of this interdisciplinary technology where knowledge, methods and expertise are required for chemistry, biochemistry, microbiology, genetics, chemical engineering and computer science. Special volumes are dedicated to selected topics which focus on new biotechnological products and new processes for their synthesis and purification. They give the state-of-the-art of a topic in a comprehensive way thus being a valuable source for the next 3–5 years. It also discusses new discoveries and applications.

In general, special volumes are edited by well known guest editors. The series editor and publisher will however always be pleased to receive suggestions and supplementary information. Manuscripts are accepted in English.

In references Advances in Biochemical Engineering/Biotechnology is abbreviated as *Adv Biochem Engin/Biotechnol* as a journal.

Visit the ABE home page at
http://www.springerlink.com

ISSN 0724-6145
ISBN 3-540-40379-5
DOI 10.1007/b 10841
Springer-Verlag Berlin Heidelberg New York

Library of Congress Catalog Card Number 72-152360

Springer-Verlag is a part of Springer Science+Business Media

springeronline.com

The use of general descriptive names, registered names, trademarks, etc. in this publication does not imply, even in the absence of a specific statement, that such names are exempt from the relevant protective laws and regulations and therefore free for general use.

Typesetting: Fotosatz-Service Köhler GmbH, Würzburg
Cover: KünkelLopka GmbH, Heidelberg; design & production GmbH, Heidelberg

Printed on acid-free paper   02/3020mh – 5 4 3 2 1 0

Advances in Biochemical Engineering/Biotechnology also Available Electronically

For all customers who have a subscription to Advances in Biochemical Engineering/Biotechnology, we offer the electronic version via SpringerLink free of charge. Please contact your librarian who can receive a password for free access to the full articles by registering at:

http://www.springerlink.com

If you do not have a subscription, you can still view the tables of contents of the volumes and the abstract of each article by going to the SpringerLink Homepage, clicking on "Browse by Online Libraries", then "Chemical Sciences", and finally choose Advances in Biochemical Engineering/Biotechnology.

You will find information about the

– Editorial Board
– Aims and Scope
– Instructions for Authors
– Sample Contribution

at http://www.springeronline.com using the search function.

Contents

Adv Biochem Engin/Biotechnol (2004) 86: 1–45
DOI 10.1007/b12439

Investigations of the Production of Cephalosporin C by *Acremonium chrysogenum*

C. Tollnick[1] · G. Seidel[1] · M. Beyer[2] · K. Schügerl[3]

[1] Aventis Pharma Deutschland GmbH, Wirkstoffe, Wirkstoffproduktion Biotechnik, Industriepark Hoechst, 65926 Frankfurt am Main, Germany
[2] Department of Biotechnology of the Technical University of Denmark, Lyngby, Denmark
[3] Institut für Technische Chemie, Universität Hannover, Callinstrasse 3, 30167 Hannover, Germany. *E-mail: Schuegerl@iftc.uni-hannover.de*

Abstract   A review is given on the morphology of *Acremonium chrysogenum* and the biosynthesis of cephalosporin C based on the published references. Investigations are presented on the comparison of cultivation media carried out by means of shake flask cultures. The process performance of a standard cultivation in well controlled bioreactor is presented and compared with other cultivations, which were executed with the same strain and bioreactor, but with various carbon-, nitrogen- and sulphur-sources keeping the concentrations of the key components at definite levels. Also the influence of dilution and enrichment of the medium on the process performance is explored. Mathematical models for the growth of *Acremonium chrysogenum* and production of cephalosporin C are reviewed and their application for control of industrial processes with complex cultivation media are discussed.

Keywords   *Acremonium chrysogenum* · Cephalosporin C · Fungal morphology · Biosynthesis · Medium composition · High density cultivation · Low density cultivation · Intracellular enzymes · Mathematical models · Process control

Abbreviations

α-AAA	α-amino adipinic acid
ACV	δ-(L-α-aminoadipyl)-L-cysteinyl-D-valine
ACVS	ACV synthetase
Arg	arginine
Asp	aspartate
ATC	Acyltransferase
ATP	adenosine triphosphate
BFM	(wet) biomass
BTM	(dry) biomass
C	carbon
CPC	cephalosporin C
CPR	CO_2 formation rate
CSL	corn steep liquor
Cys	cystein
DAC	deacetyl cephalosporin C
DAOC	deacetoxy cephalosporin C
DNA	deoxyribonucleic acid
EXP	Expandase
FT-NIR	Fourier transform near infrared
Glu	glutamate
Gln	glutamine
HPLC	high-performance liquid chromatography
HYD	Hydroxylase
Ile	isoleucine

IPNS   Isopenicillin synthetase
Lys   lysine
Met   methionine
N-   nitrogen
NADH   nicotinamide-adenine-dinucleotide (reduced form)
NADPH   nicotinamide-adenine-dinucleotide-phosphate (reduced form)
PENN   penicillin N
RNA   ribonucleic acid
rpm   revolutions per minute
S-   sulphur
SED   sediment
TGS   (dry) glucose syrup
Val   valine

1
Introduction

Cephalosporins belong together with penicillins to the family of β-lactam antibiotics. In 1945, Giuseppe Brotzu, working at an institute in Caligiari on the island of Sardinia, isolated from seawater near to a sewage outlet a fungus (*Cephalosporium acremonium brotzu*) which produced a mixture of antibiotics [1]. This mixture was effective against both Gram-positive and Gram-negative bacteria, especially against *Staphylococcus aureus* and *Salmonella typhi*. Cephalosporin C (CPC) was isolated and purified by Abraham et al. [2, 3] and its structure was elucidated by Newton and Abraham [4]. *Cephalosporium acremonium* was recently renamed as *Acremonium chrysogenum*. Cephalosporin C is also produced by *Streptomyces clavuligerus* [5] besides clavulanic acid and Cephamycin C produced by *Streptomyces* sp [6, 7]. The present review considers only the growth of *Acremonium chrysogenum* and the production of cephalosporin C by this fungus. Five distinguishable morphological forms of *A. chrysogenum* are known: filamentous hyphae, swollen hyphal fragments, arthrospores, conidia and germlings. The spores differentiate into swollen spores and germlings [8]. The presence of C-, N-, Mg-, and PO_4^{3-}-sources are necessary for germination. The hyphae grow apical and by branching. The morphological differentiation of *A. chrysogenum* was studied by several investigators. Especially the formation of strong swollen hyphal fragments was the centre of interest (Nash and Huber [9], Queener and Ellis[10], Matsumura et al. [11], Drew et al. [12], Mehta et al. [13]). According to Drew and Demain [14] the differentiation of *A. chrysogenum* into swollen hyphae and arthrospores in methionine-containing media is more distinct than in sulphate-containing media. High amounts of ammonia inhibit the sporulation and the CPC formation according to Bartoshevich [15]. Mehta et al. [13] found an effect of lysine on the morphology of this fungus. Low amounts of lysine delay the differentiation, but later enhance the fragmentation and antibiotic formation. Nash and Huber [9] showed by using density-gradient centrifugation that swollen hyphal fragment-rich fractions displayed 40% greater activity of β-lactam antibiotic production than any other morphological form.

Nash and Huber [9] and Matsumura et al. [11] denied that the formation of arthrospores has a close interrelationship with the formation of antibiotics, because they found that the fungus formed, with the same type of arthrospores, different amount of antibiotics.

Several research groups investigated the biosynthesis of CPC, which is shown in Fig. 1.

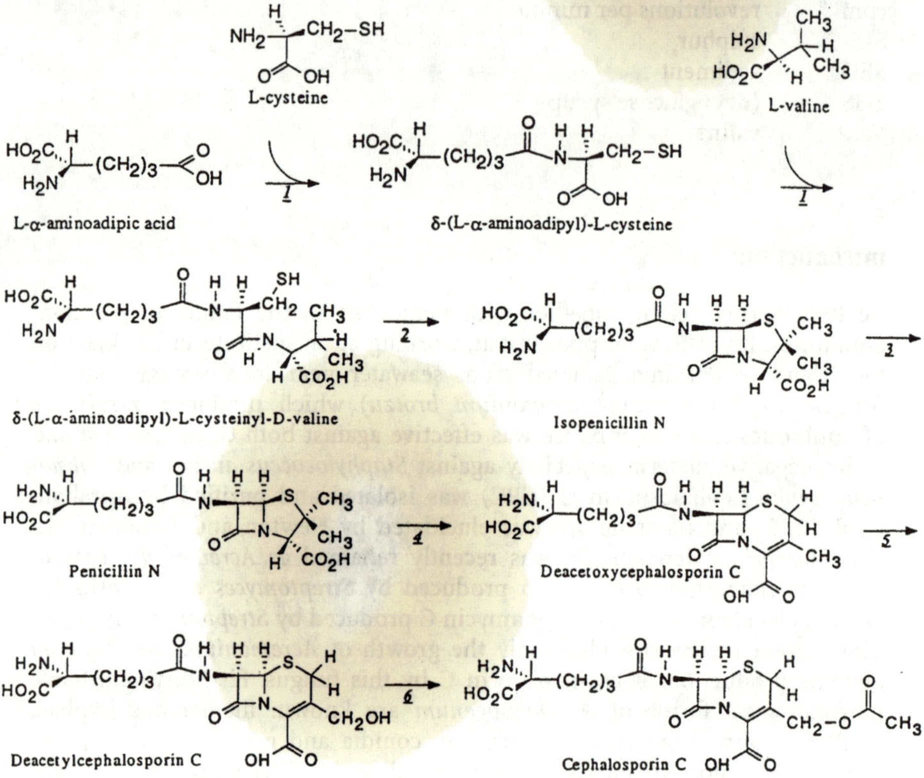

Enzymes of the Cephalosporin C biosynthesis:

1 ACV-Synthetase

2 Isopenicillin N-Synthetase

3 Isopenicillin N-Epimerase

4 Deacetoxycephalosporin C-Synthetase

5 Deacetoxycephalosporin C-Hydroxylase

6 Deacetylcephalosporin C-Acetyltransferase

Fig. 1 Biosynthesis of cephalosporin C. (Reprinted from [50] with permission from Elsevier-Science)

Because of the close relationship between penicillins and cephalosporins, their biosynthesis is usually discussed jointly. The first step of the biosynthesis is ACV synthesis from cysteine, valine and α-amino adipinic acid (α-AAA) by the enzyme ACV-synthetase (ACVS), which is the bottleneck of the cephalosporin C formation. Therefore, several researchers investigated the regulation of this enzyme. This synthesis is suppressed by ammonia [16, 17] and by phosphate [18], but is induced by methionine [19]. ATP, L-cysteine [20] and α-AAA [21] act as cofactors and have an activating effect. Phosphate, glycerolaldehyde-3-phosphate [20], ammonia [17] and Fe^{2+}-ions [18] inhibit ACVS.

The cyclisation of ACV to Isopenicillin N is oxidative step. It is catalysed by Isopenicillin *N*-synthetase (Cyclase or Dioxygenase). According to Bainbridge et al. [22] an equimolar oxygen as cofactor is necessary to this step. Other cofactors are Fe^{2+}, ascorbate and dithiothreitol [23]. Phosphate suppresses the IPNS-Synthetase (IPNS) [18].

The conversion of IPN to penicillin N (PENN) is catalysed by the enzyme Isopenicillin N-epimerase, which is extremely instable in a cell free extract [24] and not well characterized. According to Jayatilake et al. [25] the epimerase needs neither Fe^{2+} nor ascorbate as cofactors.

The ring expansion of penicillin N to deacetoxy cephalosporin C (DAOC) und its hydroxylation to desacetyl cephalosporin C (DAC) are catalysed by the bifunctional enzyme Expandase/Hydroxylase, which requires the cofactors oxygen, α-ketoglutarate [26], Fe^{2+}, ascorbate and ATP [24, 27]. Phosphate acts as inhibitor [18], the activity of the enzyme is suppressed by glucose [28], phosphate [18] and ammonia [17]. According to the investigations of Lucas [29] and Scheidegger [30] the Expandase/Hydroxylase is more strongly dependent on oxygen than the Cyclase. Therefore, Penicillin N is enriched by oxygen limitation [31].

The last step of the biosynthesis is the acetylation of DAC, which is catalysed by DAC-acyltransferase. This enzyme requires Mg^{2+} [32] and Acetyl-CoA or acetate, CoA and ATP [33] as cofactors. This acetylation is a reversible reaction. However, the back reaction, which arises at the end of the cultivation and causes the decomposition of a considerable amount of CPC, is mainly a non enzymatic reaction.

The tripeptide, δ-(L-α-aminoadipyl)-L-cysteinlyl-D-valine (ACV), is synthesised from L-α-aminoadipat (α-AAA), L-cysteine (Cys) and L-valine (Val) by ACVS. Since α-AAA is an intermediate of lysine biosynthesis, high lysine concentration causes feedback inhibition of the Homcitrate-synthatase, which was proved by Smith [34] for *Penicillium chrysogenum* and according to Mehta et al. [13] this inhibition also holds true for *A. chrysogenum*. Low lysine concentrations stimulate CPC formation, which is used for cysteine formation. Methionine plays an important role for the production of cephalosporin C. Extracellar methionine is used for cysteine synthesis. The alternative biosynthesis of cysteine through reduction of sulphate, which is preferred by *Penicillium chrysogenum*, is less important for *A. chrysogenum*. Methionine induces ACVS [19]. According to Alfonso and Luengo [35] methionine inhibits the Valine-permease. Queener et al. [36] investigated the sulphur-metabolism of *A. chrysogenum* und described sulphate dependent mutants, which are independent of methionine. However, the assimilation of sulphur by sulphate reduction has a high metabolic expenditure, because it consumes 4 moles of NADPH and 2 moles of ATP [37].

The fungus can use several nitrogen sources: amino acids, ammonia, nitrates, urea, etc.

In the industry often $(NH_4)_2SO_4$ is used. Only NH_3 can pass the plasma membrane by free diffusion. ATP consumption is required to the transport of NH_4^+-ion into the cells and to drive out the protons from the cells. Under the usual cultivation conditions (pH 5.8–6.7) the ratio NH_3/NH_4^+ is very low; therefore, diffusion only plays a role at high NH_4OH concentrations.

According to Shen et al. [16] ammonia suppresses the activity of Expandase and only slightly that of Cyclase. Zhang et al. [17] observed the suppression of ACVS and Expandase at high concentrations of ammonia. A more detailed description on the genes and enzyme of the biosynthesis of β-lactam antibiotics is given in a review by Martin [38].

The application of glucose as a C-source is only recommended during growth, because the easily consumable glucose represses the biosynthesis of cephalosporin C. To avoid this repression, slowly consumable oligo- and polysaccharides, such as (dry) glucose syrup, starch and dextrin and soy oil and lard oil, respectively, are used in the industry. They are less expensive than glucose and soy oil has higher energy content and they hinder foam formation.

2
Comparison of the Published Data on Medium Composition

The spore suspension is stored in liquid nitrogen. Agar slant cultures of the strain are used to inoculate the first preculture. The last preculture is used as inoculum for the main culture. The medium compositions of all of these stages influence the success of the production process.

The following tests, published in the literature, were performed with various medium compositions in shake flask cultures and the results were compared. The authors of these publications used various strains; therefore a direct comparison of their results is not possible. To make the results comparable, the same strain was used for all of our tests.

2.1
Composition of Agar Slants

For the tests five published agar media were used (Table 1) and 5 ml of the agar medium was poured into the slant culture tube (16×13 cm) and, after cooling and keeping it at room temperature for 24 h, it was inoculated and incubated at 28 °C.

On the Adlington agar [42] the growth was unsatisfactory. On the agars of Matsumura [43] and Fujisava [41] it was intermediate and on the of Herold/Auden medium [40] it was the highest. According to microscopic pictures (400× enlargement), in the methionine-containing media, morphological differentiation occurred and loops and budlike structures were formed.

The Herold/Auden agar was used in the following investigations.

Table 1 Composition of the tested agar media

References	Küenzi [39]	Herold and Auden [40]	Fujisawa et al. [41]	Adlington et al. [42]	Matsumura et al. [43]
Components	Concentrations (g l^{-1})				
Glucose	10.0	10.0	20.0	1.0	7.0[a]
Yeast extract	4.0	–	2.5	1.0	3.0
CSL	–	5.9	–	–	–
(NH$_4$) acetate	–	2.3	–	–	–
D,L–Methionine	2.0	2.0	–	–	–
NaCl	–	–	–	0.5	–
CaCl$_2$	–	–	–	10.0	–
Malt extract	–	–	5.0	–	–
Starch	–	–	–	–	30.0[b]
Casein hydrolysate	–	–	1.0	–	–
Agar	25.0	30.0	15.0	20.0	20.0
pH	7.7	7.4	6.8	6.8	7.0
Temperature (°C)	25	27	28	25	25
Incubation time (days)	7	7	7	10	2

[a] instead of oat flour, [b] instead of lactose.

2.2
Composition of Preculture Medium

The tests were performed with six preculture compositions published in the literature. The compositions of these media are compiled in Table 2.

2.5 ml inoculum was added to 50 ml medium in 1000 l flask and cultivated at 25 °C on a rotational shaker (HT Infors AG, Basel) at 280 rpm for

Table 2 Composition of the tested preculture media

References	Sawada et al. [44]	Demain et al. [45]	Küenzi et al. [39]	Ott et al. [46]	Feil [47]	Fujisawa et al. [41]
Components	Concentration (g l^{-1})					
CSL	30.0	32.0	23.6	–	–	5.0
Glucose	10.0	1.0	10.0	10.0	2.5	30.0
Starch	30.0	–	–	15.0	–	–
CaCO$_3$	5.0	–	–	1.5	2.5	–
(NH$_4$) acetate	–	4.4	4.5	–	–	–
CaSO$_4$	–	–	0.5	–	–	–
MgSO$_4$	–	–	0.5	–	–	–
(NH$_4$)$_2$SO$_4$	–	–	–	1.0	–	–
Malt extract	–	–	–	–	12.0	15.0
Yeast extract	–	–	–	–	16.0	–
Soy oil	–	–	–	16.0	0.2 ml	–
pH	7.0	7.2	7.2	7.0	6.4	6.8

72 h. At the beginning and at the end of the cultivation, the medium was analysed.

The best results (highest biomass and protein concentration and DNA-content) were obtained by using the Sawada medium [44]. The mycelia grow well in the preculture of Ott (46) and Feil [47], but the DNA content remained low. In the Fujisawa medium [41), only low amounts of DNA formed and only a few hyphae were produced. Therefore, the Sawada medium [44] was investigated more thoroughly. Eight parallel runs were performed and at the end of their cultivation, the main values of these cultivation were: pH 6.45 and 8.6 vol % sediment, 9.43 g l^{-1} (dry) biomass, 510 mg l^{-1} phosphate, 11.6 g l^{-1} glucose, 0.34 g l^{-1} ammonia, 182 mg l^{-1} protein and 16.49 mg l^{-1} DNA-content with an average standard deviation of 10%.

2.3
Composition of Main Cultures

The tests for the main cultures were carried out with nine different published cultivation media (Table 3).

Beside the major components given in Table 3, the media contained low amounts of minor components ((NH_4)SO_4, (NH_4)acetate, urea, Na_2SO_4, $MgSO_4$, $CaSO_4$, $CaCO_3$, $CaCl_2$, and various phosphate salts), respectively.

Except for the test with the Mehta medium, all of them were inoculated with 2 ml of the preculture of Sawada [44], which was added to 30 ml complex medium in 500-ml shake flasks and cultivated at 25 °C on a rotational shaker at 280 rpm.

The test with the Mehta medium was performed without preculture and the test was terminated after 100 h.

In the media proposed by Küenzi [39] and Demain [45] the mycelium formed swollen hyphae and arthrospores. Both of them contained CSL. The pH value increased above 8 and this was independent of the initial pH value. The biomass, protein and DNA content were higher in the medium of Küenzi than in that of Demain. In the Sawada medium [44] only thin hyphae formed without swelling.

Table 3 Main components of the tested complex media used for main cultures

Components References	Sugar (g l^{-1})	CSL (g l^{-1})	Ammonium (g l^{-1})	Soy oil (g l^{-1})	Methionine (g l^{-1})	Yeast extract (g l^{-1})
Sawada et al. [44]	46.00	–	1.92	–	3.0	–
Demain et al. [45]	27.00	25.00	0.44	–	5.0	2.0
Küenzi [39]	37.50	–	2.36	–	4.0	–
Mehta et al. [13]	60.00[a]	– 1.28	40.00	6.0	–	
Ott et al. [46]	20.00[b]	20.00	1.92	17.00	1.2	–
Telesrina et al. [48]	3.00	–	0.2	–	0.5	–
Shen et al. [28]	100.00	–	3.5	30.00	–	50.00
Feil [47]	10.0[d]	18.3	2.57	–	2.44	–
Fujiwasa et al. [41]	47.00[c]	–	–	–	5.00	–

[a] starch, [b] dry glucose syrup, [c] soybean flour was replaced by starch, [d] dextrin

Table 4 Concentration of the β-lactam products after 144 h cultivation with the media of Table 3

Medium of	Concentration (g l⁻¹)				
	PENN	DAOC	DAC	CPC	β-lactam
Demain et al. [45]	0.29	0.13	0.67	2.17	3.26
Küenzi [39]	0.23	0.20	2.17	1.58	4.18
Mehta et al. [13]	–	0.01	0.06	–	0.07
Ott et al. [46]	0.37	0.15	0.77	2.30	3.59
Telesrina et al. [48]	–	–	0.18	0.22	0.40
Shen et al. [28]	0.16	0.52	0.92	0.21	1.81
Feil [47]	1.97	0.24	0.94	2.92	6.07
Fujisawa et al. [41]	0.32	0.04	0.39	0.13	0.88

The ingredients of the medium were not consumed completely and the pH value dropped to 4.18. This medium yielded unsatisfactory results. No mycelium was formed in the Fujisawa medium [41]. Only a few mycelia were formed in the Shen medium [28].

According to Table 4, satisfactory results were obtained using the media of Demain, Ott and Feil. The best results were attained with the medium from Feil.

3
Cultivation in Shake Flask and in Bioreactors under Well-Defined Conditions

Shake cultivations of micro-organisms are used to obtain preliminary results.

Development of the cultivation medium is usually performed in shake cultures, because one can test several variations of medium compositions parallel at the same time. However, these investigations have a qualitative character, because the cultivations are performed under non-controlled conditions. Nevertheless, such investigations are useful, because we can monitor several flasks to give us an approximate idea of the progress of the process.

3.1
Shake Flask Cultivations

The tests in Sect. 2.2 were carried out in shake flask cultures. The composition of the medium was determined at the beginning and at the end of the cultivation. By means of suitable buffer salt supplement, the pH value was maintained in the range of 6.7–8.7, except of the Sawada medium in which the pH dropped to 4.18. In spite of maintaining a nearly constant pH value, the results of the cultivation were poor. Jürgens et al. [49] investigated the cultivation in shake flasks by monitoring the course of the cultivation and comparing them with those in bioreactors under well-defined conditions. The maximum concentrations of biomass, PENN and CPC, produced by the *Acremonium chrysogenum* 3/(C3) strain in a semi-synthetic medium (Table 5) and complex media are compared in Table 6.

Table 5 Composition of the semi-synthetic medium [49]

Compound	Concentration [g l^{-1}]	Compound	Concentration [g l^{-1}]
Glucose	35	L-serine	1
$(NH_4)_2SO_4$	4.2	Soy oil	20[a]
Na_2SO_4	5	Desmophen	1 ml
KH_2PO_4	3.1	(antifoam agent)	
L-Valine	1	$MgSO_4$	6.67
D,L-Alanine	2	$FeSO_4$	0.09
L-Asparagin	3.33	$MnSO_4$	0.03
L-Methionine	14.67	$ZnSO_4$	0.02
L-Arginine	3.33	$CuSO_4$	0.02
Urea	2.1		

[a] In 10 L bioreactor the soy oil concentration is 40 g l^{-1}

Table 6 Production of CPC by *Acremonium chrysogenum* 3/(C3) strain in semi-synthetic and complex media under various cultivation conditions. Comparison of maximum concentrations of biomass, cephalosporin C and penicillin N in the cell free medium filtrate [49]

Cultivation Conditions	(dry) biomass conc. [g l^{-1}]	PENN conc. [g l^{-1}]	CPC conc. [g l^{-1}]
Semi-synthetic medium without buffer shake flask culture	25	<0.1	1.0
Semi-synthetic medium 100 mM MES buffer shake flask culture	40	2.5	4.0
Semi-synthetic medium pH control, batch 10 L reactor	42	2.5	4.4
Semi-synthetic medium pH control, batch Oxygen supplement 30 L reactor	47.5	4.0	7.7
Semi-synthetic medium pH control, batch Oxygen supplement 30 L reactor	40	4.5	7.7
Complex medium pH control, batch Oxygen supplement 30 L reactor			9.5
Complex medium pH control, fed-batch oxygen supplement 30 L reactor	88	3.2	23

The product concentration in a complex medium and by fed-batch operation in a bioreactor under well-controlled conditions is considerable higher than in a semi-synthetic medium in shake flasks or in a bioreactor, but in batch operation.

Therefore, in the following, fed-batch cultivations in complex media are investigated.

3.2
Growth and Product Formation in Complex Media

In the chemical and biotechnology industry, the raw material costs usually amount to 60% of the product formation costs. Therefore, the raw material costs play an important role in the process design. In industrial production, complex media are used because they are less expensive and yield higher product concentration. Various C-, N-, and S-sources are used: dextrin, (dry) glucose syrup, starch, soy oil, peanut flour for carbon source; corn steep liquor, sulphates, methionine, cysteine, urea and other amino acids for nitrogen sources; methionine, sulphates, cysteine as sulphur sources. These medium components are used in batch and fed-batch operation, respectively. The cultivations, which are reported here, were performed in a bioreactor with a 30-L working volume with monitoring and controlling of the key process parameters.

3.2.1
Materials and Methods

Various *Acremonium chrysogenum* strains from culture collections were used in the publications of researchers. However, these strains produce only low amounts of cephalosporin C and have properties which, to some extent, do not hold for industrial production strains.

Therefore, in the following, only a high producing strain is considered, which is close to the industrial production strains. This is an *Acremonium chrysogenum* A3/2(C3) strain donated by HMR Deutschland [50]. The ampoules with 2.2 ml spore suspensions were stored in liquid nitrogen. Agar slant cultures of the strain were used to inoculate 1-L shake flasks containing a 160-ml malt- and yeast-containing medium, which had been sterilised previously at 121 °C for 20 minutes. This was cultivated on a rotary shaker at 280 rpm with 3 cm stroke at 25 °C for 72 h. The second preculture contained corn steep liquor (CSL) and glucose. It was prepared in a 10-L bioreactor (Biostat V of B. Braun Melsungen) with 6 L of the medium, which was sterilised at 121 °C for 35 min and after cooling to 25 °C it was stored for 12 h. The pH was set to 6.5; the reactor was aerated and operated with 600 rpm. After 25–30 h the main culture was inoculated with this second preculture. The amount of the inoculum was 19% of the medium volume. The main cultivations were performed in a 30-L working volume reactor (Biostat UD, B. Braun Melsungen), which was furnished with digital measuring and control units and a process guide system. (UBICON, eds, Hannover). The evaluation of the data was performed with a separate computer. The reactor was provided with instruments for measurement and control of the temperatures of the medium and reactor mantle, head pressure, dissolved oxygen (pO_2) and pH-value in the medium, and stirrer speed. A gas flow ratio control unit allowed one to prepare mixtures of pressurised air and oxygen with various compositions of aeration gas to avoid the oxygen limitation of the process. This is not unrealistic in practice, because in industrial production, the dissolved oxygen concentration can be increased by the reactor pressure, instead of being supplementing the air with oxygen.

The reactor was furnished with two on-line sampling units (Eppendorf), Autosampler (Gilson XL222. abimed) and a rack (cooled by a cryostat (Haake) to –20 °C) with 60 Eppendorf reaction tubes with a maximum sampling frequency of 10 h^{-1} and a steam-sterilised, off-line sampling unit (SV15, bbi) as well as by O_2- and CO_2-analysers in the off gas.

The concentrations of the β-lactam products (CPC, DAC, DAOC, PENN) were monitored by on-line HPLC. The concentration of urea was monitored by on-line FIA. The concentrations of amino acids were measured by off-line HPLC. The process was operated with controlled feeding of cysteine, arginine/methionine, urea and $(NH_4)_2SO_4$, respectively. The on-line control of soy oil feeding by FT-NIR spectroscopy was not installed because the on-line separation of the aqueous and oil phases by means of a hydrophobic membrane module was not successful. Therefore, the soy oil feeding was not computer controlled.

The amount of sediment, the concentrations of (wet) biomass, (dry) biomass, ammonia, phosphate, protein, sugar, organic acids, dissolved organic carbon (DOC), enzyme activities in the medium and the intracellular concentrations of DNA, RNA, amino acids and β-lactam precursors and product as well as intracellular enzyme activities of the biosynthesis were determined as well. The details of the analysis methods are described in Ref. [51]. The extracellular concentration of the dissolved medium components is given with regard to the volume of the supernatant.

3.2.2
Cultivation with a Standard Complex Medium

For these cultivations, a standard medium concentration was used and this was varied in different way [50, 52, 53]. The composition of the standard medium differs from that of Table 5 in following way: Glucose was replaced by 46.0 g l^{-1} (dry) glucose syrup (TGS). The medium was supplemented by 110 g l^{-1} corn steep liquor (CSL). The concentration of $(NH_4)_2SO_4$ was reduced to 1.7 g l^{-1}, and that of KH_2PO_4 was diminished to 2.1 g l^{-1}. Urea, valine, serine and alanine were omitted. All the runs were fed-batch cultivations with soy oil feeding. The following variations of nitrogen sources were used: $(NH_4)_2SO_4$ in batch and fed-batch operation, methionine+arginine in batch and in fed-batch, serine and alanine in batch, valine in batch and fed batch, and urea in batch and in fed-batch operation. The following sulphur sources were used: cultivation with and without methionine, Na_2SO_4 in batch, $(NH_4)_2SO_4$ in batch and fed-batch, and methionine and cysteine in fed-batch [50].

All of the cultivations were carried out at 25 °C with a dissolved oxygen concentration of 60% of its saturation value. This was maintained at a 0.5 vvm aeration rate and 600 rpm stirrer speed by varying the composition of the aeration gas mixture. The initial pH value was 6.5. The pH control was performed by 12.5% ammonium hydroxide or potassium hydroxide and concentrated sulphuric acid.

In Refs. [50, 53] a typical cultivation was described. This is considered as being a standard cultivation.

The summary of the results of these investigations is presented here.

Since, during the inoculation, the cells in the preculture were in the exponential growth phase, only a short cell adaptation was observed in the main culture. The length of this lag phase depends on the type of carbon source in the preculture and main culture.

At the start of the cultivation, pO_2 dropped within 30 h to 60% and kept at this value by increasing the oxygen content of the aeration gas. The pH value was maintained at 5.8.

During the growth phase, mainly (dry) glucose syrup (TGS), ammonia and phosphate were consumed by the fungus (Fig. 2a). At first (0 to 20 h), the free glucose used up, then the oligo- and polysaccharides (20 to 55 h) and finally the soy oil was consumed. The amylase activity in the medium started to increase at 20 h and attained its maximum at 70 h. The lipase activity started to increase as early as at 20 h and attained a broad maximum after 40–120 h [54]. According to the changes of the consumed substrate type, at least two diauxies were observed at 20 h and 60 h by means of the minima of the CO_2 production rate (CPR) and respiratory quotient (RQ) (Fig. 2b). The minima of oxygen transfer rate (OTR) were less distinct. The amount of the sediment and the (dry) biomass concentration remained constant only during the second diauxy (Fig. 2c). Protease activity and protein concentration increased during the cultivation. Some amino acids of the CSL were consumed, others produced. Methionine, alanine, asparagine and arginine were gradually consumed and exhausted at the end of the cultivation (Fig. 2d). The concentration of other amino acids also decreased. Glutamin concentration increased (Fig. 2e). The concentrations of pyruvate, succinate and acetate passed a maximum. Lactate slowly consumed and formiate produced during the cultivation (Fig. 2f).

The growth and the product formation phases could not be separated clearly, because they proceeded simultaneously. The β-lactam compounds gradually increased: CPC to 24.4 g l^{-1}, DAC to 1.94 g l^{-1}, DAOC to 0.4 g l^{-1} and PENN to 3.25 g l^{-1} (Fig. 2g).

The intracellular concentrations of enzymes of the CPC biosynthesis have very different activities. The activity of ACV synthetase is low (1.7 U g^{-1}) (Fig. 2h). IPNS activity is high (20 U g^{-1}). Activity of EXP (5 U g^{-1}) intermediate and HYD (2 U g^{-1}) low. ATC activity is very high (80 U g^{-1}) (Fig. 2i). It is well known that ACV synthetase is the bottleneck of the CPC synthesis.

In following, the influence of the variation of nitrogen and sulphur sources on the process performance is considered.

3.2.3
Cultivation with Ammonia as the Nitrogen Source

Ammonium sulphate $(NH_4)_2SO_4$ was the main nitrogen source, in addition to 14.7 g l^{-1} methionine [52]. The concentration of ammonium sulphate in the medium was increased to 11.7 g l^{-1}. NH_4OH/H_2SO_4 was used for pH control. The concentration of ammonia gradually dropped from 5 g l^{-1} at the beginning to 2 g l^{-1} at the end of the cultivation. To maintain the constant pH value the NH_3–feed was started at 40 h, the first feed rate gradually increased to 120 mg l^{-1} h^{-1} up to 60 h, it dropped at 70 h and increased again to 130 mg l^{-1} h^{-1} up to 80 h than reduced and stagnated at 50 mg l^{-1} h^{-1} during 120–140 h. At the end of the cultivation it increased again.

The methionine concentration diminished from 11 g l^{-1} to 2.5 g l^{-1} at the end. The glutamate concentration increased from 0.05 g l^{-1} to 0.19 g l^{-1} up to 100 h and

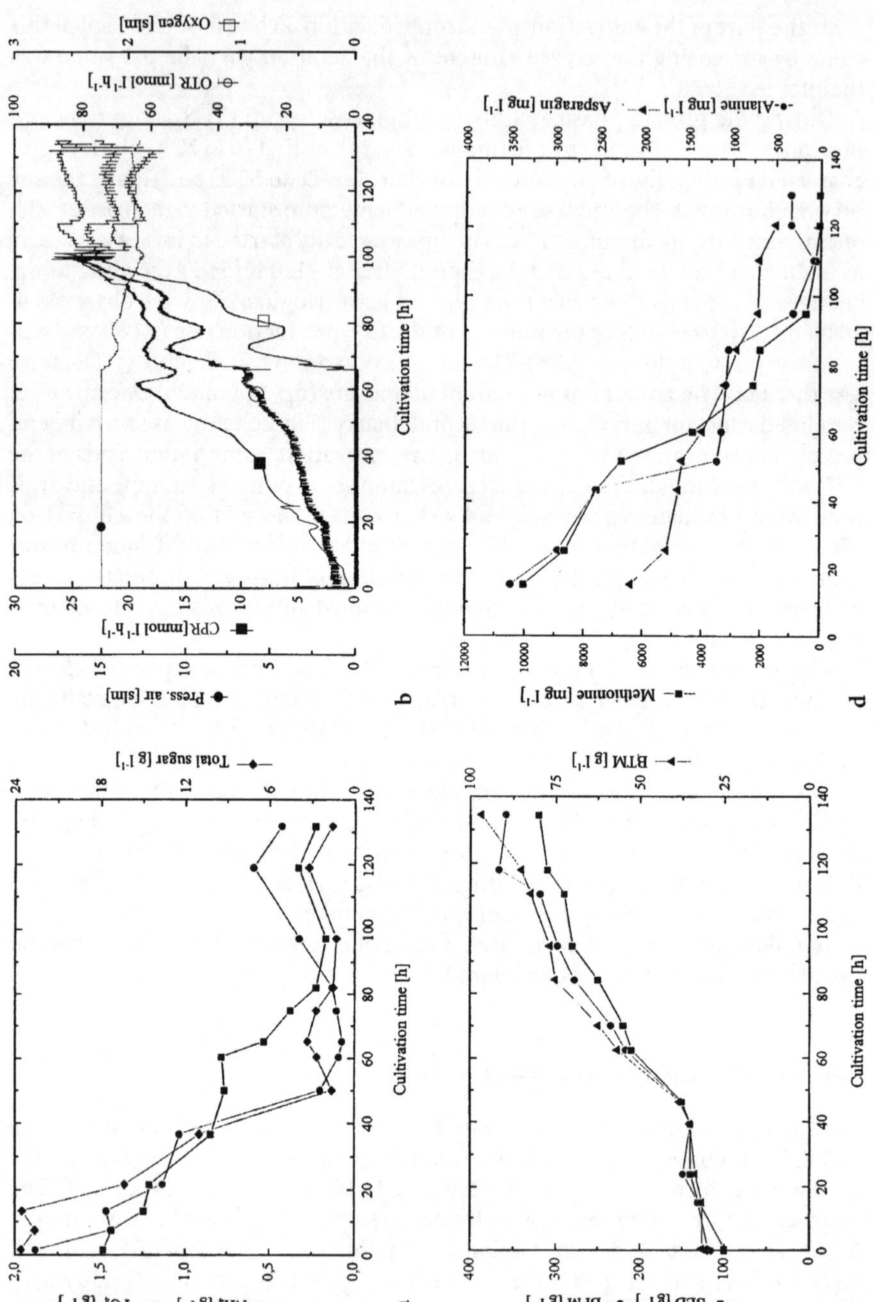

Fig. 2 a–i Standard cultivation. **a** Concentrations of ammonia, phosphate and total sugar. **b** CO_2 production rate (CPR), oxygen transfer rate (OTR), respiratory quotient (RQ). **c** Sediment, (dry) biomass. **d** Concentrations of methionine, alanine, asparagine, arginine. (Reprinted from [50] with permission from Elsevier Science)

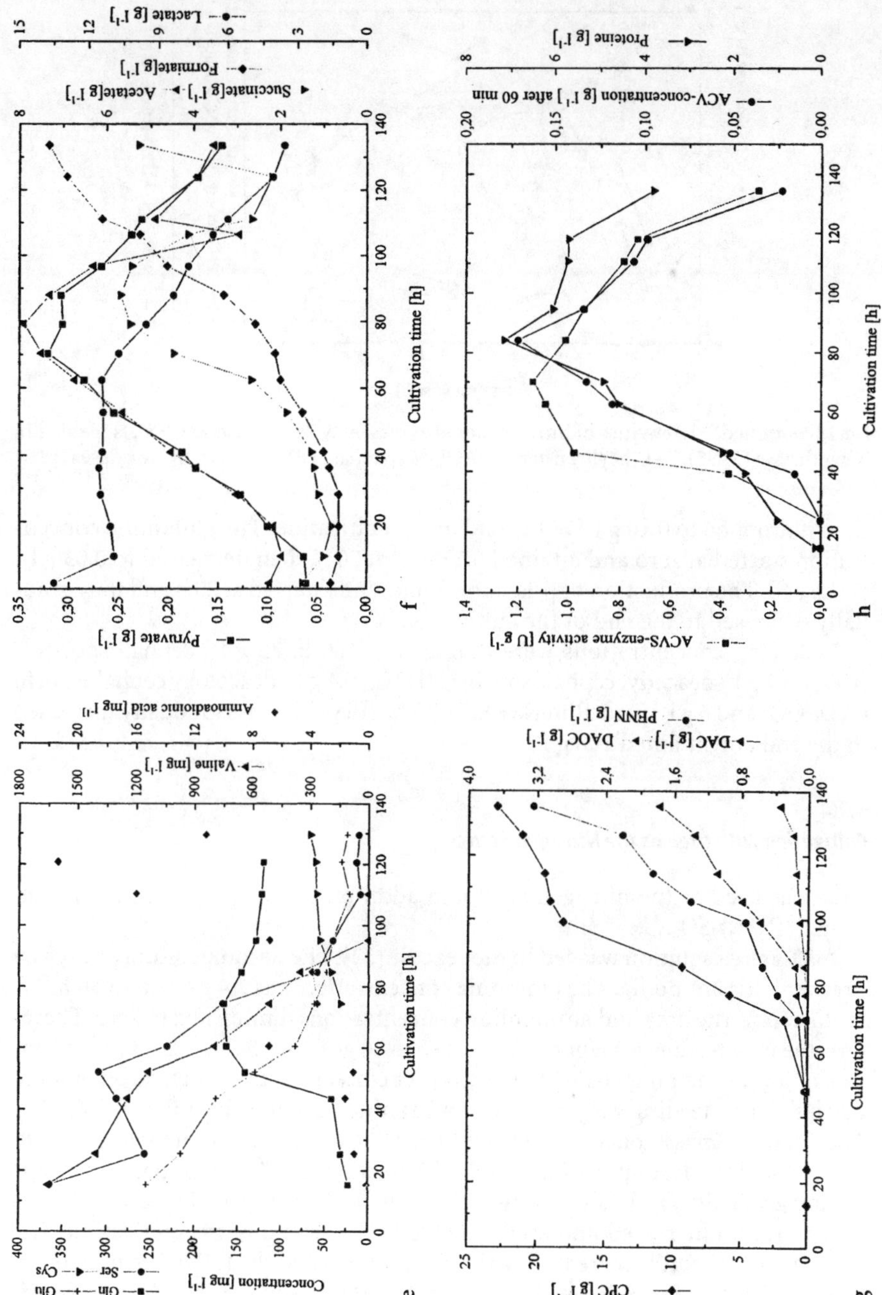

Fig. 2 (continued) **e** Glutamine, glutamate, serine, cysteine, valine and α-AAA. **f** Concentrations of pyruvate, succinate, actate, formiate and lactate. **g** Cephalosporin C (CPC), deacetyl cephalosporin C (DAC), deacetoxycephalosporin C (DAOC), penicillin N (PENN). **h** Activity of intracellular ACV synthetase, intracellular concentrations of ACV and proteins. *(continued)*

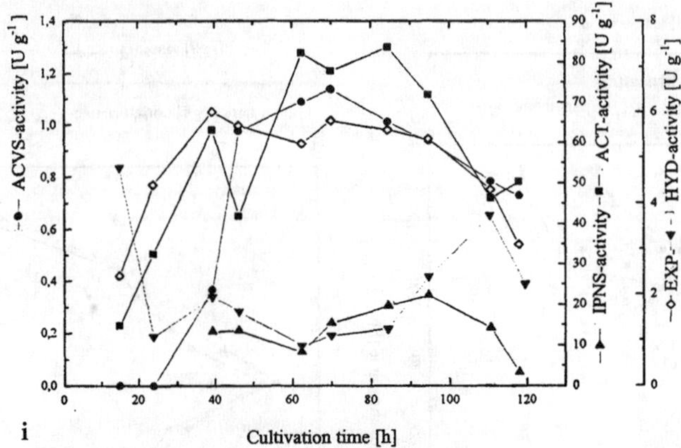

Fig. 2 (continued) **i** Activities of intracellular enzymes: ACV synthetase, (ACVS), Isopenicillin N synthetase (IPNS), DAC-acyltransferase (ACT), Expandase (EXP), DAOC hydroxylase (HYD)

than it dropped to 0.10 g l^{-1} at the end of the cultivation. The glutamine concentration started at zero and attained 0.17 g l^{-1} at 70 h, then decreased to 0.03 g l^{-1} at the end. The production of β-lactam compounds started at 20 h and they gradually increased to the end of the cultivation.

Following concentrations were obtained at 160 h: 25 g l^{-1} cephalosporin C (CPC), 5.5 g l^{-1} deacetlyl cephalosporin C (DAC), 0.4 g l^{-1} deacetoxy cephalosporin C (DAOC) and 6 g l^{-1} penicillin N (PENN). The (dry) biomass gradually increased to the end and attained 108 g l^{-1}.

3.2.4
Cultivation with Urea as the Nitrogen Source

Urea was used as the nitrogen source in addition to 14.7 g l^{-1} methionine and 3.3 g l^{-1} (NH$_4$)$_2$SO$_4$.

A 33% urea solution was fed to the reactor [52]. The decomposition of urea by urease yields ammonia. The ammonia concentration was 2.4 g l^{-1} up to 30 h. After this time, the urea and ammonium concentrations diminished to zero. Therefore, the urea feeding was increased from 5 to 20 g l^{-1} (Fig. 3a) to maintain its concentration in the range 0.5–1.0 g l^{-1} [54]. The urease activity increased at 40 h, when the urea feeding was increased and attained a high value (35 U l^{-1}) at 90 h. The (dry) biomass concentration and the CPR indicated a diauxy at 50–60 h, where the urea decomposition rate had a maximum. The methionine concentration gradually diminished from 14.7 g l^{-1} to 5 g l^{-1} at 140 h. The (dry) biomass concentration increased and attained 60 g l^{-1} at the end of the cultivation. The β-lactam compounds increased with time and reached the following maximum values: 22 g l^{-1} CPC, 3.8 g l^{-1} DAC, 0.6 g l^{-1} DAOC and 6.5 g l^{-1} PENN, which dropped to 5.6 g l^{-1} at the end.

The urea run was repeated with a 50% diluted cultivation medium. 1.7 g l^{-1} (NH$_4$)$_2$SO$_4$, 7.3 g l^{-1} methionine and urea were used as nitrogen sources. By means of on-line FIA monitoring, the urea concentration was able to be kept at 0.5 g l^{-1}

(Fig. 3b). The urea decomposition rate started from zero, attained a maximum in the range 80–100 h and diminished to zero at 150 h. The (dry) biomass concentration increased gradually to 45 g l^{-1}. The CPR and (dry) biomass concentration indicated a diauxy at 40 h. Methionine was consumed already at 100 h. The β-lactam compounds obtained following maximum values: 12. 4 g l^{-1} CPC, 3.3 g l^{-1} DAC, 0.9 g l^{-1} DAOC and 12 g l^{-1} PENN. The product concentration was very low because of the early substrate limitation.

There is no difference between the ammonia and 100% urea runs with regard to the CPC formation (Fig. 3c). Because of the high biomass production by ammonia as nitrogen source, the specific CPC product concentration or yield coefficient [CPC (g l^{-1})/(BTM (g l^{-1})] is higher with urea than with ammonia feeding. In addition the use of urea makes it possible to decouple the pH-control from the feeding of the nitrogen source. The uptake of urea does not cause a pH-shift, as does the uptake of ammonia. Therefore, the former need less correction medium. Another disadvantage of the use of ammonia is that it is not possible to recognise nitrogen limitation if the pH is not influenced by the uptake of ammonia.

3.2.5
Cultivation with Asparagine, Arginine and Ammonia

The concentration of $(NH_4)_2SO_4$ was reduced to 1 g l^{-1} in the starting medium. 3.3 g l^{-1} arginine and 3.3. g l^{-1} asparagine were used as nitrogen sources. 14.7 g l^{-1} methionine was used as the sulphur source [52]. The pH-control was provided by NH_3/H_2SO_4. The ammonia concentration gradually diminished from 2 g l^{-1} to 0.4 g l^{-1} at 150 h. The ammonia uptake rate and feed rate had sharp minima during the diauxy caused by the change from sugar to soy oil at 55 h. The ammonia uptake rate had a sharp maximum (200 mg l^{-1} h^{-1}) after the fungus started to grow on soy oil at 70 h. The CPR clearly shows two smaller diauxies caused by the change from ammonia uptake to asparagine and arginine uptakes (Fig. 4). Methionine concentration diminished from 14.7 to zero at 145 h. The (dry) biomass concentration increased stepwise and attained 84 g l^{-1} up to 140 h.

Following maximum β-lactam concentrations were obtained: 21.5 g l^{-1} CPC, 3.9 g l^{-1} DAC, 0.48 g l^{-1} DAOC and 8 g l^{-1} PENN.

3.2.6
Feeding of Methionine and Arginine

This cultivation was performed with a 50% diluted medium supplemented with 1.3 g l^{-1} $(NH_4)_2SO_4$. 1 g l^{-1} arginine, 1 g l^{-1} asparagine and 7.3 g l^{-1} methionine were fed to the medium when their concentrations in the medium dropped to zero [52]. This happened at 50 h, 75 h and 110 h. NaOH/H_2SO_4 was used for the pH control. After 20 h, the uptake rate of ammonium sulphate started to increase and attained 1.7 mmol l^{-1} h^{-1} at 30 h, dropped to 0.25 g l^{-1} h^{-1} at 45 h and increased again to 1.1 g l^{-1} h^{-1} at 65 h. After this second maximum, it dropped to zero at 80 h and remained there up to end of the cultivation (Fig. 5a). According to the CPR, a distinct diauxy appeared at 45 h and several less distinct diauxies at 70 h, at 90 and 110 h. The ammonia concentration gradually diminished from

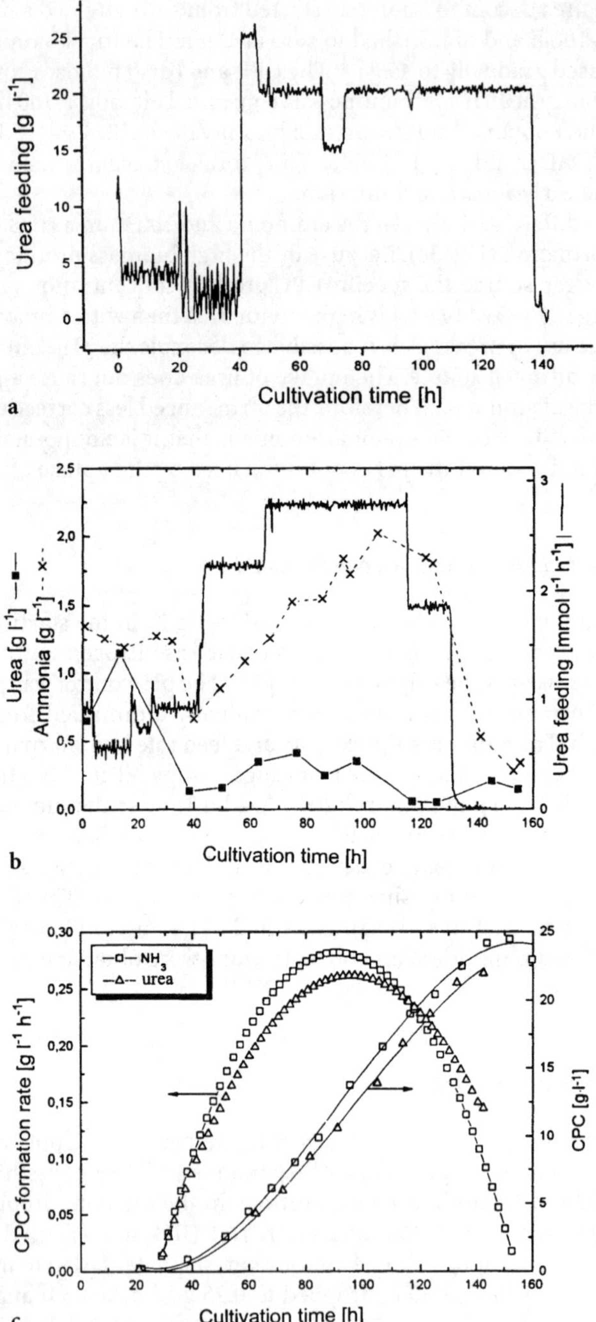

Fig. 3 a–c Alternative cultivation strategies I. Cultivation with urea as the nitrogen source. **a** Urea feeding rate, **b** Urea concentration and feeding rate with 50% medium, **c** Comparison of CPC production rate with ammonia and urea nitrogen sources

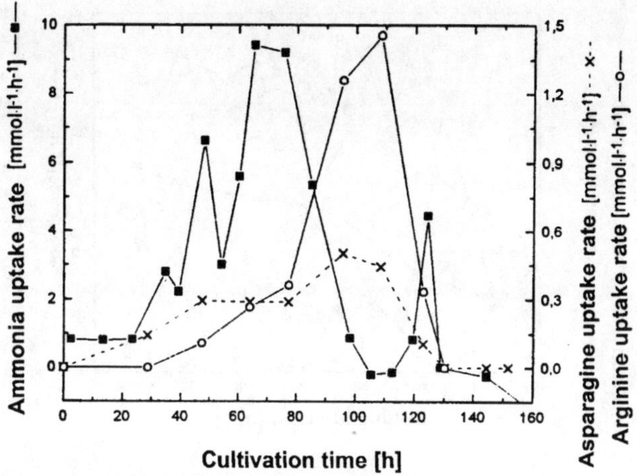

Fig. 4 Alternative cultivation strategies II. Cultivation with asparagine and arginine and ammonia. Comparison of the uptake rates of ammonia, asparagine and arginine with pH control by NH_3

1.25 g l^{-1}to 0.2 g l^{-1} during the first 80 h and remained there. Obviously the fungus was not able to utilize ammonium sulphate below this concentration. Methionine was consumed early and was exhausted at 70 h (Fig. 5b). After that, it remained at a very low value regardless of its feeding. Leucine, arginine and phenylalanine were consumed during the growth phase and exhausted up to 60 h. Only a low concentration of (dry) biomass (45 g l^{-1}) was obtained because, during the entire cultivation, the growth was limited by an insufficient amount of nitrogen compounds in spite of the feeding of arginine and methionine to the medium. For the same reason, only a low concentration of CPC (8.3 g l^{-1}) obtained which stagnated at this value after 110 h.

The concentrations of precursors increased up to the end of the cultivation and reached the following maximum values: 7 g l^{-1} PENN, 0.35 g l^{-1} DAOC and 1.25 g l^{-1} DAC.

3.2.7
Cultivation on Various Sulphur Sources

Methionine and sulphate are the most important sulphur sources. At first sulphate was used as the S-source. The methionine free 100% standard medium was supplemented by 1 g l^{-1} $(NH_4)_2SO_4$, 5 g l^{-1} Na_2SO_4 and 6.7 g l^{-1} $MgSO_4$, which are ingredients of the standard medium [52]. These were the sulphur sources. In addition, the medium was supplied with 3.3 g l^{-1} arginine and 3.3. g l^{-1} asparagine as nitrogen sources. NH_3/H_2SO_4 was used for pH control. In the first 30 h an increase of the pH value from 6.5 to 7.4 was observed, which only gradually reduced to 5.0 after the pH control was started. In the presence of methionine, the pH value remained at 6.5 up to 30 h and quickly decreased to 5.0 after the pH control was introduced. Ammonia uptake quickly increased and at 40 h attained a

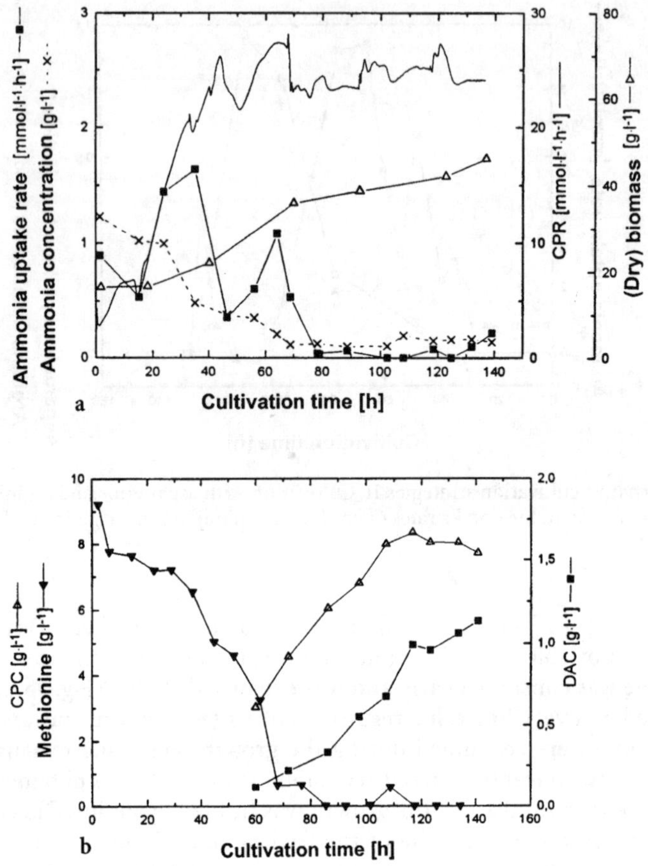

Fig. 5 a, b Alternative cultivation strategies III. Cultivation with feeding. of methionine and arginine. **a** Ammonia uptake rate, ammonia concentration and CO_2 production rate (CPR). **b** Concentrations of methionine, cephalosporin C (CPC) and deacetyl cephalosporin C (DAC)

maximum (5.5 mmol l^{-1} h^{-1}) but at 60 h dropped to zero, because ammonia was exhausted (Fig. 6a). At 50 h a distinct diauxy was observed. After this time, no growth was observed and the concentration of (dry) biomass reached 70 g l^{-1} which gradually dropped to 60 g l^{-1} by the end of the cultivation. Because of the lack of nitrogen sources at 80 h, only very low concentrations of CPC (9 g l^{-1}), DAC (1.9 g l^{-1}), DAOC (0.16 g l^{-1}) and PENN (0.74 g l^{-1}) were obtained at the end of the cultivation.

Figure 6b compares the CPC formation rates with and without methionine. In the first 50 h, the production rates were nearly the same but afterwards, the formation rate in absence of methionine quickly decreased because of the sulphur limitation.

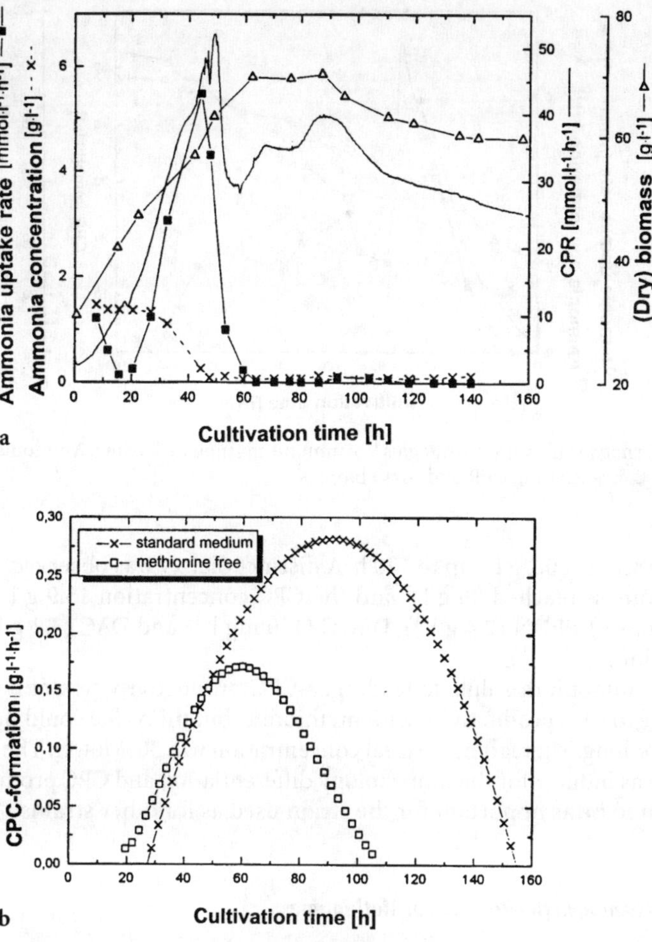

Fig. 6a, b Alternative cultivation strategies IV. Cultivation with various sulphate sources. **a** Ammonia uptake rate, ammonia concentration, (dry) biomass and CO_2 production rate (CPR) in absence of methionine. **b** Comparison of CPC production rates with standard and methionine free medium

3.2.8
Feeding of Ammonium Sulphate

A methionine-free 100% standard medium was supplemented with 1 g l⁻¹ ammonium sulphate 5.0 g l⁻¹ Na_2SO_4 and 6.7 g l⁻¹ $MgSO_4$ as ingredients of the standard medium and by 3.3 g l⁻¹ arginine and 3.3 g l⁻¹ asparagine again [52]. Ammonium sulphate was fed to the medium. Ammonia uptake rate quickly increased and at 40 h reached a maximum (6 mmol l⁻¹ h⁻¹) and afterwards decreased (Fig. 7). The ammonia concentration dropped to very low values at 55 h. To avoid the S-limitation, ammonium sulphate was fed to the system. This resulted in the uptake rate being maintained at 2 mmol l⁻¹ h⁻¹ and the ammonia

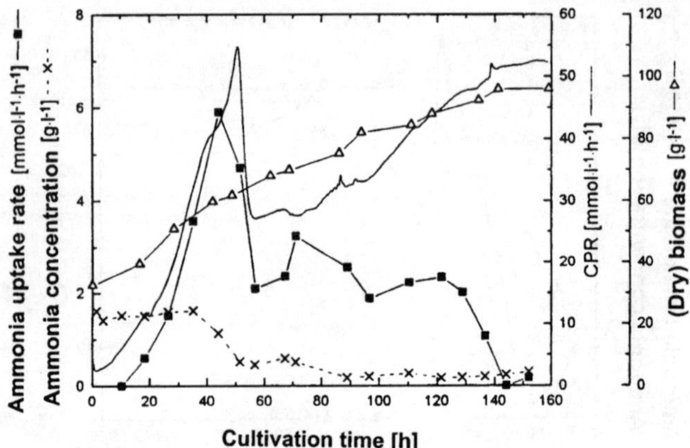

Fig. 7 Alternative cultivation strategies V. Ammonium sulphate feeding. Ammonia uptake rate, ammonia concentration, CPR and (dry) biomass

concentration at 0.2 g l⁻¹ up to 120 h. A distinct diauxy was observed at 55 h. The (dry) biomass reached 96 g l⁻¹ and the CPC concentration 15.9 g l⁻¹. The concentrations of PENN (2.4 g l⁻¹), DAOC (150 mg l⁻¹) and DAC (3.2 g l⁻¹) also remained low.

With ammonium sulphate feeding, a CPC productivity was obtained which was close to CPC productivity with methionine but this value could not be maintained for long. Thereafter, the final concentration was 30% lower. The role of methionine as inducer for the morphology differentiation and CPC production does not seem to be as important for the strain used as for other strains.

3.2.9
Cysteine Feeding in the Presence of Methionine

A 100% standard medium was supplemented by 1 g l⁻¹ $(NH_4)_2SO_4$. 3.3 g l⁻¹ arginine, 3.3 g l⁻¹ asparagine. 14.7 g l⁻¹ methionine and cysteine served as the S-source [53]. Cysteine was fed to the medium. The pH was controlled by NH_3/H_2SO_4. The methionine concentration dropped from 14.7 g l⁻¹ to 4 g l⁻¹ after 110 h and remained at this value for the rest of the time. The cysteine concentration gradually increased during this time to 0.4 g l⁻¹ and thereafter it increased to 1.4 g l⁻¹. After a lag phase, the ammonia uptake rate increased to 4.4 mmol l⁻¹ h⁻¹ at 50 h, passed a minimum at 88 h and a second maximum at 130 h (Fig. 8). The CPR closely followed this course with a distinct diauxy at 75 h, at which the sugar to soy oil change occurred. The ammonia concentration changed only slightly. At the end, a (dry) biomass concentration of 56 g l⁻¹ was obtained. The uptakes of alanine, asparagine and arginine were also delayed. The product formation started late, after 40 h and only low concentrations of β-lactam compounds were obtained: 14.1 g l⁻¹ CPC, 1.9 g l⁻¹, DAC, 0.5 g l⁻¹, DAOC and 5.4 g l⁻¹ PENN. The cysteine uptake is difficult to determine because part of the cysteine precipitates. After 110 h, the sulphate reduction seems to be the main sulphate assimilation be-

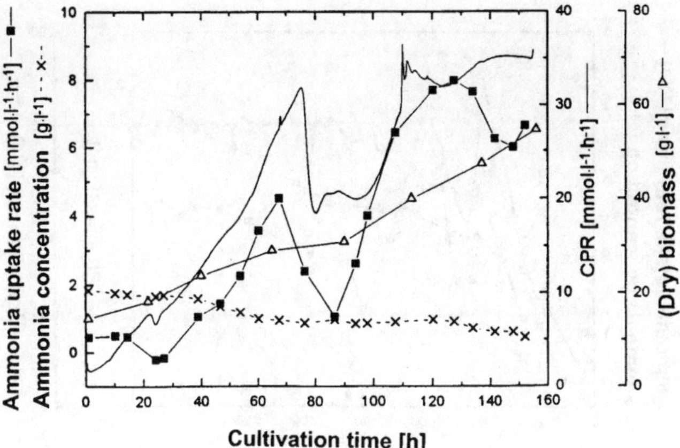

Fig. 8 Alternative cultivation strategies VI. Feeding of cysteine. Ammonia uptake rate, ammonia concentration, (dry) biomass and CPR

cause the increase of β-lactam compound was not accompanied by the uptake of methionine and cysteine. Obviously the presence of cysteine had a detrimental effect on the process.

3.2.10
Cysteine Feeding in the Absence of Methionine

The 50% standard medium was supplemented by 1 g l⁻¹ (NH₄)₂SO₄, 1.7 g l⁻¹, arginine and 1.7 g l⁻¹ asparagine. Cysteine was fed to the medium [52]. The pH was controlled by NH₃/H₂SO₄. Because the former run did not give any information on the disposal of cysteine, a second run with lower medium concentration was carried out. As Fig. 9a shows, no delay of ammonia uptake and biomass formation was observed in this case. Also the uptakes of alanine, asparagine and arginine were fast and they were exhausted as soon as at 50 h. The production of β-lactam compounds started with the uptake of methionine within 30 h. In spite of cysteine feeding, no cysteine enrichment in the medium was observed. However, the concentrations of CPC was very low and that of PENN fairly high: 7.8 g l⁻¹ CPC, 1.52 g l⁻¹ DAC, 0.48 g l⁻¹ DAOC and 6 g l⁻¹ PENN. After 60 h the CPC formation stagnated but the PENN concentration increased further because of the Expandase/Hydroxylase inhibition.

The enrichment of ACV tripeptide in the medium to 0.2 g l⁻¹ and stagnation of the PENN formation after 110 h is unusual because of the inhibition of Cyclase by the increased cysteine concentration [55].

Figure 9b shows the sulphur balance for the cultivation with methionine as the S-source and urea feeding. With increasing cultivation time, the sum of the sulphur compounds determined by analysis decreased at first. Obviously some gaseous and/or solid sulphur compounds were formed which were not measured by analysis. At the end of the cultivation, the sum of the sulphur compounds amounted to 100%. The sulphur content of the β-lactam compounds exceeded

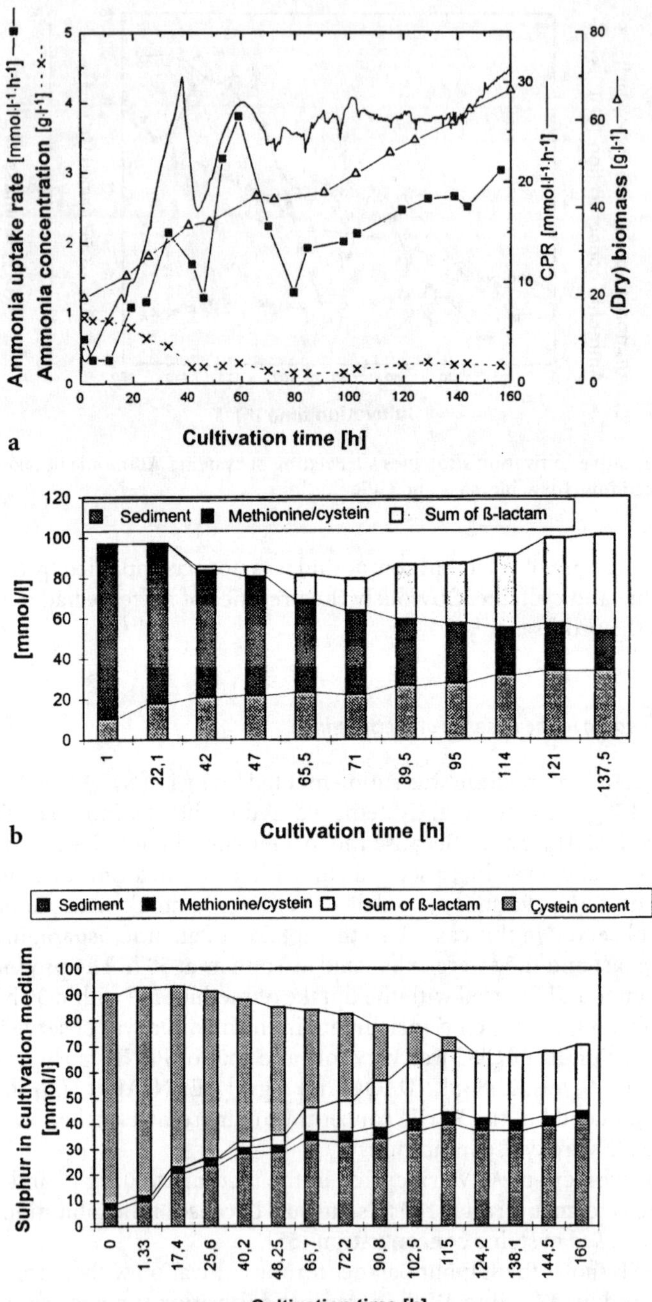

Fig. 9 a–c Alternative cultivation strategies VII. Cysteine feeding in absence of methionine. **a** Uptake rate of ammonia, concentration of ammonia and (dry) biomass and CPR. **b** Sulphur balance of a methionine free cultivation. **c** Sulphur balance of a methionine free cultivation with cysteine feeding

that of the biomass. Figure 9c shows the sulphur balance for methionine-free cultivation with cysteine feeding.

The sum of the sulphur compounds detected started to diminish after 40%. The sulphur losses are caused by non-detected compounds. The sulphur content of the biomass was considerable higher than that of the β-lactam compounds.

3.2.11
Comparison of the Process Performances with Different Medium Compositions

In Fig. 10a, cultivations with various nitrogen sources are compared.

Maximum specific productivities of CPC with regard to the (dry) biomass amount and the (dry) biomass concentrations are shown. The easily consumable substrates such as ammonia and urea stimulate the primary metabolism. The secondary metabolisms and the productivity are less influenced by the limitation of ammonia supply. This is shown by the run with amino acid feeding, which indicates that, in spite of a low growth rate, an average productivity was obtained.

The requirement of amino acids being necessary for anabolism seems to be important. In contrast to Shen et al. [56] and Zhang et al. [17], no repression of the ACV synthetase and expandase by high ammonia concentrations was found with the strain used for the present investigations. The extracellular amino acids such as leucine, alanine and arginine were quickly consumed and at 50 h their limitation occurs. As a consequence, the extracellular protease activity increased at this time and the soluble protein was decomposed. In spite of this, these amino acids could not be detected in the medium because they were consumed as fast as they were formed.

In Table 7 the concentrations of (dry) biomass and β-lactam compounds and the selectivity of CPC with regard to the sum of β-lactam compounds are compiled for the runs investigated.

The highest CPC concentration was obtained with ammonia feeding but the highest selectivity was attained with the standard cultivation. The CPC amount produced with regard to the amount of (dry) biomass formed is also high (25.1%) for the standard cultivation. A higher CPC share was only observed with cysteine feeding into the methionine free medium (25.5%), however, the CPC concentration was very low (7.9 g l^{-1}) in this cultivation.

Reference [50] reported on cultivations with the same strain on a complex medium and with glucose batch and fed-batch operations as well as with valine, alanine and serine supplement, respectively. Figure 10b shows the CPC production rates and Fig. 10c the CPC yield coefficients with regard to the (dry) biomass produced for a definite cultivation time.

The highest production rates were obtained with valine and alanine supplements and a 50% standard medium and the second highest with serine, valine and alanine supplements with a 50% standard medium. The lowest production rate was obtained with a synthetic medium and the second lowest with a complex medium, but with a glucose batch operation.

The highest CPC yield coefficient with regard to the (dry) biomass was obtained with serine, valine and alanine supplements and a 50% standard medium and the second highest with valine and alanine supplements and a 75% standard medium.

Table 7 Concentrations of BTM (dry biomass), CPC, DAC, DAOC, PENN and the selectivity of CPC CPC/Σ β-lactam [%] obtained by the investigated runs

Runs	Concentrations [g l^{-1}]					CPC Σ β-lactam
	BTM (dry biomass)	CPC	DAC	DAOC	PENN	
Standard medium	97	24.4	1.94	0.40	3.25	81
Ammonia feeding	108	25.1	5.5	0.4	6	67
Urea feeding	66	22,5	3.8	0.6	6.5	67
Urea feeding 50% medium	65	12.4	3.3	0.9	12	43
Asparagine, Arginine, Ammonia feeding	84	21.5	3.9	0.48	8	63
Methionine, Arginine, Asparagine, Feeding 50% medium	48	8.3	1.25	0.35	7	49
Methionine free, $(NH_4)_2SO4$, Na_2SO_4, $MgSO_4$	70	9	1.9	0.16	0.74	76
Ammonia sulphate feeding	96	15.9	3.2	0.15	2.4	74
Cysteine feeding	56	14.3	1.9	0.5	5.4	65
Methionine free, Cysteine feeding 50% medium	68	7.9	1.6	0.47	6.2	49

4
Cultivation in High and Low Density Media

High density cultivation is popular today. How dense should a medium be? The advantages of high density media are high biomass and product concentrations. Disadvantages are the high medium viscosity and the increased oxygen requirement and power input for operation of the reactor. Because the energy cost amounts to about 20% of the product formation costs, high power input can easily increase the production costs. At high medium concentration the yield coefficients are usually low therefore, the share of raw material costs increase as well. In addition the oxygen supplement increases the production costs.

With low density cultivation, the biomass and product concentrations are low but yields coefficients are high and oxygen supply of aerobic micro-organisms is improved and the heterogeneity of the medium in the reactor is reduced by improved mixing. These favourable effects are especially important in large industrial reactors. The enrichment and dilution of the medium can influence the process performance in another way and this was investigated and reported in Ref. [57]. In following, the results of these investigations are presented. These were carried out with the same strain and medium and in the same reactor as was used in the preceding investigations. Six cultivations were performed with different densities.

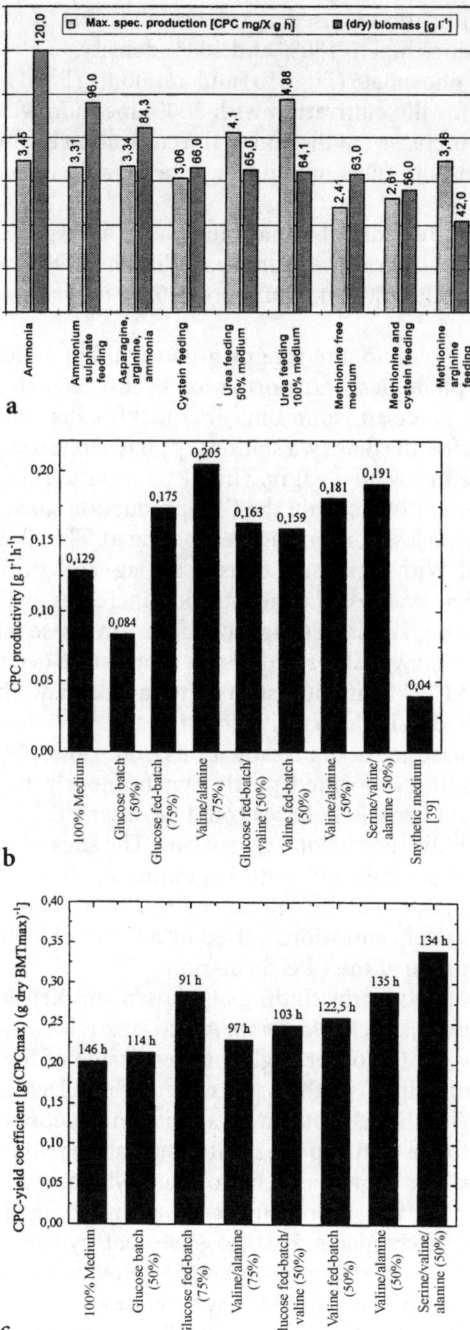

Fig. 10 **a** Comparison of the (dry) biomass concentrations and maximum specific production rates of CPC. **b** Comparison of CPC productivities [50]. **c** Comparison of CPC yield coefficients with regard to the (dry) biomass formed [50]

The standard cultivation is used as reference (100%). It was diluted to 30%, 50% and 75% and enriched to 150% and 200% density.

Sugar (Fig. 11a), phosphate (Fig. 11b) and ammonia (Fig. 11c) were consumed in parallel, except for the cultivation with 200% medium where the phosphate concentration was increased at the end of the cultivation by the lyses of the cells, and the cultivation with 50% medium, where urea was used as the nitrogen source.

The concentration of (dry) biomass increased as expected (Fig. 11d). The courses of CPC concentrations were unusual. With the 150% medium, 30 g l^{-1} was CPC produced. With the 200% medium only 15.9 g l^{-1} was produced because of oxygen limitation.

In a low density medium, the fungus grows without a lag phase und much faster and the morphological transformation occurs later than in highly dense media. Sugars, phosphate and ammonia are quickly exhausted, the first growth phase is short and after the diauxy, a stationary phase appears, before the biomass increases again due to soy oil feeding. The CPC formation starts immediately at the beginning of the cultivation, but the CPC production stops as early as 60 h after the start and only a low concentration is obtained. The CPC and biomass formation run parallel. With increasing density the lag- and growth–phases are extended. After the diauxy, the cell concentration increases again with the soy oil feeding. However at high densities the second growth phase is less significant. In a 200% medium, the growth stops and lyses a sporulation occurs because of oxygen limitation. The CPC formation starts after a delay and the maximum production rate is shifted to higher cultivation times.

With the strain used in these investigations, some unusual phenomena were observed. The product is formed partly synchronously to growth. No clear growth and product formation phases could be observed.

The carbon catabolic repression was missing. The sugar concentration varied by a factor of six without influencing the beginning and the first phase of the CPC production. The repression of the CPC formation by ammonia is not significant. The initial ammonia concentrations varied by a factor of about ten, without influencing the first phase of the CPC formation.

This can be explained by the finding of Jekosch and Krück [58, 59]. They investigated the *Acremonium chrysogenum* A3/2(C3) strain, which was used in the present work. They did not observe glucose repression of the *pcbC* gene, which is coded for Isopenicillin N synthetase, as Bremer and Demain did [60] with a wild-type strain. With the A3/2 strain no significant glucose regulation was observed in contrast to the wild-type. A similar regulation pattern was observed in the *cefEF* gene coded for Expandase/Hydroxylase, which was not regulated in the A3/2(C3) strain. The strain improvement led to mutations that suspended the glucose repression mechanisms. It is possible that by the mutation not only changed the regulations of the *pcbC* and *cefEF* genes, but also the regulation of the *pcbAB* gene, which is coded for ACV synthetase.

In Fig. 11e and Fig. 11f the maximum CPC concentrations and CPC productivities are shown. The highest CPC concentrations and productivities were obtained in the 150% medium. However, the highest yield coefficient was attained in the 50% medium (Fig. 11g). It depends on the cost structure of

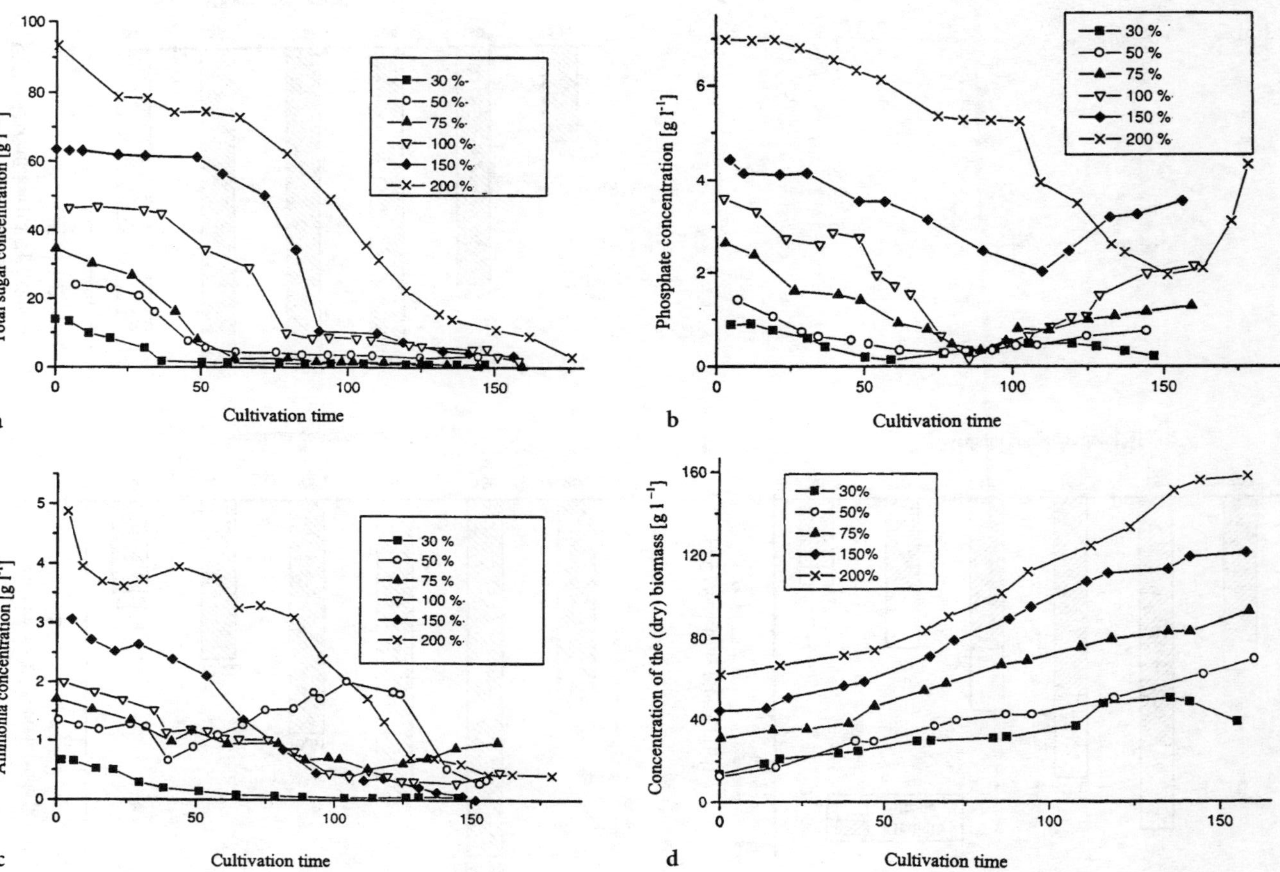

Fig. 11a–g Comparison of cultivations with different dilutions and enrichments. **a** Concentration of total sugar. **b** Concentration of phosphate **c** Concentration of ammonia. (Reprinted from [57] with permission from Elsevier Science) *(continued)*

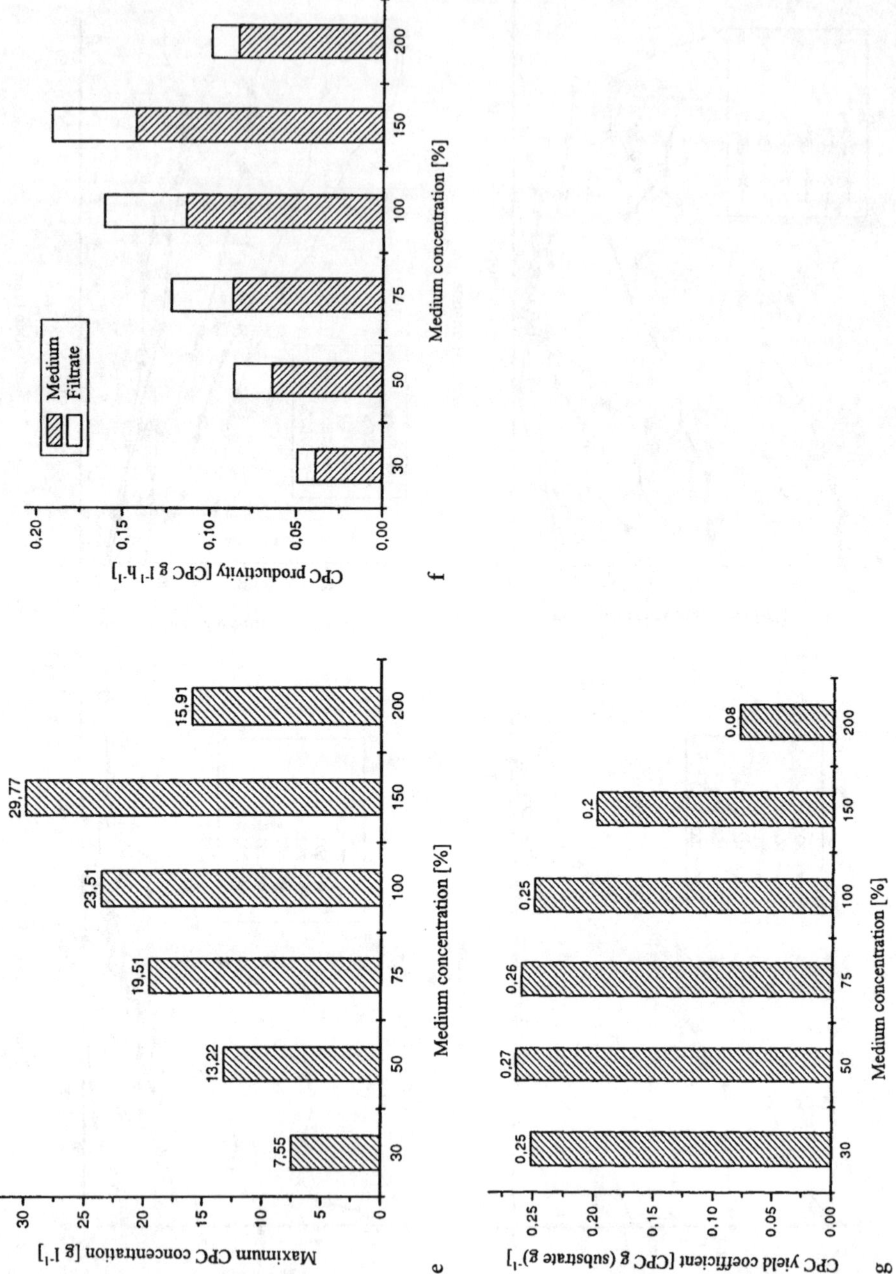

Fig. 11 (continued) **d** Concentration of (dry) biomass. **e** Maximum CPC concentration. **f** CPC productivity. **g** CPC yield

the production, whether the productivity or the yield coefficient should be optimised.

5
Precursor-Amino Acids

α-Amino adipate must have been formed by the fungus from α-keto glutarate, because it was not present in the initial cultivation medium. Its extracellular concentration gradually increased with time and reached $1.2\,\mathrm{g\,l^{-1}}$ at the end of the cultivation. The increase in the extracellular lysine concentration with time to $0.75\,\mathrm{g\,l^{-1}}$ showed that a part of the α-aminoadipate was decomposed to lysine. This indicated that the synthesis rate of the peptide AC/ACV was lower than the formation rate of α-aminoadipate. The feedback inhibition of α-aminoadipate by lysine did not cause a bottleneck in the ACV synthesis because the extracellular concentrations of α-aminoadipate and lysine continual increased. Cysteine must have been synthesised from methionine or from sulphate by the fungus because it was only in very low concentration in the initial standard medium. No extracellular cysteine uptake was observed, not even during cysteine feeding. The fast uptake of methionine indicates that the reverse transsulphuration was the preferred mechanism for the synthesis of cysteine. The cysteine synthesis by sulphate reduction was only used in absence of methionine because of its high metabolic expenditure. The enrichment of cysteine indicates a sufficient sulphur supply and it is an expression of a possible strong regulation of the intracellular cysteine pool.

Valine is an ingredient of CSL and was present in sufficient concentration $(0.5–1.0\,\mathrm{g\,l^{-1}})$ in the medium. Therefore, its biosynthesis was not necessary. The valine concentration quickly decreased from $0.5\,\mathrm{g\,l^{-1}}$ to $50\,\mathrm{mg\,l^{-1}}$ within 40 h in a methionine-free medium and within 80–90 h in a methionine-containing medium because of its high uptake rate. This difference in uptake rates indicated the inhibition of the valine permease by methionine [35].

The intracellular valine concentration was nearly constant during its uptake from the medium and increased after the valine was exhausted in the medium and therefore the biosynthesis of CPC was not limited by the lack of valine. The high intracellular valine pool indicated that it had not inhibited or suppressed the β-lactam production.

Banko et al. [74] evaluated the K_M-values of ACVS for the substrates α-aminoadipate (0.17 mM), cysteine (0.026 mM) and valine (0.34 mM). The intracellular concentrations of α-aminoadipate, cysteine and valine, which were determined by Beyer as a function of the cultivation time for the strain A3/2(C3) [75], exceeded their K_M-values. Thus theirs were high enough to saturate ACVS.

6
Morphology

Acremonium chrysogenum undergoes a morphological change during the product formation (as already discussed in the introduction). Hyphae with thin walls are formed after germination of spores of the strain A3/2(C3) and constituted a filamentous mycelium. Depending on the cultivation conditions swollen hyphae were formed and these were transformed into yeast-like, single-cell structures. These arthrospores had active metabolisms and were able to propagate as well. We were able to follow the morphology change by staining the fungus with Acridine orange. Other investigations were carried out with the DIC-method. Shortly after inoculation the fungus formed swollen hyphae. After 22 h, the increase of the biomass was accompanied by the formation of swollen hyphae and arthrospores (Fig. 12a, b). The right-hand figure shows a cell with a mean diameter of 14 µm, which exhibits a large vacuole of 7-µm diameter and several smaller ones. Most of the cells are filled with cytoplasm. After 100 h, the biomass

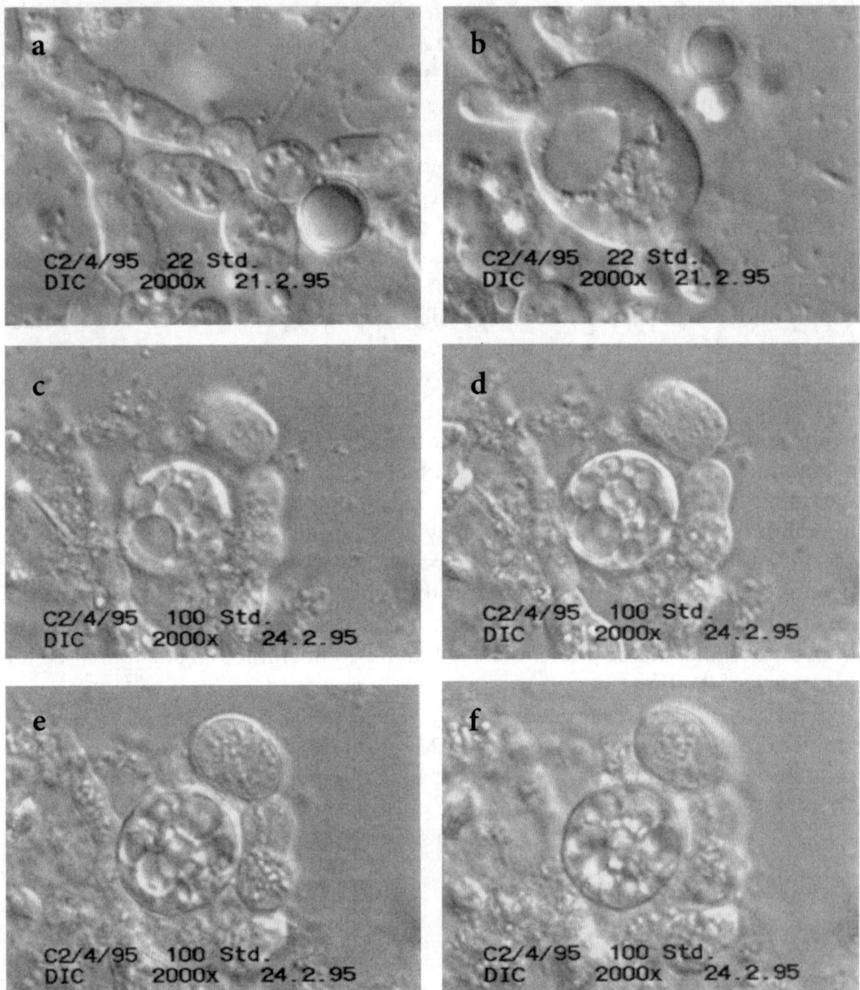

Fig. 12a–f Development of the fungus. **a, b** 22 h after inoculation. **c–f** 100 h after inoculation. (The vacuole formation was determined by DIC method) [52]

consisted mainly of arthrospores which were productive because the CPC formation rate had in this time reached its maximum. The arthrospores had a diameter of 10 µ diameter and hyphae with twelve vacuoles of diameters of 2–3 µm, which nearly filled the entire cells (Fig. 12b, d–f).

7
Mathematical Modeling and Control

7.1
Mathematical Models

The growth of *Acremonium chrysogenum* and the production of cephalosporin C were mathematically modelled first by Matsumura et al. [61]. They developed a structured-segregated model for the production of CPC and morphological differentiation, induction of CPC production by endogenous methionine and catabolite repression by glucose. Malmberg et al. [62, 63] developed a kinetic model describing the CPC biosynthesis by *Streptomyces clavuligerus* and the rate-limiting steps in the CPC biosynthesis pathway under conditions of constant dissolved oxygen concentration. The model includes six reaction steps according to Fig. 1. Cell growth was simulated according to a formal balance with fixed growth rate. Measured enzyme activities were used for the model simulations. Basak et al. [64] extended the model of Chu and Constantinides [65] for batch cultivation of *Acremonium chrysogenum* and CPC production for a fed-batch system. Cruz et al. [66] modelled CPC production and optimised the bioprocess in fed-batch cultivation with invert sugar as substrate.

A model was developed for the biosynthesis of CPC based on Fig. 1 and the in-vitro measured kinetic parameters of the enzymes of the biosynthesis by Rhee at al [67]. This model well describes the enzymatic biosynthesis of CPC.

An expert system for modelling and control design of biotechnological processes [68] was used for modelling the CPC production on (dry) sugar syrup (TGS) and soy oil [52]. The process is divided into a growth phase on sugar with parallel uptake of soy oil at a low CPC production rate and a second production phase on soy oil only with continuous feeding with a high CPC production rate. A cybernetic metabolic regulator model was developed [69, 70].

The initial lag-phase on sugar and the diauxic lag-phase for the adaptation from sugar to soy oil are modelled with metabolic regulators. A limitation term for sugar and soy oil was not considered because it was assumed that both of the concentrations stayed above the critical values.

Meyerhoff [71] developed a segregated model for cultivation on a complex medium. There was no clear separation of the growth phases on sugar and soy oil. In spite of that a diauxy was observed after sugar was exhausted. After the diauxy, soy oil was the sole carbon source.

Figure 13a shows the four-compartment model of Meyerhoff [72]. This model used the data which were evaluated with *Acremonium chrysogenum* A3/2(C3) on a complex medium. Therefore, it was the only model which could be applied for the cultivations discussed in this review. The experimental data and simulation results with this model agree well as shown in Fig. 13b. Therefore, this model was

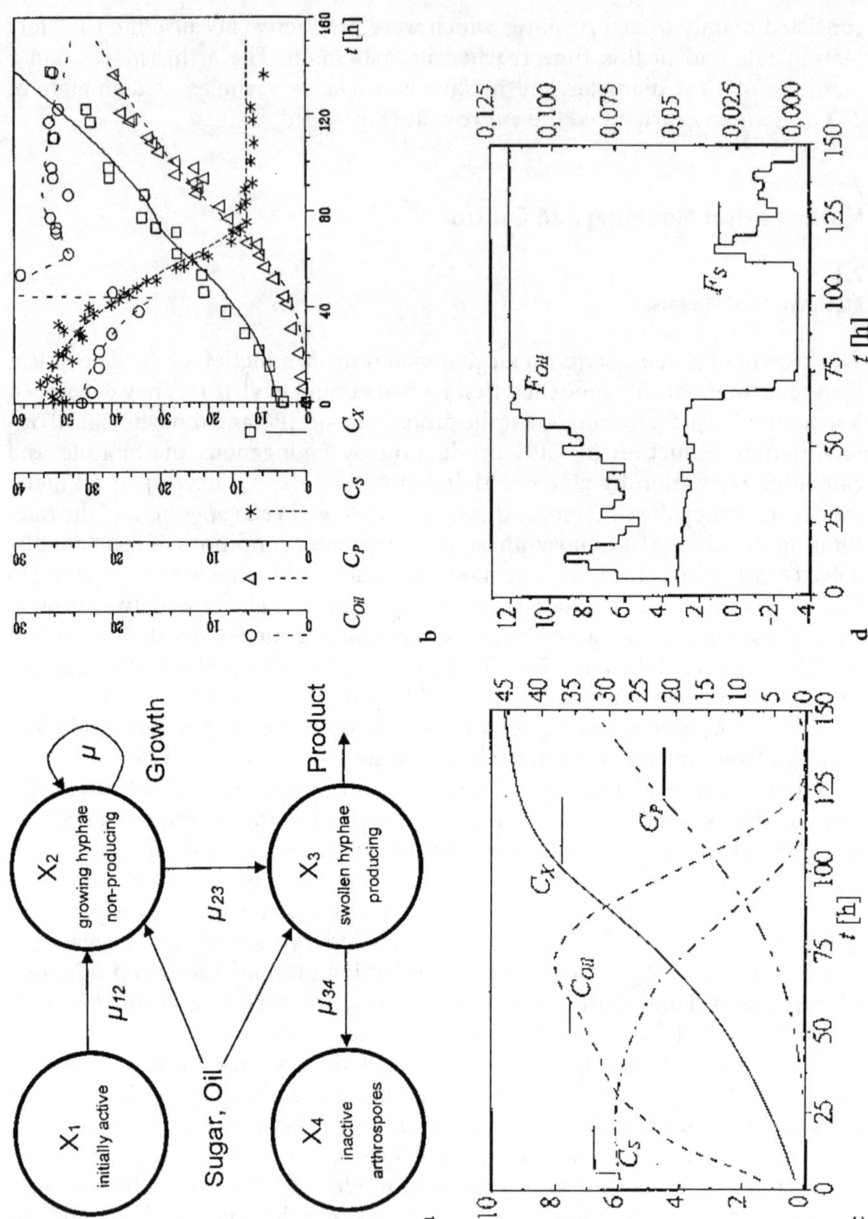

Fig. 13 **a** Structure of segregated model of Meyerhoff [71]. **b** Comparison of the measured (*symbols*) and calculated (*lines*) results for the fed-batch cultivation of *Acremonium chrysogenum*. C_X = concentration of (dry) biomass, C_S = concentration of total sugar, C_P = concentration of CPC, C_{oil} = concentration of soy oil. Concentrations in kg m^{-3}. **c** Simulation of the optimal control of sugar and soy-oil feeding for a fed batch cultivation of *Acremonium chrysogenum*. C_X = (dry) biomass, C_S = concentration of total sugar, C_P = concentration of CPC, C_{oil} = concentration of soy oil, F_S = flow rates of total sugar [l h^{-1}] and F_{oil} = flow rate of soy oil [g h^{-1}]. Concentrations in kg m^{-1}

used for developing a dynamic optimal control for the fed-batch cultivation of *Acremonium chrysogenum* and production of cephalosporin C [72, 73]. Figure 13c shows a simulation of the optimal control of sugar and soy oil feeding for a fed-batch process.

7.2
Process Control

The Meyerhoff model assumes that the substrate fed to the reactor is consumed immediately. It means that the figure does not show the feeding rates but the consumption rates of the substrates. To implement this model for the production of *Acremonium chrysogenum* and CPC production on a complex medium, it is necessary to determine the lag time between the feeding and the uptake of the substrate. Since (dry) glucose syrup and soy oil are used as carbon sources and these are not well-defined compounds, there will be a broad time-lag distribution depending on the composition of these substrates. The (dry) glucose syrup consist of glucose, maltose, trioses, oligosaccharides and polysaccharides. Glucose is immediately consumed. Maltose has to be decomposed by maltase to glucose. The oligo- and polysaccharides have to be decomposed by α-amylase to glucose before they can be consumed. With increasing degree of polymerisation the time-lag for their uptake increases. For optimal feeding, we need the time-lag distribution of the oligo/polysaccharide uptake. If we use (dry) glucose syrup in batch operation, we can observe several diauxies by means of CPR caused by the change from the consumption of free glucose to glucose from maltose, oligosaccharides and polysaccharides. Monitoring the concentrations of glucose, maltose and the difference between glucose/maltose and total sugar concentrations, which represents the oligo- and polysaccharides, by FIA-analysis and the diauxies by CPR measurement, it can be shown that these diauxies correlate with the disappearances of various ingredients of (dry) glucose syrup.

The consumption rate of (dry) glucose syrup depends on the production of necessary enzymes, maltase and α-amylase. Therefore, the activity of α-amylase, which is formed and excreted by the fungus into the cultivation medium, was monitored (Fig. 14a). No glucose can be detected in the medium as soon as the free glucose was consumed because the glucose uptake rate is higher than the glucose production rate, although oligo- and polysaccharides (measured as total sugar) are still present in the cultivation medium (Fig. 14b). According to the optimal control, the fungus requires sugar during the entire growth phase. This is the case, because it takes about 80–100 h to decompose the oligo- and polysaccharides to glucose. Thus, the fungus still has enough sugar for growth without feeding sugar to the reactor. Therefore, we can approach the glucose consumption course of the optimal control using (dry) sugar syrup in batch mode without feeding sugar to the reactor. This is not valid for other sugar sources as dextrin or starch.

The other main carbon source is soy oil. Soy oil is a mixture of unsaturated fatty acids with 18 to 22 carbon atoms. The soy oil has to be split into glycerol and unsaturated fatty acids by lipase before being consumed. The consumption of glycerol is fast, the uptake of fatty acids needs more time. The uptake rate of soy

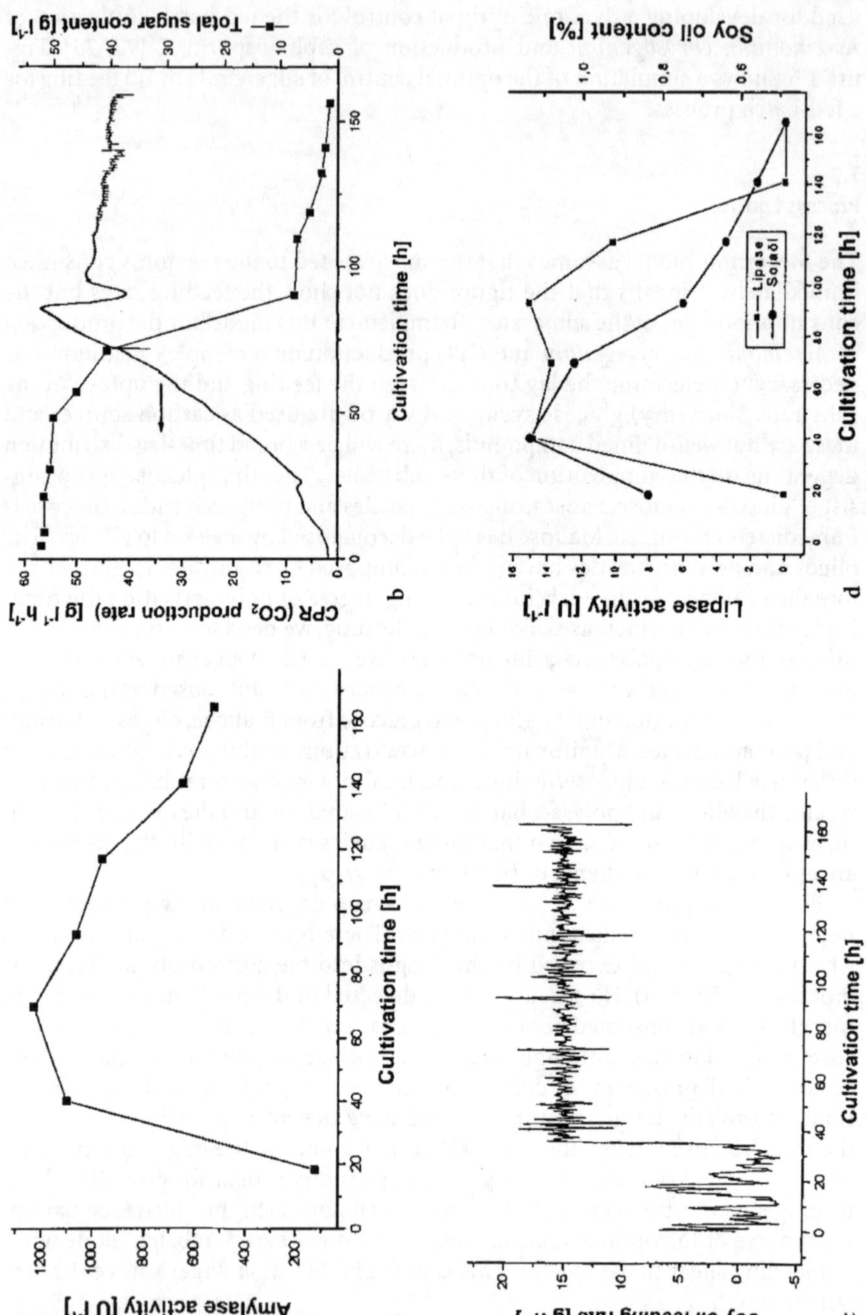

Fig. 14 **a** Amylase activity. **b** Concentration of total sugar and CPR. **c** High soy oil feeding rate during the growth phase. **d** Lipase activity and soy oil content. with high soy oil feeding rate during the growth phase.

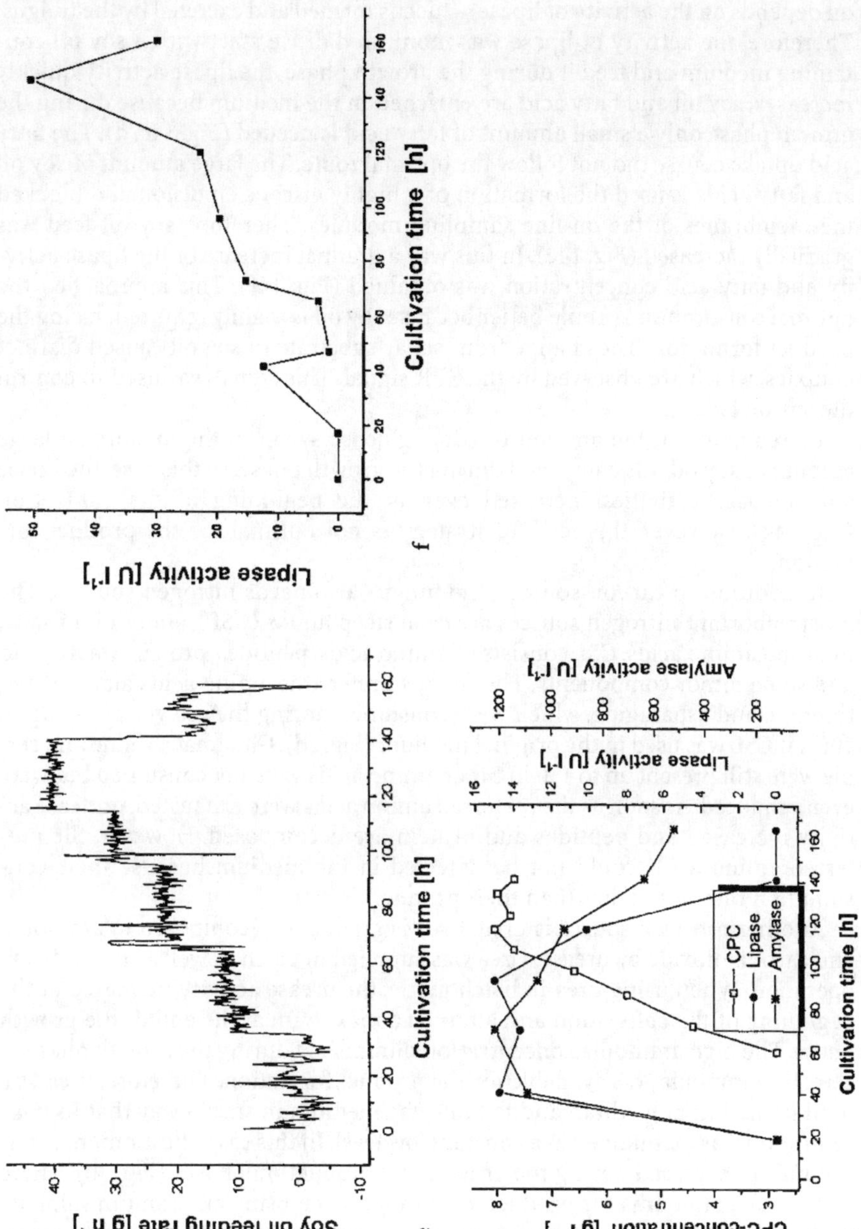

Fig. 14 e Gradually increased soy oil feeding with high feeding rate during the production phase. f Lipase activity with gradually increased soy oil feeding rate. g Activities of lipase and amylase and concentration of CPC

oil depends on the activity of lipase, which is formed and excreted by the fungus. Therefore, the activity of lipase was monitored. If we start with a soy oil containing medium and feed it during the growth phase, the lipase activity quickly increases, soy oil and fatty acid are enriched in the medium because during the growth phase only a small amount of fatty acid is needed (Fig. 14c, d). The fatty acid uptake course did not follow the optimal route. The large amount of soy oil and fatty acids caused the formation of a highly viscous emulsion and blocked the membranes of the on-line sampling modules. Therefore, soy oil feed was gradually increased (Fig. 14e). In this way a gradual increase of the lipase activity and fatty acid concentration was obtained (Fig. 14f). This approached the optimal consumption profile better because soy oil is mainly required during the product formation. The change from sugar substrate to soy oil caused distinct diauxies, which are observed by the CPR signal. This signal was used to control the soy oil feeding.

By reduction of the amount of (dry) glucose syrup in the medium, a large amount of soy oil was consumed during the growth phase. In this case, the lipase and amylase activities increased even at the beginning of the cultivation (Fig. 14g). However, this feeding strategy is not optimal for the product formation.

In addition to carbon sources, the fungus also needs nitrogen sources. The most important nitrogen sources are corn steep liquor (CSL), ammonium salts, urea and amino acids. CSL consists of amino acids, peptides, proteins, lactic acid and some minor components. The fungus prefers the amino acids alanine, methionine and asparagine, which were consumed during the growth phase up to 100 h if CSL was used in the original medium (Fig. 2d). Glutamate, valine and serine were still present up to 130 h. Other amino acids were not consumed but were even produced. As soon as the preferred amino acids were exhausted, protease activity increased and peptides and protein are decomposed. However, the preferred amino acids could not be detected in the medium because their consumption rate was higher than their production rate.

Another nitrogen source is urea. Urea is gradually decomposed to ammonia and carbon dioxide by urease. Urea was supplied in batch as well as in fed-batch operation. When using urea in batch mode, the urease activity increased at the beginning of the cultivation and attained a maximum at the end of the growth phase. The high ammonia concentration diminished during the growth phase to zero. No ammonia was available for the product formation. Therefore, urea was used in the initial medium and fed into the medium in such a way that its concentration was maintained at a constant low level. In this case, the ammonia concentration increased during the growth and production phases (Fig. 3b). There was enough ammonia for product formation. When using ammonium sulphate in the original medium and, as soon as the ammonia concentration decreased, ammonium sulphate was fed to the reactor in such a way that a high ammonia concentration was maintained during the production phase. A comparison of these three strategies indicated that the use of urea in batch mode gave the lowest productivity. Feeding urea and ammonium sulphate to the system yielded about the same productivity. However, in these cases, the preferred amino acids were consumed as early as during the growth phase, as can be seen in Fig. 5b, in which only the methionine concentration course is shown. Because methionine is not only a nitrogen and sulphur source, but also an inducer of the morphology change and the CPC formation, only a concentration of 8 g l^{-1} CPC was obtained at low methionine concentration during the production phase (Fig. 5b). Therefore, a mixture of preferred amino acids was fed to the medium in such way that the methionine concentration was maintained above 5 g l^{-1}. In this way, a 25 g l^{-1} CPC concentration was obtained. By varying the composition of the amino acids and using the proper amino acid composition for feeding, the concentration of CPC was increased to 30 g l^{-1}.

The optimal process control consists of using (dry) glucose syrup and CSL in batch mode, gradually increasing the fed-batch rate of soy oil, which is controlled by CPR, keeping the ammonia and methionine concentrations at a constant level by feeding urea and methionine to the medium. Methods of on-line monitoring of the concentrations of urea, ammonia and methionine are state of the art. On-line or in situ monitoring of soy oil and unsaturated fatty acid concentrations is not yet at our disposal. It still has to be developed probably by means of Fourier Transfer-Near Infrared (FT-NIR)-spectrometry.

A significant improvement of the productivity and yield coefficient of CPC could be obtained by the change of the regulation of the enzymes of CPC synthesis by mutation or genetic engineering [58, 59].

7.3
By-Products

During the cultivation, the concentration increase of CPC diminishes, because of the back reaction:

$$CPC + H_2O \rightarrow DAC + CH_3COOH \tag{a}$$

It is a first-order reaction with the rate constant k:

$$\frac{d[DAC]}{dt} = k[CPC] \tag{b}$$

The rate constant k [h^{-1}] was determined for several cultivations at 60 h, 80 h, 100 h, 120 h and 140 h. Following average values were evaluated: 2.9 h^{-1} for 60 h, 2.8 h^{-1} for 80 h, 2.9 h^{-1} for 100 h, 3.0 h^{-1} for 120 h and 3.25 h^{-1} for 140 h. Figure 15a shows the measured and calculated DAC concentrations for standard cultivation, with urea and cysteine feeding, respectively. Figure 15a indicates that the back reaction was active during the entire cultivation.

To prove if the back reaction is a chemical or enzymatic hydrolysis, CPC decomposition was determined in the absence and presence of the growing fungus. The decomposition was only slightly influenced by the presence of the fungus. Therefore, the back reaction is essentially an abiotic hydrolysis.

During the cultivation, the concentration of 2-(D-4-amino-4-carboxybutyl)-thiazol-4-carbonic acid (ACTCS), a decomposition product of CPC, is increased. The formation of ACTCS was investigated in the absence and presence of growing fungus and no significant difference was found. The ACTCS formation rate

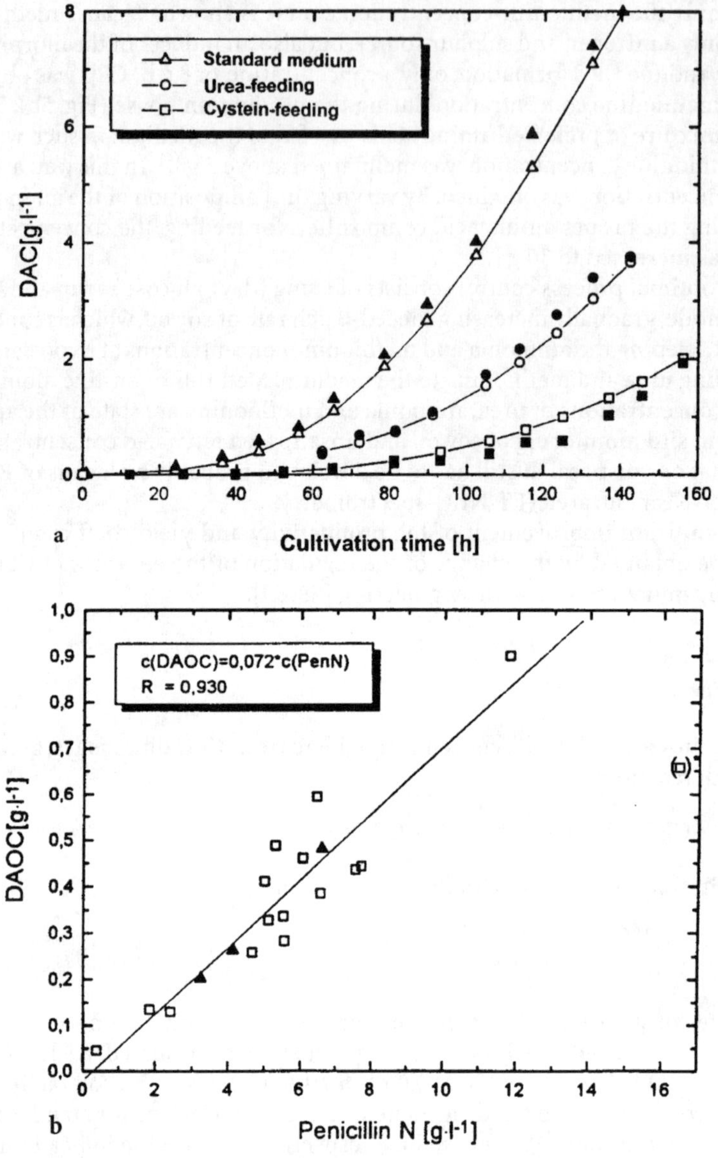

Fig. 15 a Measured (full symbols) and calculated (open symbols) DAC concentrations as a function of the cultivation time for standard cultivation and cultivations with urea feeding and cysteine feeding. **b** Relationship between the concentrations of PENN and DAOC

was in the range 20–30 mg l^{-1} h^{-1} for fungus free systems and for standard cultivation as well as for cultivations with urea and arginine asparagine. Therefore, this is an abiotic decomposition reaction.

The recovery and purification of CPC is impaired by the presence of DAOC, because the final product, 7-aminocephalopanic acid (7-ACS), is contaminated by the homologous 7-aminodesacetoxycephalospoanic acid (7-ADCS). Since the complete separation of CPC and DAOC is at the present not possible, it is important to minimise the concentration of DAOC during the production. According to the investigations of Tollnick [52], there is a close relationship between the concentrations of PENN and DAOC (Fig. 15b), which was evaluated at the time when the maximum CPC concentration was obtained. The best way to reduce the DAOC concentration is to avoid oxygen limitation, which enhances the PENN production [31]. In large industrial reactors, no complete mixing is maintained to avoid high energy costs due to a high power input through the stirrer and a high cooling capacity due to the increased heat evolution. Because of the heterogeneity of the medium, local oxygen limitations prevail and these enhance PENN formation. To understand the influence of the heterogeneity on the process performance, a combination of an advanced reactor model with a cybernetic model of metabolic regulation would be necessary. The application of Computational Fluid Dynamics (CFD) to stirred tank bioreactors could solve this problem. However, the solution of the Navier-Stokes Differential Equation (NS-DE) system for aerated stirred tank bioreactors by Euler-Langrange approach (the liquid phase is homogeneous and the gas phase is discrete) combined with the (k-ε) model for including the turbulence effects is not yet possible [76]. The solution of the NS-DE system by Euler-Lagrange is only possible at the present for bubble column bioreactors [77]. The combination of advanced reactor models with advanced metabolic regulation models is also not possible yet. It will take several years to solve this problem.

7.4
Economical Aspects

The most important pathogenic micro-organisms do not exhibit resistance to semi-synthetic cephalosporins yet, in contrast to semi-synthetic penicillins. Therefore, the demand for them is high. The sale of cephalosporin antibiotics amounted to 7 billion dollars last year [81].

8
Conclusions

Easily consumable substrates such as glucose, ammonia and urea enhanced primary metabolisms, thus increasing the growth rate. Using glucose feeding, the duration of the cultivation could be significantly shortened. However, it was not possible to influence the final CPC concentration. The catabolic repression by sugars, which was found for several other strains, was not observed for the strain A3/2 (C3). This can be explained by a mutation that suspended the glucose repression mechanisms [58, 59].

The uptake of ammonia was impeded below 0.5 g l^{-1} in the presence of amino acids. Instead of ammonia, alanine, asparagine and arginine were consumed successively. The supplementation of the medium with alanine, valine and serine enhanced the primary and the secondary metabolisms. These amino acids reduced the amount of by-products and increased the yield of CPC by 50%.

The secondary metabolisms were not influenced by ammonia limitation. The ammonia concentration did not influence the concentration of by-products either. The reduction in product formation at high ammonia concentrations was noticed by Shen et al. [16] and Zhang et al. [17] for the strain C-10, but was not observed for the strain A3/2 (C3).

The fungus had a high requirement for anabolically valuable amino acids such as alanine, phenylalanine, valine and leucine. Their exhaustion set free protease production.

The use of various sulphur sources denoted that the strain A3/2 (C3) prefers methionine.

With ammonium sulphate. a high productivity was obtained comparable to that with methionine, but this could only be maintained for a short time. Therefore the yields were 30% lower.

Cysteine feeding was detrimental to growth and product formation.

The intracellular concentrations of the substrate amino acids of ACV synthesis based on the K_M-values of this enzyme are high enough that, because of substrate limitation, no bottleneck can be expected. No extracellular ACV was observed. This indicate that the ACV conversion rate is as high as its formation rate. This is supported by the intracellular activity measurements of ACVS and Cyclase. The low activity of ACVS is the major bottleneck of the CPC synthesis. The Cyclase activity is much higher than that of ACVS. Only during the cysteine feeding was an inhibition of Cyclase observed in agreement with Jensen et al. [55]. This caused an extracellular accumulation of ACV.

Expandase exhibits a well-known oxygen dependence. Accumulation of PENN is especially high with oxygen limitation. The improvement of the oxygen supply to the fungus by dilution of the medium or by an increase in the dissolved oxygen concentration in the medium by oxygen supplementation or by increased pressure in the reactor can improve the CPC yield. An over-expression of the enzymes Expandase/Hydroxylase could also improve the CPC-yield. This would reduce the concentration of by-products PENN and DAOC, which are closely connected.

The activity of Acyltransferase was high enough that no rate limitation of the biosynthesis occurred. The extracellular accumulation of DAC was caused by extracellular decomposition of CPC. The DAC accumulation could be reduced by shortening the cultivation time.

The extra- and intracellular data collected were adequate for the development of a segregated model [71], which used to generate an optimal control strategy for the feeding of (dry) glucose syrup and soy oil [72, 73]. However, for the implementation of this control strategy for cultivations with complex substrates, it is necessary to take the time lag of the uptake of different substrate components into account. Because most of the complex medium components have to be enzymatically decomposed before they can be consumed by the fungus, they determine the uptake rate of the complex substrates. Therefore, the extracellular enzyme activities of amylase, lipase, urease and protease were monitored to evaluate the time lag of the substrate uptake. For example, the uptake of (dry) glucose syrup, which is decomposed to glucose by amylase has a very broad time lag distribution between some minutes (for glucose) to 100 h (for oligo and polysaccharides).

The distinct morphology change of the fungus during the cultivation is well known. The strong vacuolisation of the hyphae and arthrospores was unexpected. To gain more information, digital image analysis of the hyphae and arthrospores would be necessary [78].

The localization of the mRNA and the protein synthesis in the hyphae and arthrospores would help us to understand the physiology of the hyphae during the swelling and formation of arthrospores [79]. The localization of the replicating DNA could show which parts of the hyphae and arthrospores propagate [79]. The localization of the intracellular enzymes by gold immuno-assay and transmission electron microscopy (TEM) [80] would help to clear up in which parts of the swollen hyphae and arthrospores the successive steps of the biosynthesis of CPC occurs.

Acknowledgement The authors appreciate the donation of the *Acremonium chrysogenum* A3/2(C3) strain by HMR Deutschland and would like to thank Dr. B. König BC Biochemie GmbH Frankfurt, formerly with HMR Germany, Frankfurt for his support and Ulf Bäckermann, Michael Jürgens, Sylvia Feil and Yasmin Fahimi for their assistance.

9
References

1. Brotzu G (1948) Lavori dellístituto D'Igiene di Caligiari, p 1
2. Abraham EP, Newton GGF, Hale CW (1954) Biochem J 58:94
3. Newton GGF, Abraham EP (1954) Biochem J 58:103
4. Abraham EP, Newton GGF Biochem J 79:377
5. Howart TT, Brown AG, King TL (1976) J Chem Soc Chem Commun 266
6. Nagarajan RL, Boeck LD, Gorman M, Hamill RL, Higgens CE, Hoehn MM, Stork WM, Whitney JG, (1971) J Am Chem Soc 93:2308
7. Stapley EO, Jackson OM, Hernandez S, Zimmermann SB, Currie SA, Mochales S, Mahta JM, Woodruff HB, Hendin D (1972) Antimicrob Agents Chemother 2:122
8. Mora J, Davila G, Espin G, Gonzalez A, Guzma J, Hernandez G, Hummelt G, Lara M, Martinez E, Mora Y, Romero D (1980) In: Mora J Palacios R (eds) Glutamine : Metabolism, Enzymology and Regulation. Academic Press p 185
9. Nash CH, Huber FM (1971) Appl Microbiol 22:6
10. Queener SW, Ellis LF (1975) Can J Microbiol 21:1981
11. Matsumura M, Imanaka T, Yoshida T, Taguchi H (1980) J Ferment Technol 58:197
12. Drew SW, Winstanley DJ, Demain AL (1976) Appl Environ Microbiol 31:143
13. Mehta RJ, Speth JL, Nash CH (1979) European J Appl Microbiol 8:117
14. Drew SW, Demain AL (1977) Ann Rew Microbiol 31:343
15. Bartohevich YuE, Shuvalova IA (1082) Antibiotiki (Moscow) 28:8
16. Shen YQ, Heim J, Solomon NA, Wolfe S, Demain AL (1984) J Antibiot 37:503
17. Zhang J, Wolfe S, Demain AL (1987) J Antibiot 40:1746
18. Zhang J. Wolfe S, Demain AL (1988) Appl Microbiol Biotechnol 29:242
19. Zhang J, Banko G, Wolfe S, Demain AL (1987) J Ind Microbiol 2:251

20. Zhang J, Demain (1991) Biotech Adv 9:623
21. Lopez Nieto MJ, Ramos FR, Luengo JM, Martin JF (1985) Appl Microbiol Biotechnol 22:343
22. Bainbridge ZA, Scott RI, Perry D (1992) J Chem Technol Biotechnol 55:233
23. Castro JM, Liras P, Laiz L, Cortez J (1988) J Gen Microbiol 134:133
24. Martin JF, Liras P (1989) Adv Biochem Eng/Biotechnol 39:153
25. Jayatilake GS, Huddleston JA, Abraham EP (1981) Biochem J 194:645
26. Baldwin JE, Adlington RM, Schoenfield CJ, Sobey WJ, Wood ME (1989) J Chem Soc Chem Commun 15:1012
27. Shen YQ, Wolfe S, Demain AL (1984) Enzyme Microb Technol 6:402
28. Shen YQ, Wolfe S, Demain AL (1986) Bio/Technol 4:61
29. Lucas L (1993) The biosynthesis of cephalosporin C by a sulphate-dependent mutant of *Cephalosporium acremonium*. Dissertation, University of Hannover
30. Scheidegger A, Küenzi MT, Fiechter A, Nüesch J (1988) J Biotechnol 7:131
31. Zhou W, Holzhauer-Rieger K, Dors M, Schügerl K (1992) Enzyme Microb Technol 14:848
32. Fujisawa Y, Kanzaki T (1975) Agric Biol Chem 39:2043
33. Felix HR, Nüesch J, Wehrli W (1980) FEMS Microbiol Lett 8:55
34. Smith A (1985) In: Moo Young M (ed) Comprehensive Biotechnology 3 (S Drew, DIC Wang (eds), p 163
35. Aloso MJ, Luengo JM (1987) Antimicrob Agents Chemother 31:357
36. Queener SW, Wilkerson S, Tunin D, McDermott JP, Chapman JL, Nash C, Platt C, Westpheling J (1984) In: Vandamme EJ (ed) Drugs and pharmaceutical sciences, vol 22. Marcel Dekker, New York
37. Leyh TS (1993) Crit Rev Mol Biol 28:515
38. Martin JF (1998) Appl Microbiol Biotechnol 50:1
39. Küenzi MT (1980) Arch Microbiol 128:78
40. Herold Th (1985) Cephalosporin Production im Rührkessel. Dissertation. University Hannover
41. Fujisawa Y, Shirafuji H, Kida M, Nara K, Yoneda M, Kanzaki T (1975) Agr Biol Chem 39:1295
42. Adlington RM, Baldwin JE, Lopez-Nieto M, Murphy JA, Patel N (1983) Biochem J 213:573
43. Matsumura M, Imanaka T, Yoshida T, Taguchi H (1980) J Ferment Technol 58:205
44. Sawada Y, Baldwin JE, Singh PD, Solomon NA, Demain AL (1980) Antimicrob Agents Chemother 18:465
45. Demain AL, Newkirk JF (1962) Appl Microbiol 10:321
46. Ott JL, Godzeski CW, Farran JD, Horton DR (1962) Appl Microbiol 10:515
47. Feil S (1996) Mediumeinflüsse bei der Kultivierung von *Acremonium chrysogenum*. Masters' Thesis. University Hannover.
48. Telesrina GN, Bartoshevich YuE, Krakhmaleva IN, Petryushenko RM, Sazykin YuO, Navashim SM (1990) J Basic Microbiol 30:273
49. Jürgens M, Seidel G, Schügerl K (2002) Process Biochemistry 38:263
50. Seidel G, Tollnick C, Beyer M, Fahimi Y, Schügerl K (2002) Process engineering aspects of the production of cephalosporin C by *Acremonium chrysogenum* Part I. Process Biochemistry 38:229
51. Seidel G, Tollnick C, Beyer M, Schügerl K (2000) Adv Biochem Eng/Biotechnol 76:115
52. Tollnick C (1996) Kultivierungen von *Acremonium chrysogenum* in Forschung und Produktion. Medienstrategien zur Auffindung von Engpässen in der Biosynthese des Antibiotikums Cephalosporin C. Dissertation. University of Hannover
53. Seidel G (1999) Kultivierungen mit einem Hochleistungsstamm von *Acremonium chrysogenum* in komplexen und synthetischen Medien. Strategien zur Produktivitätssteigerung unter Berücksichtigung der Enzymaktivitäten der Cephalosporin c-Biosynthese. Dissertation, University of Hannover
54. Bäckermann U (1996) Untersuchung extrazellulärer Enzymaktivitäten an Kultivierungen von *Acremonium chrysogenum*. Masters' Thesis. University of Hannover
55. Jensen SE, Westlake DWS, Wolfe S (1988) FEMS Microbiol letters 49:213
56. Shen YQ, Heim J, Solomon NA, Wolfe S, Demain AL (1984) J Antibiot 37:503
57. Seidel G, Tollnick C, Beyer M, Schügerl K (2002) Process Biochemistry Part II. 38:241

58. Jekosch K, Kück U (2000) Curr Genet 37:388
59. Jekosch K, Kück U (2000) Appl Microbiol Biotechnol 54:556
60. Bremer CJ, Demain AL (1983) Curr Microbiol 8:107
61. Matsumura M, Imanaka T, Yoshida T, Taguchi H (1981) J of Ferment Technol 59:115
62. Malmberg LH, Hu WS (1991) Biotechn Bioeng 38:941
63. Malmberg LH, Sherman DH, Hu WS (1992) Ann New York Acad Sci 665:16
64. Basak S, Velayudhan A, Ladisch MR (1995) Biotechnol Progr 11:626
65. Chu WB, Constantinides A (1988) Biotechnol Bioeng 32:277
66. Cruz AJG, Silva AS, Araujo MLGC, Giordano RC, Hokka CO (1999) Chem Eng Sci 54:3137
67. Rhee JI, Seidel G, Schügerl K (2002) Chem Ing Techn 74:1171
68. Gomersall R, Grumann S, Ludewig D, Seeger M, Hitzmann B, Bellgardt KH, Munack A (1998) Automatisierungstechnik 46:368
69. Ludewig D (1999) Expoertensystem zur Entwicklung von Prozessmodellen für biotechnische Prozesse. Fortschritt-Berichte 20 Nr 289. VDI Verlag, Düsseldorf
70. Ludewig D, Bellgardt KH, (1997) BioTech 4:31
71. Meyerhoff J (1995) PhD Thesis, University of Hannover
72. Bellgardt KH (2000) In: Schügerl K, Bellgardt KH (eds) Bioreaction Engineering. Modeling and Control, Springer Verlag, p 391
73. Bellgardt KH (1998) Adv Biochem Eng Biotechnol 60:153
74. Banko G, Demain AL, Wolfe S (1987) J Am Chem Soc 109:2858
75. Beyer M (1976) Untersuchung des Engpasses der Cephalosporin C-Biosynthese in Hochleistungsstämmen von *Acremonium chrysogenum*. Dissertation. University of Hannover
76. Reuss M, Schmalzried S, Jenne M (2000) In: Schügerl K, Bellgardt KH (eds) Bioreaction Engineering. Modeling and Control. Application of Computational Fluiddynamics (CFD) to modeling Stirred Tank Bioreactors. Springer, Berlin Heidelberg New York, p 207
77. Lübbert A (2000) In: Schügerl K, Bellgardt KH (eds) Bioreaction Engineering. Modeling and Control. Bubble column bioreactors. Springer, Berlin Heidelberg New York, p 247
78. Paul GC, Thomas CR (1998) Adv in Biochem Engineering/Biotechnolgy 60:1
79. Freudenberg S, Fasold KI, Müller SR, Siedenberg D, Kretzmer G, Schügerl K, Giuseppin M (1996) J Biotechnol 46:265
80. Adolph S, Müller S, Siedenberg D, Jäger K, Lehmann H, Schügerl K, Giuseppin M (1996) J of Biotechnol 46:221
81. Schmidt FR (2002) In: Osiewacz HD (ed) Aspects of Manufacture and Therapy in the Myotica. Industrial Applications. Springer, Berlin Heidelberg New York, p 69

Received: June 2002

Adv Biochem Engin/Biotechnol (2004) 86: 47–82
DOI 10.1007/b12440

Plasmid Copy Number and Plasmid Stability

Karl Friehs

Universität Bielefeld AG, Fermentationstechnik, Technische Fakultät, 33501 Bielefeld,
Germany. *E-mail:* karl.friehs@uni-bielefeld.de

Abstract Many expression systems in research and industry use plasmids as vectors for the
production of recombinant proteins or non-proteinous recombinant substances. Plasmids
have an essential impact on productivity. Related factors are plasmid copy number, structural
plasmid stability and segregational plasmid stability. Plasmid copy number determines the
gene dosage accessible for expression and many plasmids lead generally to a high productivity. To analyze an expression system the quantification of plasmid copy number is very helpful. Therefore, different methods for the determination of plasmid copy number are described. Structural plasmid stability exists, when all generated plasmids have the correct base
sequence. The analysis of structural instabilities is not trivial and some methods are reported. When all daughter cells get at least one plasmid during cell division, the culture is segregational stable. The development of plasmid free cells can lead to a significant loss in productivity. Different methods for lab scale and industrial scale help to avoid segregational instability. Since plasmids are used as pharmaceuticals, additional aspects of stability have to be
taken into account. These include stability during downstream processing, stability after application and stability during storage and shipping.

Keywords Plasmid copy number · Structural plasmid stability · Segregational plasmid stability · Expression systems · Recombinant proteins · Gene therapy – vaccination

List of Abbreviations and Symbols

Abbreviations for some proteins may be found in small letters (gene names)

λ_{PL}	Promoter of phage Lambda
μ	Specific growth rate
^{3}H	Tritium
AGE	Agarose Gel Electrophoresis
AIDS	Acquired Immune Deficiency syndrome
asd	aspartate-semialdehyde dehydrogenase
Bla	β-Lactamase
bla	β-Lactamase
bp	base pair
BstEII	Restriction enzyme from *Bacillus stearothermophilus*
ccd	coupled cell division
cer	ColE1 replication

CGE	Capillary Gel Electrophoresis
cGMP	current Good Manufacturing Practice
cI857	Mutated repressor of phage Lambda
cop	Copy number
DnaA	DNA biosynthesis; initiation binding protein
DnaB	DNA biosynthesis; helicase
doc	Death on curing
EcoRI	Restriction enzyme of *E. coli*
EDTA	Ethylenediaminetetraacetic Acid
FIA	Flow Injection Analysis
FIP	Flow Injection Processing
fmol	femto-mol
FtsH	Filamentation temperature sensitive, a protease
GFP	Green Fluorescent Protein
HindIII	Restriction enzyme of *Haemophilus influenzae*
hok	host killing
HPLC	High Performance Liquid Chromatography
ILA	Immuno Ligand Assay
IPTG	Isopropylthiogalactoside
IS	Insertion Sequence
kbp	kilo base pair
kid	killing determinant
kis	killing suppressor
lacZ	β-Galactosidase
Lux	Luciferase
mur	murein
mvp	maintenance of virulence plasmid
OD	Optical Density
ori	origin of replication
par	partitioning
pas	plasmid addiction system
PCR	Polymerase Chain Reaction
pem	plasmid emergency maintenance
phd	prevent host death
pnd	Promoting nucleic acid degradation
pro	Proline
RepA	Replication, helicase
res	Resolution site
RNA I	RNA involved in replication control
RNA II	RNA involved in replication control
Rom	RNA one inhibitor modulator
SDS	Sodium Dodecyl Sulfate
sok	Suppressor of killing
sop	Stabilization of plasmid
SOS	Related to stress and repair systems
ssb	Single strand binding protein
tac	Hybrid of *trp* and *lac*

TE	Tris(ethylthio)methane-EDTA
tet	Tetracycline resistance
TGGE	Temperature Gradient Gel Electrophoresis
TN	Transposon
tnp	Transposition
trf	Transacting replication functions
trp	Tryptophan
umu	Ultra violet mutator
uvr	Ultra violet resistance
X-gal	5-Bromo-4-chloro-3-indolyl-beta-D-galactoside
YO-PRO	Quinolinium, 4-[(3-methyl-2(3*H*)-benzoxazolylidene)methyl]-1-[3-(trimethylammonio)propyl]- diiodide
YO-YO	Quinolinium, 1,1'-[1,3-propanediylbis[(dimethyliminio)-3,1-propanediyl]]bis[4-[(3-methyl-2(3*H*)-benzoxazolylidene)methyl]]-tetraiodide

1
Plasmids in Biotechnology

Much has been written about the discovery, the naming, the biological fundamentals and the "scientific career" of plasmids. Instead of writing more words on these topics a book should be recommended here, which allows a very good overview on all important subjects of plasmid biology including discovery, structure, classification, replication, horizontal gene transfer and the spreading of antibiotic resistance. The book was written by David K. Summers, has the clear title "The Biology of Plasmids" and was published by Blackwell Science Ltd 1996. Those, who are able to read German, can find a lot of valuable information in the book "Biologie Bakterieller Plasmide" by Wolfgang Schumann, published by Vieweg, Braunschweig 1990.

Natural plasmids and their derivatives are very helpful to understand for example DNA replication. Vectors based on plasmids are important tools for genetic engineering and one could say that genetic engineering would not have developed in it's amazing way without plasmids. Thousands of different plasmids constructed in laboratories all over the world, enable the scientific community to do research in life science and plasmids allow the production of interesting proteins in sufficient amounts.

However, not only life science research depends to a significant part on plasmids. In addition, biotechnology as an industrial line of production uses plasmids to produce high value proteins such as insulin or interferon and fine chemicals such as enzymes, to give just a few examples.

Since experiments in the fields of gene therapy and genetic vaccination using naked plasmid DNA, as therapeutic vectors were successful, plasmids themselves became a more and more important medical and economical product.

Success and failure in all these fields of plasmid application rely on some basic factors mainly plasmid copy number and plasmid stability. Plasmid copy number, which depends on the mode of replication resp. replication control, plays as gene dosage an important role in expression systems and of course determines the amount of plasmid as a product. Plasmid stability can be divided in structural and segregational stability. Structural stability describes the correct replication during cultivation of plasmid bearing cells leading to a non-altered base sequence of the plasmid. When plasmids are passed on to the daughter cells, so that each daughter cell has at least one copy, the system shows segregational plasmid stability. In medical applications other stabilities for plasmids may be of interest: the stability during downstream processing and purification, the pharmacokinetic stability in men or animals and the stability as a drug during storage and shipping.

This chapter discusses fundamental factors with emphasis on *E. coli* (not exclusively), but basic conclusions can be applied to all plasmid bearing expression and production systems.

2
Plasmid Copy Number

What is the meaning of plasmid copy number? The most used definitions are "copies per cell" [1] and "copies per chromosome" [2]. Additionally one can find in the literature "fmol plasmid DNA" or "mg plasmid DNA per mg cell dry weight" as indications of quantity [3, 4]. Sometimes comparisons with other plasmid quantities, for example relatively to the copy number of a wild type plasmid serve as specifications for the plasmid copy number [5].

Plasmids can be classified as low-copy, medium-copy or high-copy plasmids. The ranges of different classes fluctuate in the literature. For low-copy plasmids the range lies between 1 and 10, for medium-copy between 10 to 20 and high-copy plasmids can reach 700 copies and more [6]. Other classifications define 1–2 plasmids as low-copy, 6–9 as "moderate" and up to 100 as high copy [7]. Different copy numbers occur due to different mechanisms of replication or different genes at the origin of replication. The *ori*-sequence of the plasmid P15A leads to the low-copy type, whereas plasmids with the *ori* of pBR322 show a medium copy number. A single point mutation in the gene for RNA I of the replication control of pBR322 leads to the high-copy type [8]. Other mutations increase the copy-number [9] and some models describe the factors influencing copy numbers [10]. For a comparison of such models, see the article by Patnaik [11].

Normally the copy number is controlled by the regulation of replication. To reach the "normal copy number" after cell division, the replication has to be controlled depending on the actual copy number. That means, that the frequency of replication per plasmid corresponds with the total copy number in the cell [12].

2.1
Gene Dosage and Copy Number

The gene dosage describes the number of genes that are available for the expression of proteins. In recombinant expression systems with plasmids as vectors, the gene dosage depends on the number of the specific gene to be expressed on the plasmid and the plasmid copy number. A special case would occur if during replication multimeric concatenates are formed, which means that two or more copies of a plasmid would be on one circled molecule. Additionally genes can be integrated as single or multiple copies into the chromosome. Such integrations into the chromosome do not play a significant role in *E. coli* expression systems so far, but are important in other systems (e.g. yeast).

In most cases, only one copy of the specific gene is placed on a plasmid and therefore the gene dosage depends only on the plasmid copy number. For the production of recombinant proteins high-copy number plasmids are favored because normally a high gene dosage leads to high expression levels [13]. There can be some cases where an increase in the copy number does not lead to an increase in gene product [14]. This may be due to the limited transcription and translation capacity of cells [15]. The metabolic burden imposed by plasmid replication and recombinant gene expression may decrease the growth rate leading to a lower productivity [16].

Under special circumstances plasmids with a low copy number are chosen as vectors. Some mammalian genes show a higher stability in vectors with small copy numbers [17]. To express proteins, which have toxic effects on the host, low-copy plasmids are more suited [18]. The stabilization of a plasmid at a low copy number can improve recombinant protein production due to significant reduction of the metabolic burden [19].

2.2
Determination of Plasmid Copy Number

Various methods for the determination of plasmid copy numbers have been developed. Many of them include the quantification of DNA. No matter which of these methods is used, a DNA standard is essential.

2.2.1
Quantification of DNA in Solutions

The exact DNA content has to be measured to prepare DNA standards. Different methods show different problems. Table 1 gives examples of methods and their problems.

Table 1 Methods to quantify DNA insolutions

Method	Problems	Ref.
Absorption at 260 nm	RNA and proteins can lead to significant errors	[20]
Fluorescence after staining	Fluorescence may not be quantitative	[21]
Quantification of deoxyribose	Errors due to hydrolysis and reaction, labor intensive	[22]
Quantification of phosphate after complete hydrolysis	Errors due to hydrolysis and reaction, labor intensive	[20]

Table 2 Factors influencing the quantification of DNA by absorption at 260 nm

Factor	Influence	Error
Chromosomal DNA	High	Less plasmid DNA than measured
RNA	High	Less plasmid DNA than measured
RNA/proteins	High	Often less plasmid DNA than measured
Plasmid forms	Low	Linear forms show decreased absorption
pH	Low	Higher absorption when pH high

The fastest and most widely used method is to measure the absorption at 260 nm. Proteins, also absorbing at this wavelength, lead to errors. Therefore, the quotient of the absorption at 260 nm/280 nm is calculated and should lie between 1.6 and 1.8 to give reliable results. However, RNA can cover up proteins.

Especially after preparations of plasmid DNA, the errors through fractions of chromosomal DNA can be significant. Some of the factors influencing the quantification of plasmid DNA by absorption at 260 nm are shown in the Table 2.

The method where phosphate is determined after hydrolysis could be a good approach. Another method uses a special fluorescent nucleic acid stain and claims that the presence of up to 30% total DNA as single stranded DNA does not interfere with the assay [23].

2.2.2
Methods to Measure Plasmid Copy Number

An overview of methods to determine the plasmid copy number is shown in Fig. 1.

The methods to analyze the plasmid copy number can be divided into two main categories: the direct methods and the non-direct methods. The non-direct methods are related to the phenotype and quantify plasmid located proteins. A well-known method is to measure the activity of plasmid based β-lactamase, an enzyme that leads to the resistance of cells against ampicillin [24]. It is expected, that a high copy number leads to a high activity and that this activity is directly proportional to the copy number. Related to this method is an assay called "rate of segregation" [25]. This assay is used with plasmids, which are randomly distributed to the daughter cells at cell division. The random distribution is connected to the plasmid copy number. If a culture of cells, containing a plasmid coding for the resistance against an antibiotic, is successively diluted with fresh medium without antibiotic, cells will loose their plasmids. Depending on the rate of loss and the model applied, plasmid copy number can be calculated. Other indirect methods use luciferase [26] or GFP, the green fluorescent protein [27] as a reporter protein.

But the expression of an enzyme or any other protein and the activity/function of that protein depends on much more factors than plasmid copy number, like promoter strength, mRNA stability, proteolysis and protein folding. They may vary in experiments although using the same expression system and

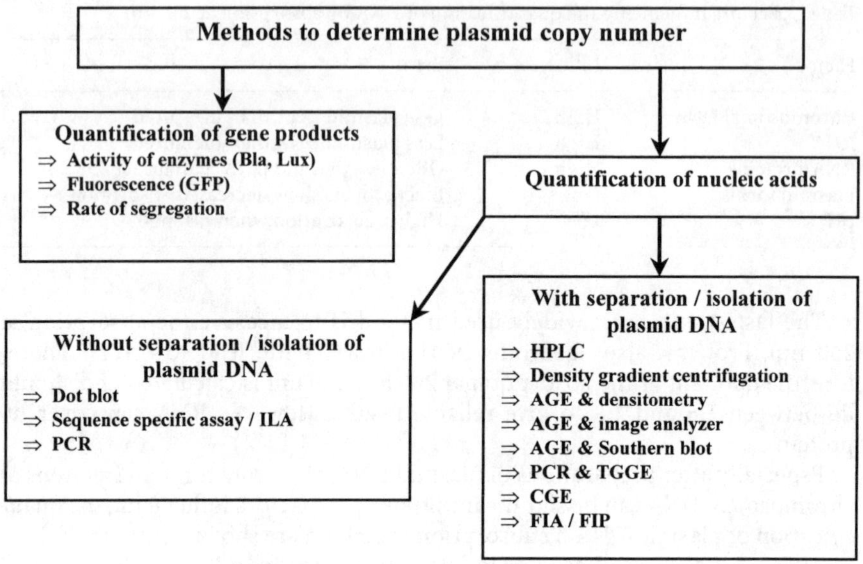

Fig. 1 Methods for the determination of plasmid copy number. Bla: β-lactamase, Lux: luciferase, GFP: green fluorescent protein, ILA: immuno ligand assay, PCR: polymerase chain reaction, HPLC: high performance liquid chromatography, AGE: agarose gel electrophoresis, TGGE: temperature gradient gel electrophoresis, CGE: capillary gel electrophoresis, FIA: flow injection analysis, FIP: flow injection processing

conditions. Therefore, these methods have to be used wisely and with restrictions [28].

Direct methods are measuring the quantity of nucleic acids. All of these methods start with taking a sample and lysis of cells. The procedures can be split into methods without plasmid separation and methods, where the plasmid DNA is separated from chromosomal DNA and RNA. Often the plasmid DNA is isolated in an extra step.

Methods without separation or isolation of plasmid DNA include the dot blot hybridization assay and the sequence-specific DNA assay. The dot-blot method starts with binding the nucleic acids of a cell lysate on a membrane. With special gene probes, often marked with radionuclides, plasmids are hybridized. The intensity of blackening of an x-ray film due to the attached radioactivity is proportional to the copy number [29]. The sequence specific assay, an immunoassay method, uses two pairs of two oligonucleotide probes. One pair is for the quantification of chromosomal DNA, the other pair works with the plasmid DNA. One probe of a pair is biotinylated and the other fluoresceinated [30]. The probes are chosen in a way, that they both anneal to the same strand of the target DNA in direct neighborhood. The probe target hybrids are captured by the biotinylated probe and measured with an anti-fluorescein immuno ligand assay. Chromosomal DNA and plasmid DNA have to be cut with a restriction enzyme to give reliable results. Therefore, total DNA from a cell sample has to be prepared before analysis. Such samples can also be analyzed using real-time

PCR and fluorescence tagged primers specific for plasmid and chromosomal sequences.

Methods with separation of plasmid DNA from chromosomal DNA can be subdivided into those without and those with plasmid isolation. Without plasmid isolation means, that the plasmids are separated during the assay, but not isolated before in an extra step. Such methods often measure the copy number per chromosome and use HPLC, agarose gel electrophoresis followed by densitometry and PCR followed by Temperature Gradient Gel Electrophoresis, TGGE.

Samples of cleared cell lysates can be placed directly on an HPLC column and the plasmid DNA and chromosomal DNA may be quantified from the corresponding peaks of an UV detector, provided that RNA is separated from DNA and plasmid DNA separates from chromosomal DNA [31].

The cleared lysates can also be placed on an agarose gel. After electrophoresis, the nucleic acids are quantified using x-ray films like in the dot-blot method. Both, gel electrophoresis and pulsed field gel electrophoresis can be used [32, 33]. After transfer of the nucleic acids on a membrane during the Southern-blot method, non-radioactive samples can be marked with radioactive probes and quantified [34].

A different agarose gel assay uses total bacterial DNA and ethidium bromide. A negative photo film of the gel is used to determine the ratio of chromosomal DNA to plasmid DNA via densitometry [35]. Although the method seems to lead to reliable results, the described calculations are not very clear and the method itself may have some pitfalls as will be described later, when showing own results in the next section.

A combination of PCR and Temperature Gradient Gel Electrophoresis (TGGE) can be used to determine plasmid copy number [36]. In this method a specific amount of a standard, which is homologous to a region on the plasmid except in one base, is mixed with the sample of isolated DNA. After PCR, a probe is added which forms homoduplexes with the plasmid DNA and heteroduplexes with the standard DNA. They can be separated with TGGE and the bands can be quantified, e.g. using marked probes. The relatively intensity of the bands, related to the amount of added standard, gives the number of copies in the sample which can be used to calculate the number of copies per cell.

In some cases plasmid DNA is isolated and purified before quantification. The classical method uses density gradient centrifugation. Cells are grown in media containing [^{3}H]-thymidine. After centrifugation the plasmid and the chromosomal fractions are isolated and counted in a liquid scintillation counter [37] and the ratio of plasmids to chromosomes is determined using the chromosomal fraction as internal standard [25].

All methods where plasmid DNA is isolated are using one of the established isolation assays followed by a purification step before quantification. The purification can be achieved by agarose gel electrophoresis, capillary gel electrophoresis or chromatography.

2.2.3
Determination of Plasmid Copy Number Using Agarose Gel Electrophoresis

Agarose gel electrophoresis is used to purify and separate plasmid DNA from chromosomal DNA and RNA after plasmid isolation. Following electrophoresis, the bands are stained with ethidium bromide. The quantification of these bands can be achieved by various methods.

One method is taking photographs from the gel and evaluating the bands on the negative with densitometers [1, 35, 38]. Although this method is often used, results have to be handled with care. The following example should illustrate the problems, which may occur.

In a series of experiments, the strain *E. coli* JM103 with the plasmid pUC12 was cultivated in shake flasks. Cells were counted in a Neubauer chamber under the microscope and plasmid DNA was isolated by the alkaline SDS-method. Plasmid DNA was linearized using EcoRI and placed together with λ-DNA-HindIII standard on an agarose gel in TE-buffer. After 4 h at 4 V cm^{-1} the DNA was stained with ethidium bromide.

The bands became visible on a 254 nm transilluminator. Photographs were taken using Polaroid films with or without a negative and exposure times ranged form 4 to 25 min. A Mamiya C3 6×6 camera and black and white films (ILFORD HP5) were used with exposure times between 7 and 11 min.

For the quantification of the bands, a densitometer (TLC Scanner II, Camag) was connected to an integrator (SP 4270 Spectra Physics) and transmission of the bands at 450 nm was measured.

A direct analysis of the gels with this apparative set up was not possible due to the high noise. Therefore, the photonegatives were used to quantify the blackening of the bands. The Polaroid 665 films showed many inhomogeneities within a charge or even on one negative. The ILFORD HP5 gave homogenous results. The exposure time of films has to be chosen carefully. If it is to short or to long the blackening is not linear to the concentration.

Cutting λ-DNA with HindIII creates eight fragments of different sizes. Knowing the quantity of the total amount of λ-DNA applied on the gel, one can calculate the DNA content of each band formed by these fragments. To quantify an unknown band the blackening of a standard band has to be compared with the blackening of that unknown. Does it make a difference which of the λ-DNA bands is used as the standard? The simple answer is: yes. This can be seen in Table 3.

Table 3 Influence of the chosen standard band on the plasmid copy number

Standard band	Plasmid Copy number
23130 bp	17
9416 bp	18
6557 bp	19
4361 bp	22

Depending on the standard band the number of plasmid copy number may differ up to 30%.

Digitally images of the gels can be obtained and analyzed. One system of hardware and software is E.A.S.Y. by Herolab, Germany. With this system, a linear plasmid DNA band was compared with 12 bands of two standard lanes (λ-DNA cut by BstEII). Plasmid copy numbers were calculated using the evaluation tool of the software to determine the blackening volume of the bands and correlating the volumes of the linear plasmid band to the blackening volumes of the standard bands. Figure 2 shows the results.

The plasmid copy number varies when correlated to different standard bands. It looks if the error increases when the large fragments 8454 bp and 7242 bp and the small fragment 702 bp are used for calculation. Using standard fragments between 6369 bp and 1264 bp, the copy number was calculated from 52 and 59 (average 56) per cell, which is in an acceptable range. The same applies to the second standard lane. Here the copy number lies between 33 and 39 (average 36). The range is acceptable but the number is clearly lower (36%!) than with the first standard lane. This may be due to errors when pipetting microliters of standard on the gel. The conclusions of these experiments are twice. The handling of the standard has to be very carefully and errors when pipetting such small volumes have to be kept as low as possible. The standard bands, which are chosen to calculate the DNA content of an unknown plasmid band, should represent fragments of a similar size as the plasmid.

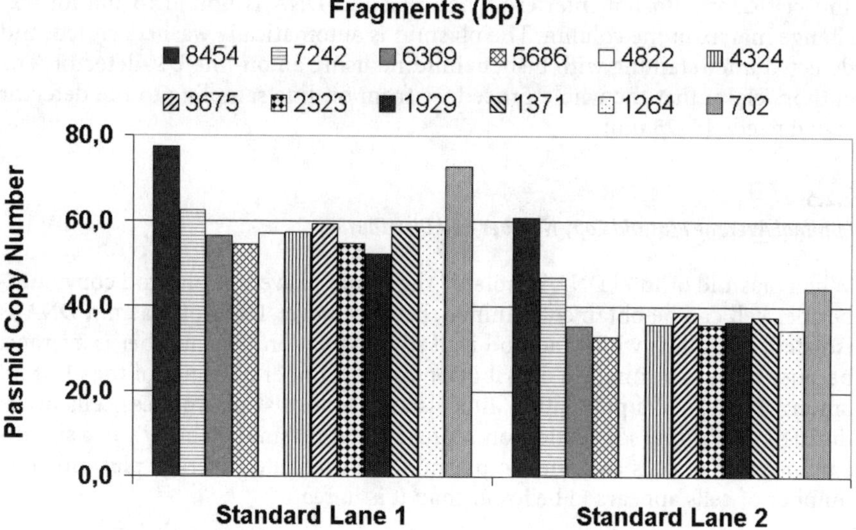

Fig. 2 Different plasmid copy numbers from one linear plasmid band correlated to different standard bands of two standard lanes

2.2.4
Measurement of Plasmid Copy Number During Cultivations

Many of the methods to determine plasmid copy number are labor intensive, may deal with radioactivity, need a lot of time or show significant errors. When using *E. coli* strains for the production of recombinant proteins the plasmid copy number is an important process variable. It would be an advantage to be able to determine the plasmid copy number in an acceptable time like within the doubling time of cells, which is usually around 30 min for *E. coli*. All methods described so far are not able to determine the plasmid copy number within 30 min. Possible approaches to fit into this time limit are assays using fluorescence spectroscopy after plasmid isolation and staining with bisbenzimidine (Hoechst 33258) [39]. Another method uses capillary gel electrophoresis (CGE), which gives the opportunity to perform gel electrophoresis very rapidly. CGE uses UV-detection or induced fluorescence detection [4, 40] for plasmid DNA samples, the latter with a much higher sensitivity.

Schmidt and colleagues developed a method were the plasmid DNA is isolated very fast, quickly linearized by an adequate dosage of restriction enzyme and quantified using a liquid gel of hydroxy propyl methyl cellulose with YO-YO or YO-PRO as fluorescence dyes. The result is available in about 20 minutes, short enough to be used in process control [4].

For process control, on-line measurements of variables are necessary and welcome. A flow-injection processing system uses a miniaturized expanded bed column to catch plasmid DNA from automatically lysed samples [41, 42]. Cells and cell debris do not interfere when plasmid DNA is bound to the ion-exchange matrix in the column. The plasmid is automatically washed, eluted, and detected after staining with bisbenzimidine using an on-line UV-detector. The authors claim that the whole procedure from on-line sampling to the detector signal needs 15–25 min.

2.2.5
Minimal Average Plasmid Copy Number and Distribution

When plasmid or total DNA is isolated the minimal average plasmid copy number per cell can be obtained. Minimal, because never 100% of plasmid DNA is isolated, regardless which method is used for isolation. The number is average, because there is normally a distribution in the number of copies of the plasmid among the cells of a population. It is also possible, that plasmidless cells occur during cultivation (see section on segregational plasmid stability). If a significant number of cells contains no plasmid, the calculated average plasmid copy number of cells appears to be lower than it is in reality.

2.3
Changing the Plasmid Copy Number

It makes sense to increase the copy number in one or the other way, if a higher plasmid copy number leads to a higher productivity.

Every change of the copy number is based on a change in the replication of the plasmid. Special sequences on the plasmid (*cis*-acting elements), like the origin of replication (*ori*), are necessary for duplication. Other genes, which might be located on the chromosome (*trans*-acting elements) are coding for specific proteins for the replication. Both elements can be targets for changing the copy number [43].

Many plasmids have at least the gene for a protein, which is regulating the initiation of replication, but all plasmids need special sequences for replication. Vectors used in *E. coli* expression systems possess such sequences and/or genes for replication of natural plasmids like ColE1, pSC101, R6K and F-factor. These plasmids show different mechanisms and regulations of replication [44]. Therefore, changes in plasmid copy number cannot be generalized. However, a rough classification could be achieved by dividing them into three categories: (1) mutations of replication genes and sequences, (2) extracellular changes causing intracellular effects and (3) runaway-replication plasmids.

2.3.1
Mutations Influencing Plasmid Copy Number

Modifications in the *cop* gene or its regulation (as in pSC101) can lead to an increase of the plasmid copy number. If the *cop* coded protein RepA for replication changes its binding property on DNA through a point mutation, the plasmid copy number increases [45]. Mutations in *cop* can lead to an amplification of plasmids [46]. In addition, alterations in the gene regions of replication initiation have similar effects [47, 48]. With plasmid RK2 mutations in the gene *trfA*, coding for an initiation protein could be detected as the reason for a higher plasmid copy number [49].

Other targets for mutations are sequences of ribonucleic acid RNAI and RNAII respectively, both involved in the regulation of replication [50, 51]. Modifications in *trans* elements of replication can lead to an amplification of pBR322 as seen in strains of *E. coli* CP79 and CP143 [52]. In reverse cases, mutation or deletions can also produce lower copy numbers or stop replication completely [53].

2.3.2
Extracellular Changes and Copy Number

Nutrient limitations in the cultivation medium can increase plasmid copy number. This could be shown for limitations in amino acids [54, 55], in phosphate [56] or in iron [57]. It can be assumed that the observed effects are similar to those during amplification of plasmids with chloramphenicol [58]. In this case growth rates are decreased and protein biosynthesis is partly stopped. The *oriC* related chromosomal DNA replication is interrupted, while plasmid production, due to another *ori* sequence, is still running.

An increase of plasmid copy numbers could also be observed after the overexpression of a plasmid located recombinant gene. After induction of overexpression, translation of not plasmid located genes was hindered, which stopped growth and the replication of the chromosome of *E. coli* [59]. These effects are also very similar to effects, triggered by chloramphenicol and gave a six times higher plasmid copy number [60].

The addition of ethanol can influence plasmid copy number. It is assumed, that changes in the cell membrane have an effect on replication of specific plasmid, because DNA is bound to the membrane during duplication (see later) [61]. As a result treatment of cells with ethanol leads to an increase of plasmid copy number [62]. It was reported that special composition of media also might lead to higher plasmid copy numbers [63]. This was seen, when using special amino acids were employed [64].

Not only the cultivation conditions for the growth of microorganisms show influences on the copy number, vice versa, the copy number can influence the growth. To keep a high copy number the cells have to invest energy and metabolites in the replication machinery. This can decrease the growth rate [65, 66]. On the other hand, a low growth rate, achieved deliberately by specific cultivation conditions, could increase the copy number [67].

2.3.3
Runaway Replication Plasmids

Runaway replication plasmids are very special. Under non-induced conditions, just one copy is present per cell. After induction of the replication, the copy number can increase up to 800 copies and more [68, 69]. This can be an advantage for the cultivation of recombinant cells. A high plasmid copy number always means a lot of additional synthesizing work for the cells and is a real metabolic burden [16]. With runaway replication plasmids, the copy number during growth can be kept at one copy per cell and the metabolic burden is consequently low. After reaching a sufficient biomass concentration, the production of plasmid located proteins is started by induction of gene expression. At the same time the gene dosage is increased, by induction of the runaway replication.

Temperature shifts can start the amplification. In this case, the genes for replication are controlled by the temperature inducible promoter of phage Lambda, λ_{PL} [70]. The promoter is controlled by constitutive expression of λ repressor cI857. This is a mutation of the wild type repressor. After a shift from 28 °C to 42 °C the repressor cI857 is irreversibly denatured and the strong λ_{PL} promoter leads to a high expression of replication factor RepA. That increases the plasmid copy number dramatically up to 1000 copies per cell [1, 71].

A similar system regulates plasmid replication via the *tetA* promoter induced by autoclaved chlortetracycline [72]. Another possibility to increase the copy number is the induction of the production of RNAI or RNAII. This can be reached by using a synthetic promoter, repressible by the repressor of the *lac*-operon and therefore inducible by adding IPTG [73]. A further way for higher plasmid copy numbers uses the co-transcription of plasmid located protein genes and replication elements under the control of a *trp* promoter [74]. Strong promoters like the *tac* promoter, can lead to a through reading of replication genes and hereby increasing the copy number [69]. Amplification can also be started through adding a new C-source when using an appropriate promoter [75].

The production of many plasmids due to runaway replication can be lethal for cells and decreases the overall productivity. A new kind of runway replication vectors, which allow precise controlled replication, can reduce these effects [76].

3
Structural Plasmid Stability

Changes of the base sequence of a plasmid by insertions, deletions or point mutations are summarized under the term "structural plasmid instability". As it was mentioned before, this might lead to a change in the copy number. However, if alterations in the sequence occur for regulation or in structural genes of recombinant proteins, heavy losses in productivity might be the consequence. Either less protein is expressed or only protein fragments may be produced. Even proteins with absolutely wrong amino acid sequences and maybe unknown activities, can occur, due to frame shift mutations.

The reasons for structural instabilities are manifold [77]. Mutagenes or free oxygen radicals can lead to changes in the sequence. Errors of DNA polymerase or insufficient repair mechanisms are further explanations for mutations. Recombination events between sequences on the plasmid and sequences on the chromosome can cause significant alterations [78]. Sometimes no other reasons are described, than the expression of genes [79, 80]. IS-elements can lead to structural instabilities [81]. Unfortunately, especially shuttle vectors, which allow the use of one vector in different species or genus, show a lower structural stability, than specific host vectors.

3.1
Determination of Structural Plasmid Stability

The analysis of structural changes of plasmids depends on the kind of alteration. Deletions and insertions can be seen quite easily by restriction analysis and comparison of the length of fragments by agarose or polyacrylamide gel electrophoresis. With these methods, differences in one base can be determined [82]. Exact analysis of fragment size distribution can also be achieved by mass spectrometry [83]. When working with plasmid based expression systems the correct pattern of fragments should be verified from time to time by restriction analysis. But structural instabilities can also lead to some strange observations as the following example will show [84].

The plasmid pJMC40 (pBR322 with a large insert of chromosomal DNA of *E. coli*) was checked before some experiments by restriction analysis. Plasmid DNA was transformed into *E. coli* DH5α, isolated and cut using the restriction enzyme EcoRI. The expected fragments were 2.5 kbp, 4.9 kbp and 6.7 kbp. But an additional fragment of 8.0 kbp occurred on the gel. This band could not be explained with partial digestion and after repeated restriction analysis; the only explanation was an extra fragment. When using other restriction enzymes this hypothesis failed, because the sum of the fragments shows discrepancies. DNA from the first plasmid isolation was used for a new transformation of *E. coli* DH5α. Plasmids from different clones were prepared and analyzed and surprisingly some of the plasmids showed the original expected pattern of 2.5–4.9–6.7 kbp and some showed the pattern 2.5–4.9–8.0 kbp. Now the case was obvious. The first plasmid isolation consisted of two different plasmids, a 14.1 kbp and a 15.4 kbp plasmid. After transformation of the original pJMC40 plasmid sample into *E. coli* DH5α a structural instability, a 1.3 kbp insertion, had occurred in one or more plasmids in the beginning of the cultivation. The explanation could be that the chromosomal DNA fragment on the plasmid interfered with the same chromosomal DNA on *E. coli* and led to a recombination and a second different plasmid. The insertion must have happened quite early after the transformation, leading to an equal distribution of cells with one or the other plasmid.

The determination of point mutation is much more complicated. Due to increasing importance of gene analysis in medicine, some new methods are under development. One is using glass fiber DNA sensors, which show alterations of the DNA by changes in hybridization reactions [85].

With PCR followed by gel electrophoresis point mutations can be detected in known gene segments [86]. Single point mutations can be detected with TGGE. This method is based on altered melting temperatures of DNA fragments even if they differ in just one base [87]. 2D electrophoreses can be used for such examinations too[88]. Capillary electrophoresis is also able to separate and verify point mutations [89, 90]. Unknown fragments, with an attached fluorescence dye, are mixed with defined sequences, marked with a different fluorescence dye. Under denaturing conditions the defined sequences hybridize with the unknown sequences. Now different point mutations show up as single peaks. The use of several defined sequences with specific markers allow the simultaneous determination of point mutations in different regions of long DNA fragments [91].

3.2
Methods to Increase Structural Plasmid Stability

To avoid structural instability the already described conditions which may lead to such instabilities should be avoided. Sometimes growth conditions have to be changed to increase structural stability [92]. Most important is the choice of the host strain. Strains should have a significant decreased capability of recombination [93]. Many of the *E. coli* strains used in genetic engineering and for the expression of heterologous proteins, show a lowered recombination frequency.

Especially when eukaryotic DNA is placed on a plasmid, structural instability may become a problem. Inverted repeats or specific secondary structures, such as Z-DNA, are targets for DNA repair systems. The UV repair system (*uvr*) and the SOS repair pathway (*umu*) rearrange or delete DNA fragments. Strains, with these genes removed, show a better structural plasmid stability [94].

4
Segregational Plasmid Stability

In most cases, the phrase "plasmid stability" stands for segregational plasmid stability. The successful partitioning of plasmids on to the daughter cells is a central problem in plasmid dependent expression systems. Plasmids are always a metabolic burden for the cells, which leads to a preference of plasmid free cells during cultivations [16]. The simple formula is: the appearing of plasmid free cells means a reduced amount of recombinant product [95]. Therefore different efforts are made to make sure that every daughter cell has at least one plasmid to start with and that the amount of plasmid free cells in a cultivations is as low as possible, at best zero.

The mechanism of plasmid replication has an important influence on segregational plasmid stability [96, 97]. Different models describe and calculate plasmid maintenance during cultivations [98–102].

Modeling the segregational plasmid stability is complex and depends on different types of plasmids and biological properties. Many factors can influence the segregation of plasmids on daughter cells. But it makes sense to establish simple models, which allow the simulation of plasmid stability during cultivations.

Non-segregated models only distinguish between cells with plasmid(s) and cells without plasmid(s) and are mainly based on differences in the specific growth rate of plasmid bearing and plasmid free cells. A segregated model would take more differences into account like different plasmid copy numbers in cells [103].

One problem during cultivations is the regularly higher specific growth rate of plasmid free cells, compared with the specific growth rate of plasmid harboring cells [104]. If plasmid free cells arise early at the beginning of cultivation, they can overgrow the whole population. At the end, there might be a lot of biomass but no plasmid and subsequently no recombinant product. That makes it clear, why different approaches are made to enhance segregational plasmid stability. This is easily to achieve in tubes and shake flasks, but is a true challenge in bioreactors or industrial scale cultivations.

4.1
Analysis of Segregational Plasmid Stability

The basic method to determine the segregational plasmid stability of a system is the labor- and time-consuming technique of parallel plating of samples [105]. Traditional plasmid vectors used in biotechnology have a selection marker, often an antibiotic resistance operon. Same amounts of diluted samples are plated on selective and non selective plates. The ratio of colonies on the plates presents the plasmid stability. To overcome plating errors the good old stamp method by Lederberg can be used. First, a sample is plated on a non selective plate and after colonies have occurred, the exact pattern is transformed on selective plates. A faster, but more complex method is using flow cytometry together with a reporter protein as for example β-galactosidase [106]. The well-known green fluorescent protein GFP was used in combination with a cell sorter to determine the segregational plasmid stability in prokaryotes [108], yeast [108] and mammalian cells [27, 109]. All these methods depend on the expression of a plasmid sited protein and therefore underlie errors due to expression and maturation process.

4.2
Enhancing Segregational Plasmid Stability

Segregational plasmid stability can have a significant influence on the production of heterologous proteins or plasmids themselves. A variety of methods, many of them based on stabilization machineries found on natural plasmids, were developed to ensure the partition of plasmids on to the daughter cells. According to these various mechanisms, these methods can be divided in different types as listed in Table 4.

Examples of these different mechanisms will be presented in the following sections.

4.2.1
Resistance Against Toxic Substances

One of the simplest method to enhance segregational plasmid stability and widely used in laboratories is to add an antibiotic to the medium, against which the cells develop resistance due to plasmid located genes [110]. All cells without a plasmid would be killed or at least prevented from growth.

This method is not applicable on an industrial scale. The costs of adding antibiotics could be enormous. For example, the usual concentration of ampicillin to stabilize a plasmid carrying the β-lactamase gene is 100–150 mg l^{-1}. For a 30 m^3 cultivation 3–4.5 kg ampicillin would be needed. Additionally any antibiotic must be removed during down stream processing or at least should not be present in a detectable concentration in all pharmaceutical and food production processes.

In particular the use of the stabilization system ampicillin and β-lactamase shows other significant disadvantages [111, 112]. Due to historical reasons,

Table 4 Mechanisms of segregational plasmid stability

Resistance against toxic substances
Complementation of chromosomal mutations
Post segregational killing of plasmid free cells
Influences of plasmid size and form
Active plasmid partitioning
High copy number and plasmid distribution
Lowering the difference in specific growth rates by cell internal factors
Influences of cultivation conditions
Separation of plasmid free and plasmid bearing cells
Integration into the chromosome

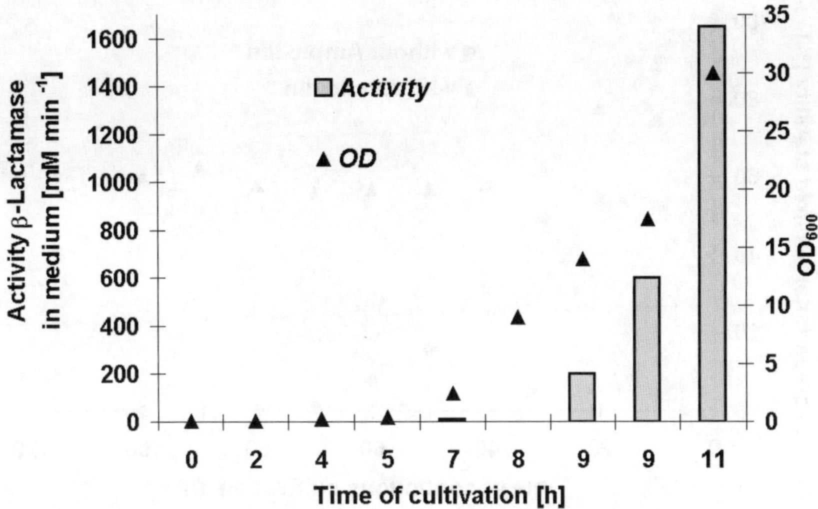

Fig. 3 Activity of β-lactamase in the medium during cultivation of plasmid bearing cells with the *bla*-gene on the plasmid

many plasmid vectors, also many commercially available vectors, carry the gene for β-lactamase *bla*. β-lactamase, expressed in *E. coli* is transported into the periplasm and overexpression leads to secretion of the enzyme into the medium [113]. Ampicillin may have a stabilizing effect during lag-phase and in the beginning of the exponential phase but is destroyed soon. A special problem are the inocula. They may contain a lot of β-lactamase so that the whole ampicillin of the main culture would be destroyed so fast, that no stabilization effect is achieved.

During experiments with *E. coli* carrying the plasmid pAA182, the activity of β-lactamase in the medium was measured. At a moderate biomass concentration, the enzyme activity was so high that a 500-µl sample of the cell free cultivation broth would destroy the ampicillin of a 40-l culture within 1 min (Fig. 3).

For all cultivations using ampicillin for plasmid stabilizations washing of the biomass of the preculture is highly recommended. In addition, a combination of methicillin and ampicillin alleviated segregational plasmid stability [112].

In continuous cultures the permanent input of fresh medium containing ampicillin and the continuous output of β-lactamase could lead to a better stabilization via ampicillin.

The strain E. coli JM103 with plasmid pOU79 was cultivated in continuous cultures with and without ampicillin as shown in Fig 4.

After 60 h of cultivation in Luria Bertani medium without ampicillin the plasmid stability decreased to 0%. The whole culture contained only plasmid free cells. Using medium with 100 mg l⁻¹ ampicillin the segregational plasmid stability dropped to 60% but did not decrease further through 100 h cultivation [114].

Plasmids without antibiotic resistance genes are especially preferred for gene therapy or gene vaccination [115, 116]. One reason is that if the resistance genes

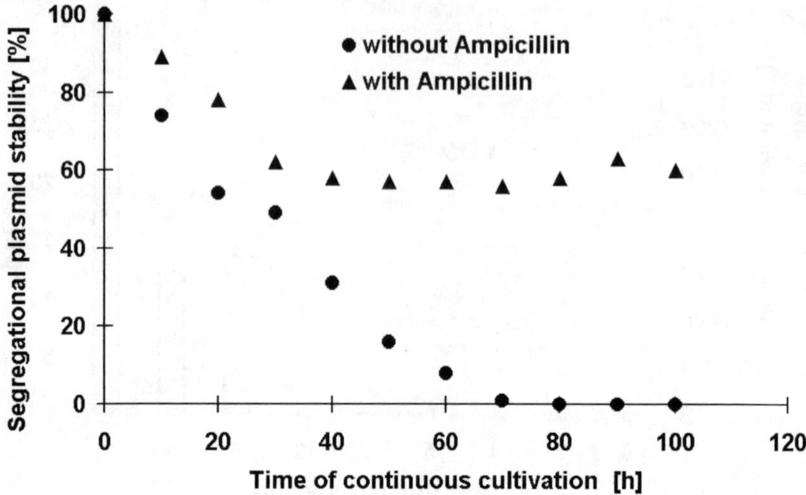

Fig. 4 Stabilization of plasmid segregation by ampicillin during continuous cultivation

are expressed in a human body, the resistance proteins may lead to unwanted side effects. Another reason is that resistance genes may be transformed via horizontal gene transfer onto pathogenic microorganism disabling antibiotics to be effective [117, 118]. This is a growing problem in hospitals. Another growing problem are allergic reactions, which could be observed in animals when plasmid DNA was used for vaccination [115].

One way to produce plasmids without a resistance gene but still by adding an antibiotic is the repressor titration method by Williams and coauthors [119, 120]. In this system, the gene for resistance against kanamycin is located on the chromosome and is controlled by the *lac* promoter. The gene for the *lac* repressor is also located on the chromosome. Under normal conditions, the expression of the resistance gene is repressed and cells will not grow if kanamycin is added to the medium. As soon as a high copy plasmid is brought into the cell, which carries one or more copies of the *lac* promoter, the number of repressor molecules is too low to occupy all promoter regions and therefore the expression of the chromosomal resistance gene can be high enough for the survival of plasmid bearing cells in kanamycin medium.

4.2.2
Complementation of Chromosomal Mutations

For industrial purposes special systems were developed. Strains were mutated so that they were not able to produce an essential substance or protein. The correct gene for the production of this essential substance or protein was placed on a plasmid and therefore only those cells could survive and divide, which carried the plasmid.

Two kind of mutations can be distinguished: a mutation where the correct phenotype can be established by adding missing metabolites into the medium and mutations, where the correct phenotype can only be reached by the appropriate gene.

By choosing a medium, which does not contain the missing substance, the plasmid stability can be enhanced in the first case. This can be seen with a system, where the adenylosuccinate synthetase was knocked out from the chromosome and placed on a plasmid [121]. This enzyme catalysis the synthesis of adenylosuccinate from inosine-monophosphate, which is an essential step in the pathway of the adenine nucleotide synthesis. Adding adenine nucleotides into the medium would allow plasmid free cells to grow.

Other systems of this type of plasmid stabilization operate with the plasmid located complementation of chromosomal mutations in the gene for aspartate semialdehyde dehydrogenase, *asd* [122], in *Salmonella typhimurium* or in the gene for D-alanine racemase in *Bacillus subtilis* [121].

In addition, essential genes involved in the synthesis of tryptophan [123, 124] or threonine [125] can stabilize plasmids, but show normal phenotypes if these amino acids were added.

The other type of complementation systems cannot establish correct phenotype by adding specific substances to the medium. This has the advantage that complex media can be used, which are sometimes preferred in biotechnology industry.

The ValS-system uses strains, which have a chromosomal mutation in the valyl-tRNA-synthetase gene. This enzyme is necessary for binding the amino acid valine to its corresponding tRNA. The gene for this enzyme was placed on a plasmid vector and established normal growth of plasmid bearing cells [126]. All appearing cells without the plasmid are hindered in protein synthesis and die.

The gene *ssb* also codes for an essential protein, the single strand binding protein, which is necessary for DNA replication [127] and is used in a stabilization system. Other methods are based on genes for enzymes involved in the synthesis of the murein sacculus. These enzymes are responsible for the building of murein tetra peptides. The corresponding genes of the *mur* operon are made available by a plasmid and even high doses of the enzymes due to high copy numbers are not lethal for the host [128].

The proline auxotrophy of the widely used host *E. coli* JM83 is used to stabilize a plasmid harboring the *proBA* genes [129]. No plasmid loss was observed during growth and induction phases.

In all these systems, essential genes are knocked out on the chromosome. Revertants, where the knocked out gene function is reestablished by another mutation, are loosing this stabilizing effect. This is a real problem for an industrial system.

4.2.3
Post Segregational Killing of Plasmid Free Cells

A special type of enhancing the segregational plasmid stability is found on natural plasmids. They are based on intracellular mechanisms, which kill plasmid

Table 5 Post-segregational killing mechanism

Plasmid/prophage	Gene(s)	Toxin	Antidote	References
F-factor	*ccd*	CcdB	Protein CcdA	[131]
RP4 (RK2)	*parDE*	ParE	Protein ParD	[131]
R1, R100	*parD (pem)*	PemK	Protein PemI	[131]
R1, R100	*parB (hok/sok=kis/kid)*	Hok	Antisense *sok* RNA	[130, 132]
P1	*phd/doc*	Doc	Antisense *phd* RNA	[133, 134]
R483	*pnd/pndB*	PndB	Antisense RNA	[135, 136]
PTF-FC2	*pasABC*	PasB	PasA	[137]

free cells and are called post segregational killing. Such functions were found on the natural plasmids or phages F, R1, P1 and RP4 (identically with RK2) [130]. The mechanisms consist of two components: a chromosomal stable anchored gene for a toxic substance and a plasmid located antidote, which neutralizes the lethal function of the toxin.

Two different mechanisms are known. In one mechanism, the antidote is a protein, a real antitoxin [131]. This is the case with the plasmid R1. The second mechanism has an antisense RNA as antidote. This so called *hok/sok* stabilization mechanism consists of two genes: the *hok* gene (*host* killing) and the *sok* gene (suppression of killing). The *hok* gene codes for the toxic Hok protein which consists of 52 amino acids. The *sok* gene codes for an antisense RNA, which is complementary to the mRNA of the Hok protein and prevents the translation of the protein. A weak constitutive promoter controls the expression of the hok-mRNA and the mRNA itself is very stable. The sok-antisense-RNA is expressed through a strong constitutive promoter. However, it has a short half-life. That means, that after loosing the plasmid and therefore the continuous production of antisense RNA, existing antisense RNA is degraded and the hok-mRNA is translated. The created Hok protein kills the cell [130]. Table 5 summarizes post segregational killing mechanisms.

Postsegregative killing systems are found on plasmids or prophages in many microorganisms [138]. Some of them can be used to stabilize plasmids in *E. coli*. For example the *mvpA/mvpT* genes of *Shigella flexneri* can stabilize otherwise unstable plasmids in *E. coli* [139].

However not all authors believe in plasmid stabilization through postsegregational killing. They found that such systems are more involved in the competition between plasmids which gives completely new sights on evolution of antibiotic resistance [140].

4.2.4
Influences of Plasmid Size and Form

The size of a plasmid has an influence on its segregational stability. Experiments with *Bacillus subtilis* showed that with increasing size the segregative plasmid stability is decreasing [141].

Multimerization of plasmids is also the reason for segregational instability [142, 143]. Genes, leading to the monomerization of multimers, enhance plasmid stability, as *res-tnpR* system in the transposon TN100 (the [144]. The plasmid RP4 (identically with RK2) carries a gene for a resolvase, which monomerizes multimers [145]. The gene for the resolvase is called *parA* (*par*titioning). The enzyme can be used to remove parts of the plasmid lying between distinct regions [146, 147]. The system is called *parCBA* [148] and stabilizes the plasmid in different Gram-negative bacteria [149]. The resolvase site lies within a section of 100 base pairs between the two promoters of *parDE* (the post segregational killing function) and *parCBA*. The gene *parB* codes for a nuclease [150]. The function of *parC* is still unclear.

Multimers can out-replicate monomers and accumulate within a culture, creating a subpopulation with significant lower plasmid stability. The theory behind this phenomenon describes a confusion of replication control by multiple origins of replication on multimers, leading to a copy number depression, which causes a lower statistical distribution probability of plasmids [151]. Multimers of multicopy plasmids are resolved to monomers by the site-specific recombination system Xer-cer. The *cer*-region placed on a multicopy plasmid leads to a high segregational plasmid stability.

The stabilization of the plasmid pSC101 is based on a gene for a gyrase, located on the par-region of the plasmid [152]. This gyrase is responsible for the formation of supercoiled plasmid forms. Due to this forms partition supporting proteins can interact with the plasmid DNA. Par-regions for active plasmid partitioning are described in the following section.

A special point to consider is the size of the fragment, which is cloned into the plasmid. Clearly, a correlation was found between fragment size and plasmid stability. Clones transformed with the largest plasmid from a series of different sized plasmids have a significantly higher rate of plasmid loss [153]. This may be due to a lower copy number. Beside these effects, interestingly the lag phase increased with increasing fragment size.

4.2.5
Active Plasmid Partitioning

Specific regions, also called *par*-regions (which leads sometimes to confusions), are responsible for a correct segregation of plasmids onto daughter cells. These *par* regions consist of three components: a *cis*-acting centromere-like site and two *trans*-acting proteins that form a nucleoprotein complex at the centromere [154].

During the partitioning of the prophage P1, the DNA is pinned on specific places of the cytoplasm membrane due to a sequence called *parS*. Proteins involved in the binding of P1 to the membrane are also located within the par region. The protein belonging to the gene *parB* interacts with the DNA and is probably responsible for the exposition of the binding region *parS* [155]. A high affinity nucleoprotein complex is formed by ParB and the *E. coli* integration host factor [156]. This complex binds to the *parS* region.

Plasmid R1 has two par-regions: *parA* and *parB*. The region *parB* stabilizes the plasmid via post segregational killing through the *hok/sok* mechanism. The

Table 6. Active partitioning systems

Plasmid/Phage	Par-region	Gene product/function	Ref.
R1	*parA*	ParM: ATPase ParR: binding protein *parC*: binding region	[159, 160]
P1	*par*	ParA: ATPase ParB: binding proteins *parS*: binding region	[161, 162]
P7	*par*	ParA: ATPase ParB: binding proteins *parS*: binding region	[163]
F-factor	*sop*	SopA: ATPase SopB: binding proteins *opC*: binding region	[164, 165]

region *parA* consists of the segments *parM, parR* and *parC*. The ParM protein is an ATPase, which interacts with the ParR protein. The ParR protein binds to the centromere-like *cis*-active *parC* region [157]. The sequence *parC* lies in front of *parM* and *parR*. It is supposed, that the combined functions of the *parM, parR* and *parC* segments lead to a pair to pair arrangement of the replication regions during plasmid replication, through which a controlled segregation of plasmids onto the daughter cells can be reached [158]. Table 6 summarizes some active partitioning systems.

Replication proteins DnaA and DnaB have effects on plasmid partitioning [166]. DnaA plays a significant role during replication of both the chromosome and plasmid pSC101. DnaB shows helicase activity that unwinds strands during replication. Using immunofluorescence microscopy it could be shown that replication proteins are involved in moving plasmids successively in younger zones [167]. Since the active plasmid partitioning may be coupled with the attaching of the plasmids to the membrane, enzymes involved in the biosynthesis of major membrane components influences active partitioning. For example leads a mutation in the protease FtsH to a defective plasmid partition in *E. coli* [168]. Plasmids and bacterial chromosomes are partitioned in a mitotic-like way in prokaryotes. This could be observed using fluorescence in situ hybridization with F plasmid and plasmid R1 and chromosomes of *B. subtilis, E. coli* and *Caulobacter crescentus* [169].

Par– regions can be used to stabilize expression plasmids. This was demonstrated with the runaway replication plasmid pOU130 [170] shown in Fig. 5.

The disadvantage of stabilizing plasmids with *par* regions is the significant increase of the plasmid size, which may influence structural and segregational stability.

Active partitioning systems could be found in *Lactococcus lactis* [171], *Coxiella burnetii* [172] and *Agrobacterium tumifaciens* [173]. It can be assumed that *par*– systems can be found in many natural plasmids of different species.

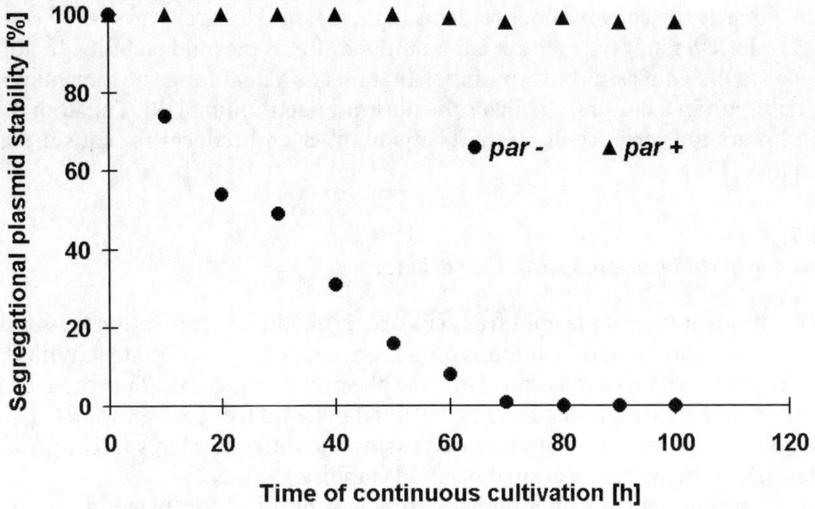

Fig. 5 Comparison of segregational plasmid stability with and without active partitioning regions and genes on the plasmid. *E. coli* JM103 pOU130 (*par+*) and *E. coli* JM103 pOU79 (*par–*) were cultivated continuously in LB-medium

4.2.6
High Copy Number and Plasmid Distribution

The correct distribution of plasmids, that means that at least one plasmid is moved to each daughter cell, depends directly on the plasmid copy number. If the copy number is high, the chances are good, that at least one plasmid is present in each of the two halves of a dividing cell. Simple models predict that a plasmid copy number of at least 20 plasmids per cell lead to a 100% segregational plasmid stability [174].

Such simple models are non-segregated and are based on different specific growth rates of plasmid free and plasmid bearing cells. However, especially if the segregative plasmid stability is not supported by genetically determined functions like a *par* region or a resistance gene and depends only on the plasmid copy number, it is important to know in which range the copy number differs inside the population. Therefore, good models have to consider more factors. Such factors would be the number of plasmids in a cell at cell division, the extent of plasmid multimerization and the distribution of plasmids between the two daughter cells [175]. In such a model all factors influencing the plasmid copy number like the metabolic burden according to a distinct copy number or the whole copy number control machinery in the cell would play important roles. From such models one can derive that an average medium copy number, where all cells have a very similar medium copy number, would show a higher stability, that an average high copy number where the distribution allows the development of cells with low copy numbers. Such cells show a high loss rate and would lead to plasmid free cells, which might overgrowth the population [175].

As long as the copy number remains high, plasmid free cells arise only rarely [147]. However, not only a high copy number affects plasmid stability. In addition a stabilized plasmid copy number, that means a near range of copy number distribution in a culture, stabilizes the plasmid partitioning [19]. The Rom protein lowers the variance in plasmid copy number and reduces the rate of plasmid loss [176].

4.2.7
Lowering the Difference in Specific Growth Rates

A big disadvantage of plasmid free cells occurring during cultivation lays in the fact, that plasmid free cells often show a higher specific growth rate μ. With this higher μ, plasmid free cells can overgrow plasmid bearing cells. This can lead to significant losses in productivity especially if plasmid free cells show up early in the cultivation process. Therefore decreasing the differences in specific growth rates may increase segregational plasmid stability.

One system depends on a higher expression burden after plasmid loss. The chromosomal gene for the making of pili is set under the control of the *tac* promoter. The plasmid carries the *lacI* gene, responsible for the production of the repressor of the promoter. After plasmid loss and therefore loss of the repressor, plasmid free cells will grow more slowly than plasmid bearing cells due to the higher metabolic burden triggered by the pili synthesis [177].

On the other hand, the growth rate is influenced by the expression of plasmid located genes. An overexpression of genes located on plasmids leads to a lowered segregational plasmid stability due to a decreased specific growth rate [178, 179]. In addition, maintenance is increased by overexpression and this also destabilizes plasmids [180]. Therefore, the use of tightly regulated promoters is very useful. They allow separating a process into two parts: a growth part, without expression and therefore higher plasmid stability and a production part. During a short production phase new plasmid free cells have no significant influence on productivity of plasmid located proteins [181].

4.2.8
Influences of Cultivation Conditions

In this section influences of cultivation conditions are described as found in the literature. Often authors gave not biological explanation and just noticed the phenomenon.

Cultivation conditions can influence segregational plasmid stability. A decrease of the dissolved oxygen concentration in the cultivation medium lead to a decrease in plasmid stability in *Streptomyces lividans* [182]. A similar influence was observed when *E. coli* was cultivated [183]. In addition, changes in the concentration of phosphate or shifts in pH influenced segregational plasmid stability. A shift to low pH increased stability in *Yersinia enterocolitica* [184]. It is supposed that analogous to a pH optimum for growth rate there is also a pH optimum for plasmid stability. This was shown for *Lactococcus lactis* [185]. Plasmid stability was reduced at higher temperatures [179]. Stress triggered by a higher temperature may have an influence on stability. This has to be distinguished from special systems, where temperature sensitive replication proteins lead to a total plasmid loss with increasing temperature [186]. In another case, stability also was lower when the temperature of cultivation was lowered [187]. This seems to be astonishing because normally a decrease in temperature would reduce the difference in specific growth rate and this should increase plasmid stability.

Media show different effects. Defined glucose media show higher segregational plasmid stability than complex media with glucose as the limiting substrate [188]. Minimal media lead to higher plasmid stability than adding of amino acids [189]. Changing the ratio of carbon- to nitrogen in media to a higher carbon content reduces plasmid stability [190]. The addition of yeast extract influenced plasmid stability in *Saccharomyces cerevisiae* [191]. Increasing feed rate of yeast extract decreased plasmid stability. The use of anti foams, which is often necessary in industrial processes, showed no influence on plasmid stability [192] but the quality of the inocula can have a significant impact on segregational plasmid stability.

Continuous cultures can be used to determine the effect of different cultivation conditions on plasmid stability. With *Pseudomonas sp.* a higher rate of plasmid loss was observed with increasing dilution rate [193]. Continuous cultures can also be used to isolate highly stable plasmids without knowing the exact reasons for this stability [194]. Changes in the dilution rate during continuous cultures lead to a higher stability [195]. Feeding cycles can increase stability [196]. To get higher cell densities often fed batch cultivations are chosen. Plasmid stability may depend on the type of feeding strategy. Exponential feeding of *E. coli* harboring a shuttle plasmid resulted in a plasmid stability of 45% [197]. In the case of other feeding strategies such as constant, linear or intermittent feeding, the plasmid stability varied between 20% and 60% [197].

Very interesting is the enhancement of stability through immobilization. This could be observed in *E. coli* [198], *Bacillus subtilis* [199] and *Corynebacterium acetoacidophilum* [200]. Plasmid stability was improved up to 50% when *Saccharomyces cerevisiae* was immobilized in calcium alginate gels [201].

One other aspect is very important when plasmid based expression systems are used. The induction of plasmid located genes always decreases plasmid stability. This maybe due to toxic effects of the recombinant protein as shown with pectate lyase overexpression [202] or an antigen [203]. Therefore, the usage of regulatable promoters can have significant advantages. Such promoters allow cultivation without expression and a small metabolic burden and a production phase after induction, in which plasmid stability no longer plays a role [204]. Basal transcription may lead to stability problems but can be avoided by catabolite repression [205]. The mode of induction of one promoter, for example the induction of *lac*-operator based promoters can have an influence on stability. Induction with IPTG leads to a higher rate of plasmid loss than induction with lactose [206].

4.2.9
Separation of Plasmid Free and Plasmid Bearing Cells

A completely different method is the separation of plasmid bearing and plasmid free cells. This can be achieved using flocculation. Cells bearing the plasmid pORN 108, which is responsible for the overproduction of pili, can be separated by flocculation from plasmid free cells, which do not form flocks. In an integrated process, the plasmid free cells are removed from the cultivation by a special separator [207].

4.2.10
Integration into the Chromosome

If one defines segregational stability as the passing of recombinant genes responsible for the expression of heterologous proteins, onto all daughter cells, the integration of these recombinant genes into the host chromosome could be a way for stabilization. For this purpose a series of vectors were developed for *E. coli* [208]. Such vectors exist also for many other species like *Rhizobium meliloti* [209]. It is important where on the chromosome the integration will appear and that this can be controlled. Uncontrolled integration could lead to lethal or at least negative changes in the chromosome.

The plasmid pBRINT, a derivate of pBR322, was constructed in such a way, that the integration place lies in the *lacZ* gene of the host chromosome. An additional advantage of this system is the simple selection of positive clones by blue color formation using x-gal [210]. Blue colonies still have the intact *lacZ* gene to express β-galactosidase. The white colored colonies have a high probability of an insert.

The attachment site for the phage λ on the chromosome is another place for a directed insertion. Using λ based vectors, recombinant DNA can be inserted into this site [211, 212].

A disadvantage could be the high structural instability if more than one copy is inserted. On the other hand, only one stable copy on the chromosome means a low gene dosage. Therefore, the productivity of clones with chromosomal inserts are normally significant lower, than plasmid based expression systems and the structural stability of the constructs must be watched carefully.

4.3
A Comment on Segregational Plasmid Stability in Scientific Publications

Often the plasmid copy number and plasmid stability have a more or less important influences on specific results presented in scientific publications. If authors for example, discuss factors influencing the production of a recombinant protein, segregational plasmid stability would be an important information. If the determination of the fraction of plasmid free cells is not described, the scientific conclusions should be interpreted with distrust.

5
Stabilities when Using Plasmids as Therapeutics

5.1
Plasmids as Vectors in Human and Veterinary Medicine

Plasmids can be used as so-called non viral vector systems in gene therapy and genetic vaccination [213, 214]. Gene therapy will become a very fundamental method in medicine as a safe and efficient way to treat or prevent diseases [215]. Targets of gene therapy are more and more not monogenetic diseases, like cystic fibrosis. A main target for gene therapy today is cancer [216–220]. Other targets are cardiovascular diseases [221] and cerebrovascular disorders [222] among others. A special area for plasmid vectors in medicine is tissue engineering [223, 224].

Another very promising field of plasmids as therapeutics is genetic vaccination [225–228]. Specific targets of genetic vaccination treatments are AIDS [229], influenza [230], hepatitis C [231], herpes simplex [232] or specific animal infections like canine *Leishmania* [233].

When using plasmids in medicine some other aspects of stability arise:
– Stability during production
– Stability during and after application in men and animals
– Stability during storage and shipping.

5.2
Stability During Plasmid Production

Large amounts of plasmid DNA in the scale of grams to kilograms have to be produced to use plasmids as a therapeutic. Such an industrial plasmid production leads to a number of downstream problems [234]. One of them is the stability of plasmid forms.

Plasmids can occur in several distinct forms: supercoiled, relaxed open circled and linear [235]. As a drug plasmids have to be of a special purity [236] and one aspect of purity depends on the distribution of these forms [237]. Therefore, a reliable analysis method is necessary to guarantee quality and to optimize downstream processing. The usual method to determine plasmid form distribution in DNA samples is agarose gel electrophoresis. A more reliable and faster method uses capillary gel electrophoresis [238]. Process validation and production according to cGMP are major challenges when scaling up plasmid isolation methods to an industrial scale [239–242].

5.3
Stability During and After Application in Men and Animals

One basic method to transfer plasmids into mammalian cells is the calcium-phosphate precipitation technique [109]. This process can be applied for ex-vivo gene therapy, where cells are taken from the body and genetically modified. After selection and cultivation, the cells are retransferred into the body. For in vivo therapies, plasmids can be used directly as naked DNA [243]. Plasmid DNA is taken up by isolated cells but also after injection by tumors or normal tissues like skeletal muscles or liver [244, 245]. The mechanism of plasmid uptake by mammalian cells is unclear [215] and the efficiency is low [246]. Better results can be achieved when using plasmids together with liposomes or histidine rich peptides [247–249]. Using these methods, the stability of forms and activity has to be taken into concern and methods have to be developed that nicking of supercoiled DNA is minimized [250].

When plasmids are placed into men or animals by one or the other method the distribution of DNA and the pharmacokinetic stability of the plasmids are important factors of the treatment [251]. The metabolic instability of plasmid DNA in the cytosol is a potential problem using gene drugs [252]. The accumulation, degradation and excretion of plasmid DNA, that means the plasmid stability after application as a drug, determines the usefulness of plasmids as therapeutics [253, 254].

5.4
Stability as a Drug During Storage and Shipping

To use plasmids as a drug it is necessary to find pharmaceutical formulations where the DNA remains stable during shipping and storage. The choice of buffer is of some importance [255]. Free radical oxidation may be a major degradative process for plasmid DNA in pharmaceutical formulations [256]. The addition of free radical scavengers and specific metal ion chelators decrease the damage by free radicals. Ethanol was also found to enhance plasmid DNA stability and extra addition of EDTA showed a synergistic stabilizing effect [256].

The instability in liquid formulations is a major problem on the way to good gene drugs. Freezing of plasmids would stabilize the DNA but would also complicate the application and would increase the costs. A good and reliable way to stabilize plasmid DNA is dehydration. Although dehydration procedures may also stress the DNA, freeze drying together with sugars lead to highly stable formulations and make plasmid based gene drugs easy to handle[257].

6
Conclusions

The aim of this article is to show that specific aspects occur when plasmids are used as vectors for genetically modified microorganisms. Plasmid copy number and plasmid stabilities determine the quality of an expression system. More and more theories and models describe factors influencing plasmid copy number and plasmid stability but mechanisms are too complex to generate a global model, which predicts all plasmid related quantities. Therefore, it is necessary to get more empirical data regarding copy number and stability from every expression system. This requires appropriate analytical methods. Without determining plasmid copy number or at least segregational plasmid stability, no real conclusions on recombinant productivity can be made honestly. Basic research on plasmid replication and segregation will lead to improved expression systems with optimized plasmid copy numbers and plasmid stabilities. The use of plasmids in gene therapy and genetic vaccination will bring new challenges for downstream processing and pharmacokinetics.

7
References

1. Leipold RJ, Krewson CE, Dhurjati P (1994) Plasmid 32:131
2. Kiewiet RJ, Kok JF, Seegers ML, Venema G, Bron S (1993) Appl Environ Microbiol 59:358
3. Schendel FJ, Baude EJ, Flickinger MC (1989) Biotechnol Bioeng 34:1023
4. Schmidt T, Friehs K, Flaschel E (1996) J Biotechnol 49:219
5. Austin SJ, Eichorn BG (1992) J Bacteriol 174:5190
6. Mayer MP (1995) Gene 163:41
7. Agophonov MO, Trushkina PM, Sohn JH, Choi ES, Rhee SK, Ter-Avanesyan MD (1999) Yeast 15:541
8. Boros I, Pósfai G, Venetianer P (1984) Gene 30:257
9. Müller AK, Rojo F, Alonso JC (1995) Nucleic Acids Res 23:1894
10. Kuo H, Keasling JD (1996) Biotechnol Bioeng 52:633
11. Patnaik PR (2000) Biotechnol Lett 22:1719
12. Paulsson J, Ehrenberg M (2000) J Mol Biol 297:179
13. French C, Ward JM (1996) Ann NY Acad Sci 799:11
14. Nacken V, Achstetter T, Degryse E (1996) Gene 175:253
15. Kim JY, Ryu DDY (1991) Biotechnol Bioeng 38:1271
16. Yazdani SS, Mukherjee KJ (2002) Bioprocess Biosyst Eng 24:341
17. La Fontaine S, Firth SD, Lockhart PJ, Paynter JA, Mercer FB (1998) Plasmid 39:245
18. Müller J, Van Dijl JM, Venema G, Bron S (1996) Mol Gen Genet 252:207
19. Grabherr R, Nilsson E, Striedner G, Bayer K (2002) Biotechnol Bioeng 77:142
20. Voß C (1999) Master thesis, University Bielefeld
21. Rye HS, Glazer A (1995) Nucleic Acids Res 23:1215
22. Dische Z (1955) The nucleic acid, vol 1. Academic Press, New York
23. Noites IS, O'Kenney RDO, Levy MS, Abidi N, Keshavarz-Moore E (1999) Biotechnol Bioeng 66:195
24. Miller CA, Cohen SN (1993) Mol Microbiol 9:695
25. Norström K (1993) In: Hardy KG (ed) Plasmids, 2nd edn. Oxford University Press, Oxford, p 1
26. Coronado C, Vazquez ME, Cebolla A, Palomares AJ (1994) Plasmid 32:336
27. Vogel M, Wittmann K, Endl E, Glaser G, Knüchel R, Wolf H, Niller HH (1998) Biotechniques 24:540
28. Betenbaugh MJ, di Pasquantonio VM, Dhurjati P (1987) Biotechnol Bioeng 29:1164
29. Hinnebusch J, Barbour AG (1992) J Bacteriol 174:5251
30. Mermelstein L (1997) Engineering Foundation Conference on Biochemical Engineering X, May 18–23, 1997, Kananaskis, Alberta, Canada
31. Coppella SJ, Acheson CM, Dhurjati P (1987) Biotechnol Bioeng 29:646
32. Projan SJ, Carleton S, Novick RP (1983) Plasmid 9:182
33. Ward AC, Hillier AJ, Davidson BE, Powell IB (1993) Plasmid 29:70
34. Olsson T, KEkwall, T Rusla (1993) Nucleic Acids Res 21:855
35. Pushnova EA, Geier M, Zhu YS (2000) Anal Biochem 284:70
36. Kang J, Immelmann A, Welters S, Henco K (1991) Biotech Forum Eur 91(10):590
37. Weaver KE, Clewell DB, An F (1993) J Bacteriol 175:1900
38. Wrigley-Jones CA, Richards H, Thomas CR, Ward JM (1992) J Microbiol Methods 16:69
39. Argyropoulos D, Savva D (1997) Biotechnol Tech 11:605
40. Breuer S, Marzban G, Cserjan-Puschman M, Durrschmid E, Bayer K (1998) Electrophoresis 19:2474

41. Nandakumar MP, Nordberg Karlsson E, Mattiasson B (2001) Biotechnol Lett 23:1135
42. Nandakumar MP, Nordberg Karlsson E, Mattiasson B (2001) Biotechnol Bioeng 73:406
43. Del Solar G, Espionsa M (2000) Mol Microbiol 37:492
44. Schumann W (1990) Biologie bakterieller Plasmide, Vieweg Braunschweig
45. Xia G, ManeD, Yu Y, Caro L (1993) J Bacteriol 175:4165
46. Panayotatos N (1998) US Patent 5,716,803
47. Acebo P, Alda MT, Espinosa M, Del Solar G (1996) FEMS Microbiol Lett 140:85
48. Nesvera J, Patek M, Hochmannova J, Abrhamova Z, Becvarova V, Jelinkova M, Vohradsky J (1997) J Bacteriol 179:1525
49. Fang FC, Durland RH, Helinski DR (1993) Gene 133:1
50. Moser DR, Campbell JL (1983) J Bacteriol 154:809
51. Lin-Chao S, Chen WT, Wong TT (1992) Mol Microbiol 6:3385
52. Hofmann KH, Neubauer P, Riethdorf S, Hecker M (1990) J Basic Microbiol 30:37
53. Phillips GJ (1999) Plasmid 41:78
54. Wrobel B, Wegrzyn G (1997) J Biotechnol 58:205
55. Wegrzyn G (1999) Plasmid 41:1
56. Horn U, Krug M, Sawistowski J (1990) Biotechnol Lett 12:191
57. Angelov I, Ivanov I (1989) Plasmid 22:160
58. Clewell LC (1972) J Bacteriol 110:667
59. Rinas U (1996) Biotechnol Prog 12:196
60. Teich A, Lin HY, Andersson L, Meyer S, Neubauer P (1998) J Biotechnol 64:197
61. Niki H, Hiraga S (1998) Genes Dev 12:1036
62. Basu T, Poddar RK (1997) Biochem Mol Biot Int 41:1093
63. Wan NC, Goodrick JC (1996) US Patent 5,487,986
64. Neubauer P, Hofmann K, Riethdorf S, Hecker M (1990) Acta Biotechnol 10:303
65. Mason CA, JE Bailey (1989) Appl Microbiol Biotechnol 32:54
66. Lee JH, Lee KJ (1994) J Biotechnol 33:195
67. Kramer W, Mattanovitch D, Elmecker G, Weik R, Luettich C, Bayer K, Katinger H (1994) Prog Biotechnol 9:827
68. Nordström K, Uhlin BE (1992) Bio/Technology 10:661
69. Togna AP, Shuler ML, Wilson DB (1993) Biotechnol Prog 9:31
70. Larsen JEL, Gerdes K, Light J, Molin S (1984) Gene 28:45
71. Goetting C, Thierbach G, Pühler A, Kalinowski J (1998) Biotechniques 24:362
72. Herman-Antosiewicz A, Obuchowski M, Wegrzyn G (2001) Mol Biotechnol 17:193
73. Bachvarov D, Jay E, Ivanov I (1990) Folia Microbiol 35:177
74. Chew LC, Tacon WC (1990) J Biotechnol 13:47
75. Compagno C, Tura A, Ranzi BM, Alberghina L, Martegani E (1993) Biotechnol Prog 9:594
76. Burian J, Kay WW (1998) WO Patent 98/41,636
77. Summers DK, Beton CWH, Withers HL (1993) Mol Microbiol 8:1031
78. Cheng KC, Loeb LA (1997) Curr Topics Microbiol Immunol 221:5
79. Kaprálek F, Jeämen, P (1992) Biotechnol Lett 14:251
80. Cordes C, Meima R, Twiest B, Kazemier B, Venema G, Van Dijl JM (1996) J Bacteriol 178:5235
81. Chawla M, DasGupta SK (1999) Plasmid 41:135
82. Martin R (1996) Elektrophorese von Nucleinsäuren, Spektrum Akademischer Verlag, Heidelberg
83. Hung KC, Ding H, Guo B (1999) Anal Chem 71:518
84. Friehs K, Bailey JE (1989) J Biotechnol 9:305
85. Healey BG, Matson RS, Walt DR (1997) Anal Biochem 251:270
86. Prockop DJ, Rock MJ, Ganguly A (1999) US Patent 5,874,212
87. Wartell RM, Hosseini S, Powell S, Zhu J (1998) J Chromatogr A 806:169
88. Vijg J (1995) Bio/Technology 13:137
89. Righetti PG, Gelfi C (1997) Biochem Soc Trans 25:267
90. Mitchelson KR, Cheng J, Kricka LJ (1997) Trends Biotechnol 15:448

91. Khrapko K, Coller HA, Hanekamp JS, Thilly WG (1998) Nucleic Acids Res 26:5738
92. Jain RK, Samanta SK, Rani M (1998) Lett Appl Microbiol 26:265
93. Brena-Valle M, Serment-Guerrero J (1998) Mutagenesis 13:637
94. Greener AL (1996) US Patent 5,552,314
95. Ollis DF (1982) Phil Trans R Soc London B297:617
96. Sharma UK (1993) Curr Sci 64:283
97. Bingle LEH, Thomas CM (2001) Curr Opin Microbiol 4:194
98. Dunn A, Davidson AM, Day MJ, Randerson PF (1995) Microbiology 141:63
99. Bhattacharya P, Roy D (1995) J Ferment Bioeng 80:520
100. Löser C, Ray P (1996) Acta Biotechnol 16:271
101. Satish G, Kavitha S, Chindambaram M (1998) Bioprocess Eng 18:343
102. Espinosa M, Cohen S, Couturier M, Del Solar G, Diaz-Orejas R, Giraldo R, Janniere L, Miller C, Osborn M, Thomas CXM (2000) Plasmid replication and copy number control. In: Thomas CM (ed) Horizontal gene pool. Harwood Academic Publishers, Amsterdam, Netherlands, p 1
103. Boudrant J, Le B, Fournier F, Fonteix C (2001) Modelling of segregational plasmid insta-bility of a recombinant strain suspension of *Escherichia coli*. In: Merten O-W, Mat-tanovich D, Lang C, Larsson G, Neubauer P, Porro D, Postma P, Teixeira de Mattos J, Cole JA (eds) Recombinant protein production with prokaryotic and eukaryotic cells. Kluwer Academic Publishers, New York, p 125
104. Imanaka T, Aiba S (1981) Ann NY Acad Sci 369:1
105. Weber AE, San KY (1989) Biotechnol Tech 3:397
106. Miao FT, Kompala DS (1993) Biotechnol Bioeng 42:708
107. Siegele D, Campbell L, Hu JC (2000) Methods Enzymol 305:499
108. Hegemann JH, Klein S, Heck S, Guldener U, Niedenthal RK, Fleig U (1999) Yeast 15:1009
109. Batard P, Jordan M, Wurm F (2001) Gene 270 :61
110. Lee DC, Kim HS (1996) Biotechnol Tech 10:937
111. Stiegelmeier C (1995) PhD thesis, University Bielefeld
112. Kim CH, Lee JY, Kim MG, Song KB, Seo JW, Chung BH, Chang SJ, Rhee SK (1998) J Ferment Bioeng 86:391
113. Georgiou G, Shuler ML, Wilson DB (1988) Biotechnol Bioeng 32:741
114. Friehs K, Schügerl K (1990) In: DECHEMA Biotechnology Conferences 4B, VCH Wein-heim, p 705
115. Morsey MA (1999) US Patent 5,922,583
116. Carriere M, Neves C, Wils P, Soubrier F, Mahfoudi, Scherman D (2001) Pharma Sciences 11:5
117. Actis LA, Tolmasky ME, Crosa JH (1999) Front Biosci 3:D43
118. Reisinger EC, Allerberger F (1999) Wien Klin Wochenschr 111:537
119. William SG, Cranenburgh RM, Weiss AME, Wrighton CJ, Sherratt DJ, Hanak JAJ (1998) Nucleic Acids Res 26:2120
120. Sherratt DJ, Williams SG, Hanak JAJ (1999) US Patent 5,972,708
121. Brey RN, Fulginiti JP, Anilionis A (1995) US Patent 5,919,663
122. Nakayama K, Kelly SM, Curtiss R III (1988) Bio/Technology 6:693
123. Matsui T, Sato H, Sato S, Mukataka S, Takahashi J (1990) Agric Biol Chem 54:619
124. Sakoda H, Imanaka T (1990) J Ferment Bioeng 69:75
125. Nudel BC, Pueyo MG, Judewicz ND, Guilietti AM (1989) Antonie Leeuwenhoek J Micro-biol 56:273
126. Skogman SG, Nilsson J (1984) Gene 31:117
127. Porter RD, Black S, Pannuri S, Carlson A (1990) Bio/Technology 8:47
128. Morsey MA (1997) WO Patent 9714,805
129. Fiedler M (2001) Gene 274:111
130. Gerdes K, Jacobsen JS, Franch T (1997) Plasmid stabilization by post-segregational killing. In: Setlow JK (ed) Genetic engineering: principles and methods, vol 19. Kluwer Academic Publishers, New York, p 49
131. Jensen RB, Gerdes K (1995) Mol Microbiol 17:205

132. Easter CL, Sobecky PA, Helinski DR (1997) J Bacteriol 179:6472
133. Lehnherr H, Maguin E, Jafri S, Yarmolinsky MB (1993) J Mol Biol 233:414
134. Galen J (2000) WO Patent 2,000,032,047
135. Pecota DC, Kim CS, Wu K, Gerdes K, Wood T (1997) Appl Environ Microbiol 63:1917
136. Nagel JHA, Gultyaev AP, Gerdes K, Pleij CWA (1999) RNA 5:1408
137. Rawlings DE (1999) FEMS Microbiol Lett 176:269
138. Weaver K, Greenfield T, Ehli E, Kirschenmann T, Heine M, Walz K (2001) Plasmid 45:
     156
139. Sayeed S, Reaves L, Radnedge L, Austin S (2000) J Bacteriol 182:2416
140. Cooper TF, Heinemann JA (2000) Proc Natl Acad Sci USA 97:12,643
141. Wojcik K, Wieckiewicz J, Kuczma M, Porwit-Bobr Z (1993) Acta Microbiol Pol 42:127
142. Summers DK, Sherratt DJ (1984) Cell 36:1097
143. Kim BG, Shuler ML (1991) Biotechnol Bioeng 37:1076
144. Bellani M, Nudel C, Rivas CS (1997) Biotechnol Lett 19:331
145. Gerlitz M, Hrabak O, Schwab H (1990) J Bacteriol 172:6194
146. Kristensen CS, Eberl L, Sanchez-Romero JM, Givskov M, Molin S, De Lorenzo V (1995)
     J Bacteriol 177:52
147. Rosche TM, Siddique A, Larsen MH, Figurski DH (2000) J Bacteriol 182:6014
148. Easter CL, Schwab H, Helinski DR (1998) J Bacteriol 180:6023
149. Eberl L, Kristensen CS, Givskov M, Grohmann E, Gerlitz M, Schwab H (1994) Mol
     Microbiol 12:131
150. Grohmann E, Stanzer T, Schwab H (1997) Microbiology 143:3889
151. Summers D (1998) Mol Microbiol 29:1137
152. Kim J-Y, Kang HA (1996) J Ferment Bioeng 82:495
153. Smith MA, Bidochka MJ (1998) Can J Microbiol 44:351
154. Gerdes K, Moller-Jensen J, Bugge Jensen R (2000) Mol Microbiol 37:455
155. Hayes F, Austin S (1994) J Mol Biol 243:190
156. Surtees JA, Funnell BE (2001) J Biol Chem 276:12,385
157. Jensen RB, Gerdes K (1997) J Mol Biol 269:505
158. Jensen RB, Lurz R, Gerdes K (1998) Proc Natl Acad Sci USA 95:8550
159. Jensen RB, Grohmann E, Schwab H, Diáz-Orejas R, Gerdes K (1995) Mol Microbiol
     17:211
160. Weitao T, Dasgupta S, Nordstrom K (2000) Mol Microbiol 38:392
161. Davis MA, Martin KA, Austin SJ (1992) Mol Microbiol 6:1141
162. Bouet JY, Surtees JA, Funnell BE (2000) J Biol Chem 275:8213
163. Radnedge L, Youngren B, Davis M, Austin S (1998) EMBO J 17:6076
164. Motallebi-Veshareh M, Rouch DA, Thomas CM (1990) Annu Rev Genet 23:37
165. Lemonnier M, Bouet JY, Libante V, Lande D (2000) Mol Microbiol 38:493
166. Miller C, Cohen SN (1999) J Bacteriol 181:7552
167. Bignell CR, Haines AS, Khare D, Thomas CM (1999) Mol Microbiol 34:205
168. Inagawa T, Kato J, Niki H, Karata K, Ogura T (2001) Mol Gen Genom 265:755
169. Moller-Jensen J, Jensen RB, Gerdes K (2000) Trends Microbiol 8:313
170. Edler C, Friehs K, Schügerl K (1989) DECHEMA Biotechnol Conf 3A:549
171. Kearny K, Fitzgerald GF, Seegers JF (2000) J Bacteriol 182:30
172. Lin Z, Mallavia LP (1999) J Bacteriol 181:1947
173. Kalnin K, Stegalkina S, Yarmolinsky M (2000) J Bacteriol 182:1889
174. Bentely WE, Kompala DS (1989) Biotechnol Bioeng 33:49
175. Paulsson J, Ehrenberg M (2001) Q Rev Biophys 34:1
176. Goss PJ, Peccoud J (1999) Pacific Symposium on Biocomputing 65
177. Ogden KL, Davis RH, Taylor AL (1992) Biotechnol Bioeng 40:1027
178. Glick BR (1995) Biotechnol Adv 13:247
179. Corchero JL, Villaverde A (1998) Biotechnol Bioeng 58:625
180. Bhattacharay SK, Dubey AK (1995) Biotechnol Lett 17:1155
181. Miao F, Kompala DS (1989) Ann Biochem Eng Symp 19 Meeting 47
182. Wrigley-Jones CA, Richards H, Thomas CR, Ward JM (1993) Biotechnol Bioeng 41:148

183. Ryan W, Parulekar SJ, Stark BC (1989) Biotechnol Bioeng 34:345
184. Li H, Bhaduri SA, Magee WE (1998) Appl Environ Micorbiol 64:1812
185. Beal C, D'Angio C, Corrieu G (1998) Biotechnol Lett 20:679
186. Masakazu S, Hiroyuki K, Akiko T, Hiroshi M, Katsuaki S, Tsuyoshi N (1998) US Patent 5,756,347
187. Gupta R, Sharma P, Vyas VV (1995) J Biotechnol 41:29
188. O'Kennedy R, Houghton CJ, Patching JW (1995) Appl Microbiol Biotechnol 44:126
189. Shu S, Shuler ML (1992) Biotechnol Bioeng 40:1197
190. Huang CT, Peretti SW, Bryers JD (1994) Biotechnol Bioeng 44:329
191. Gupta JC, Pandey G, Mukherjee KJ (2001) Enzyme Microb Technol 28:89
192. Koch V, Rüffer H-M, Schügerl K, Innertsberger E, Menzel, H, Weis J (1995) Process Biochem 30:435
193. Hempel C, Erb RW, Deckwer WD, Hecht V (1998) Biotechnol Bioeng 57:62
194. Vyas VV, Gupta S, Sharma P (1994) Enzyme Microb Technol 16:240
195. Stephens ML, Christensen C, Lyberatos G (1992) Biotechnol Prog 8:1
196. Patnaik PR (1994) Can J Chem Eng 72:929
197. Nayak DP, Vyas VV (1999) J Microbiol Biotechnol 15:73
198. Barbotin JN (1994) Ann NY Acad Sci 721:303
199. Craynest M, Mater D, Barbotin JN, Truffaut N, Thomas D (1996) Prog Biotechnol 11:452
200. Mukherjee KJ, Deb JK, Ramanchandran KB (1996) Enzyme Microb Technol 18:555
201. Lu CT, Guillan A, Roca E, Nunez MJ, Lema JM (2000) Biotechnol Lett 22:1247
202. Tierny Y, Hounsa CG, Hornez JP (1999) Microbios 97:39
203. Chaves AC, Abath FGC, Lima-Filho JL, Cabral JMS, Lucena-Silva N (1999) Bioprocess Eng 21:355
204. Choi S-J, Park D-H, Chung S-I, Jung K-H (2000) J Microbiol Biotechnol 10:321
205. Pan SH, Malclom BA (2000) Biotechniques 29:1234
206. Mari YM, Espinosa AE, Ubieta R, Batista OF, Fernandez ML (1999) Biochem Biophys Res Com 258:29
207. Henry KL, Davis RH, Taylor AL (1990) Biotechnol Prog 6:7
208. Peredelchuk MY, Bennett GN (1997) Gene 187:231
209. Elo P, Semy S, Kereszt A, Nagy T, Papp P, Orosz L (1998) FEMS Microbiol Lett 159:7
210. Balbas P, Alvarado X, Bivar F, Valle F (1993) Gene 136:211
211. Diederich L, Rasmussen LJ, Messer W (1992) Plasmid 28:14
212. Boyd D, Weiss DS, Chen JC, Beckwith J (2000) J Bacteriol 182:842
213. Mahato RI, Smith LC, Rolland A (1999) Adv Genet 41:95
214. Nordstrom JL (1999) Drug Target Deliv 10:15
215. MacLaughlin FC, Li S, Li Y, Fewell J, Rolland A, Smith LC (2000) NATO Science Series, Series A: Life Sciences 323:81
216. Butterfield LH, Ribas A, Economou JS (1999) DNA and dendritic cell-based genetic immunization against cancer. In: Lattime EC, Gerson SL (eds) Gene therapy of cancer: translational approaches from preclinical studies to clinical implementation. Academic Press, San Diego, p 285
217. Cooper MJ (1999) Non-infectious gene transfer and expression systems for cancer gene therapy. In: Lattime EC, Gerson SL (eds) Gene therapy of cancer: translational approaches from preclinical studies to clinical implementation. Academic Press, San Diego, p 77
218. Maecker HT, Syrengelas A, Levy R (2000) Methods Mol Med 29:221
219. Hersh EM (1999) Cancer gene therapy by direct transfer of plasmid DNA in cationic lipids. In: Lattime EC, Gerson SL (eds) Gene therapy of cancer: translational approaches from preclinical studies to clinical implementation. Academic Press, San Diego, p 319
220. Schatzlein AG (2001) Anti-Cancer Drugs 12:275
221. Baumgartner I, Isner JM (2001) Annu Rev Physiol 63:427
222. Weihl C, Macdonald RL, Stoodley M, Luders J, Lin G (1999) Neurosurgery 44:239
223. Bonadio J (2000) Adv Drug Delivery Rev 44:185

224. Bonadio J (2000) J Mol Med 78:303
225. Lee DJ, Takabayashi K, Corr M, Raz E (2001) Principles of genetic immunization. In: Hengge UR, Volc-Platzer B (eds) Thes skin and gene therapy. Springer, Berlin Heidelberg New York, p 177
226. Abdelnoor AM (2001) Curr Drug Targets Immune Endocr Metab Disord 1:79
227. Mor G, Eliza M (2001) Mol Biotechnol 19:245
228. Gregersen JP (2001) Naturwissenschaften 88:504
229. Pachuk C, McCallus DE, Weiner DB, Satishchandran C (2000) Curr Opin Mol Ther 2:188
230. Chen Z, Kurata T, Tamura S-I (2000) Jpn J Infect Dis 53:219
231. Inchauspe G (1999) J Hepatol 30:339
232. Eo SK, Pack C, Kumaraguru U, Rouse BT (2001) Expert Opin Biol Ther 1:213
233. Gradoni L (2001) Vet Parasitol 100:87
234. Prazeres DMF, Ferreira GNM, Monteiro GA, Cooney CL, Cabral JMS (1999) Trends Biotechnol 17 169
235. Schleef M, Schmidt T, Flaschel E (2000) Dev Biol 104:25
236. Levy MS, O'Kennedy RD, Ayazi-Shamlou P, Dunnill P (2000) Trends Biotechnol 18:296
237. Ferreira GN, Monteiro GA, Prazeres DM, Cabral JM (2000) Trends Biotechnol 18:380
238. Schmidt T, Friehs K, Schleef M, Voss C, Flaschel E (1999) Anal Biochem 274:235
239. Breul A, Muller M, Hebel H, Vu J, Schorr J (2001) Cell 33:232
240. Horn NA, Marquet M, Meek JA (2000) Biotechnol Bioprocess 25:329
241. Sullivan SM (1999) Curr Opin Drug Discovery Dev 2:129
242. Clark DW, Ciccarelli RB (2000) Biotechnol Bioprocess 25:347
243. Lockie T, Herweijer H, Zhang G, Budker V, Wolff JA (1999) Drug Target Deliv 10:235
244. Horton HM, Parker SE (2001) Methods Mol Med 65:175
245. Horton HM, Parker SE (2001) Methods Mol Med 65:185
246. Hengge UR, Tschakarjan E, Mirmohammdsadegh A, Goos M, Meyer HE (2001) Uptake of DNA by keratinocytes. In: Hengge UR, Volc-Platzer B (eds) The skin and gene therapy. Springer, Berlin Heidelberg New York, p 81
247. Tang F, Hughes JA (2001) Methods Mol Med 65:79
248. Pichon C, Goncalves C, Midoux P (2001) Adv Drug Delivery Rev 53:75
249. Gregoriadis G (1999) Curr Opin Mol Ther 1:39
250. Ando S, Putnam D, Pack DW, Langer R (1999) J Pharm Sci 88:126
251. Ciccarelli RB, Pachuk CJ, Samuel M, Winter LA, Satishchandran C (2000) Methods Mol Med 29:473
252. Lechardeur D, Sohn KJ, Haardt M, Joshi PB, Monck M, Graham RW, Beatty B, Squire J, O'Brodovich H, Lukacs GL (1999) Gene Ther 6:482
253. Hengge UR, Dexling B, Udvardi A, Volc-Platzer B, Mirmohammdsadegh A (2001) Safety and pharmacokinetics of naked plasmid DNA: studies on dissemination and ectopic expression. In: Hengge UR, Volc-Platzer B (eds) The skin and gene therapy. Springer, Berlin Heidelberg New York, p 67
254. Han S-O, Mahato RI, Sung YK, Kim SW (2000) Mol Ther 2:302
255. Poxon SW, Hughes JA (1999) J Pharm Sci Technol 53:314
256. Evans RK, Xu Z, Bohannon KE, Wang B, Bruner MW, Volkin DB, (2000) J Pharm Sci 89:76
257. Anchordoquy TJ, Koe GS (2000) J Pharm Sci 89:289

Received: October 2002

Adv Biochem Engin/Biotechnol (2004) 86: 83–158
DOI 10.1007/b12441

Bioprocesses for the Manufacture of Ingredients for Foods and Cosmetics

Peter S. J. Cheetham

Zylepsis Ltd, 6 Highpoint, Henwood Business Estate, Ashford, Kent TN24 8DH, UK
E-mail: p.cheetham@zylepsis.co.uk

Abstract This chapter describes the rapid growth in the use of biotransformation processes to manufacture food and cosmetic ingredients. Newly introduced processes include those to produce both ingredients not previously available, and also improved processes to make ingredients previously produced by chemical synthesis, fermentation or extraction from natural sources.

What can we learn from what has already been achieved that will help us in the future? This question is very important because the key challenge is not just to do innovative research, but also to develop such research into cost-effective industrial scale processes that deliver products of proven utility to the end customer, and at prices that give good returns on R&D investments to the manufacturer. Therefore biotransformation R&D information is presented in the real-life context of manufacturing, IP, regulatory and safety, product costs and quality etc.

Examples of new ingredients and processes are given that illustrate the great variety of ingredients produced biochemically, and the range of raw materials and enzyme reactions used to make them. These examples also illustrate how the technical advances made in developing these processes and products are inseparable from economic factors, especially production costs and functional benefits that create the demand and set the performance, cost, and quality standards for the product. Many of the key factors necessary for the translation of research into commercial successes are identified and described, as well as some special features that have helped individual processes to become successful. This review also proves that a key factor for success is management that can successfully integrate and implement the large number of different technical and commercial factors involved. The wide range of examples provided also prove that very many of the technical advances made in the field of applied biocatalysis have actually been made while developing processes for food and cosmetic ingredients; rather than for pharmaceuticals as is very often assumed. Consequently it is hoped that this review will help future research efforts, and especially aid the development of research into commercial success in this field.

List of Abbreviations

5′-AMP	5′-Adenosine monophosphate
5′-GMP	Guanosine monophosphate
5′-IMP	5′-Inosine monophosphate
ADH	Alcohol dehydrogenase
ATC	Amino-Δ2-thiazoline-4-carboxylate
cDNA	Copy DNA
CO_2	Carbon dioxide
CoA	Coenzyme A
DHA	Docosahexaenoic acid
DHAA	Dehydroascorbic acid
DNA	Deoxyribonucleic acid
DSP	Downstream processing
EC	European community
EPA	Eicosapentaenoic acid
G	Glucose
GI	Glucose isomerase
HFCS	High fructose corn syrup
IP	Intellectual property
ISPR	In situ product recovery
LBG	Locust bean gum
l	Litres
M	Million
MSG	Monosodium glutamate
PCR	Polymerase chain reaction
R&D	Research and development
RNA	Ribonucleic acid
SWOT	Strengths: Weaknesses: Opportunities: Threats
TM	Trademark
tpa	tonnes per annum
UV	Ultraviolet

This chapter is dedicated to the late Professor Malcolm D. Lilly F.R.S., F. Eng. whose intellect, vision and enthusiasm made him a Founder of Biochemical Engineering, as a teacher, administrator and researcher, resulting in a number of industrial successes such as the first use of an immobilised enzyme for the manufacture of semi-synthetic penicillin antibiotics. His efforts resulted in the first University Department of Biochemical Engineering in the UK and influenced developments throughout the world, particularly in Europe [95]. However, perhaps his most important memorial is the great number of students and co-workers who have benefited immeasurably from working with 'MDL'.

1
Introduction and Objectives

This chapter has several objectives. The first is to demonstrate the rapid growth and extensive use of enzymes and microorganisms as the biocatalysts for the manufacture of functional ingredients for use in food and personal care consumer products. These include processes for both traditional and new and emerging products. It describes the wide range of different enzyme reactions used, but only when there is good evidence that they are already used on a manufacturing scale. The only exception to this exclusive use of large-scale bioprocesses as examples, is for those few aspects considered to limit new process and product development; such as new biocatalyst discovery and development.

The chapter also demonstrates how technical advances are indivisible from economic factors in determining success or failure, especially the efficacy and ease of use of ingredients in end products that creates value for consumers. Examples range from long established products such as ascorbic acid and high fructose corn syrup, through to more recently established speciality ingredients such as L-carnitine and isopropylmyristate/palmitate, and then to new and emerging materials such as theanine and vanillin; that altogether create a large and growing economic value.

Many of the challenges that have to be overcome are illustrated, and also the key factors required for the development of commercially successful processes and products. These include the identification of new active materials as potential new products; and also the discovery and development of new biocatalysts and biocatalyst based processes that can produce products that meet the cost and quality targets set by customers. Success relies on several factors. These include processing options, such as in situ product recovery methods, the importance of process integration approaches, the requirement for identifying good targets by market research, and then effective new product development. Success involves the careful integration of new science with many different technical, engineering, financial and legal aspects; such as safety, labelling and regulatory requirements, IP and patenting, raw material sourcing, process and product costing, and product formulation, and packaging. All of these aspects are absolutely essential if successful new products are to be developed that will create sustained consumer demands for them.

Finally, this chapter shows that much very interesting science and technology has been developed as a result of applying biocatalysis technology to food ingredients. In fact, surprisingly to many, a high proportion of the new advances in biocatalysis and its commercial applications have actually arisen from R&D aimed at food ingredients, rather than from pharmaceutical and healthcare R&D as is generally assumed. More recently applications of biocatalysts for he manufacture of cosmetic and personal care ingredients have also begun to make a significant contribution.

2
Economic Factors

Whereas most of us only occasionally require treatment with drugs, we all eat and drink processed foods and beverages several thousand times a year, very many of which contain ingredients produced using biocatalysts. Therefore food ingredients, especially for processed foods and beverages, have a big social value as well as creating big business. Ingredient values depend on their 'in-product' functionality when formulated into complete foods and beverages. This depends a lot on how an ingredient interacts with the other ingredients, which can affect not only their activities, but also oxidative, microbiological and physical stability and other factors. Therefore assessment of flavour, mouthfeel, visual appearance, microbial and oxidative stability, emulsifying, foaming and phase stabilising activities, and nutritional properties is essential. This is done by means of manufacturing trials, and shelf-life and taste panel tests etc.

Real and sustained commercial success is only achieved once an ingredient has established a strong advantageous market position over its competitors, both direct and indirect competitors. This advantage can be in one or more of price, functional performance, quality, consumer perception etc. (Table 1). Therefore in food, beverage and most other industries it is increasingly accepted that most innovations arise directly from 'market pull' incentives rather than from 'technical push' forces, and that the need for biocatalysts and bio-processes is actually a derived-demand arising from customers needs. This means that the market value of the products made by biocatalysts, the costs of production; and the patentability, safety, and consumer acceptability of the product must all be acceptable for the new product to succeed. The importance of 'market pull' forces also emphasises the importance of discovering new functional ingredients and of applications studies to maximise their use in end-products. Some of the important commercial features and driving forces in operation are summarised in Table 1.

So how does a company progress from either a request from a customer company, or from a research report describing the discovery of a new material, probably in very small quantities, in low yield, and only partially pure form, through to a product that can be reliably manufactured, cheaply and in a pure form, and on a multi-tonne scale and then sold world-wide, safely and legally, and that provides significant benefits to customers, including the manufacturers of major branded consumer goods? So although it is useful to ask the standard question "What technical characteristics make for a good biocatalysts?", the most useful questions to ask are "What are the desirable characteristics of successful bioproducts that create high added values by meeting customers

Table 1 Commercial features and driving forces

- Research often begins by analysing natural and traditional products so as to identify what are the active materials that give them their valuable and distinctive functional characteristics
- Very many natural raw materials are chiral, and are highly oxygenated, making them very suitable as precursors for the production of chiral products
- Demand is increasing for ingredients with new and improved functional characteristics, especially related to healthy and environmentally friendly properties and suitable for use in processed and convenience products. For instance polycyclic musks are the largest volume fragrance materials
- Ingredient performance and product efficacy is very dependant on how compatible the various ingredients are. They not only have to have the physical and chemical characteristics to allow then to be easily formulated into the end-product; but also have to be sufficiently stable to pH, heat etc, both during processing and when over long periods of storage to allow long shelf-lives for the products they are used in
- The food, beverage and cosmetic ingredient industries are not as technically sophisticated as the pharmaceuticals industry, and so processes have to be relatively easy to scale up, integrate into existing operations, and commission as manufacturing processes
- There is a strong incentive to continuously improve processes, especially because of cost pressures from competitors, particularly if these are fully optimised and depreciated chemical processes
- Any one ingredient may be used in a wide range of products, so that very adaptable marketing and technical support is required
- Ingredients often have to be stable in hostile environments when formulated into end products. An extreme example is fragrance materials used in household bleach products. Other examples are detergents where enzymes are used in encapsulated form
- The value of food and personal care products, and hence of ingredients for them depends very much on consumer satisfaction features such as flavour and mouth feel for a food, moisturising activity and skin feel for a skin cream. However these can vary a lot with the age and culture etc of the consumer
- Consumers valuation of food and personal care products also depends on their perception of risk. So genetically engineered foods are often viewed with suspicion, which may limit the uptake of this technology; although cloned food enzymes, such as chysomin for cheese making are well acceptable as food processing aids.
- The high cost of obtaining regulatory approval for new molecules as food additives, such has been done for xanthan gum and Aspartame, is one of the limits on innovation
- Increasing globalisation in terms of sources of raw materials, product ranges and companies 'interests'. This means that a proposed new ingredient needs to be suitable for use in European, American and Asian products
- Opportunities for innovations are as likely to occur from changes in regulatory, fiscal and patent and raw material supply positions as from basic scientific advances, which can lead to new products that offers outstandingly new functional benefits, or new processes which allow manufacture at much lower costs
- Success is more likely when a new product fits into a companies existing product range, so that it can benefit from the existing distribution, marketing and sales networks, and also from the customer credibility already established by the pre-existing products

needs and solving their problems?", and then "Can we make such a product using a well integrated and cost-effective bioprocess?".

I have tried to answer these questions from experience gained in both large and small companies, of applying enzyme and microbial technologies to making functional ingredients for foods, beverages and personal care products [1, 2]. Therefore rather than refer to academic research that concentrates mostly on scientific novelty, I have drawn heavily on recent industrial R&D as this is largely driven by market needs and also addresses many of the commercial factors needed for success, especially market needs, manufacturing requirements, product safety, functional efficacy and cost.

The starting point for any process are ingredients that have been traditionally consumed, either because they naturally occur in fruit, cereals or vegetables etc., or are produced in traditional processed foods and beverages. Originally active ingredients such as flavours, colour or antioxidants were only produced in a dilute form, as just one part of the whole food or beverage, such as cheese, butter, yoghurt, beer or wine, depending on the use of Saccharomyces, Lactobacillus strains etc; and in the East products such as miso, soy sauce and natto using strains such as *Aspergillus oryzae* and *sojai* and *Bacillus natto*. These key ingredients greatly enhance consumer satisfaction and hence the value of these ingredients. Food and cosmetics ingredients include a wide range of different functional materials produced as pure active chemicals (Table 2). From these traditional products improvements have been made by introducing enzyme steps, particularly for the hydrolysis of proteins, oils and lipids, polysaccharides such as pectins etc, in a wide range of products including brewing, baking, starch processing, dairy and fruit processing etc. This scientifically-based approach began over 100 years ago, as the first enzyme patent dates from 1894, invented by Takamine for the use of diastatic amylases. The scale of economic activity generated by food and cosmetic ingredients is indicated in Table 3 and Fig. 1. Table 3 shows the approximate production volumes and sales of food ingredients, and also the inverse relationship between the price of a product and

Table 2 What are food and cosmetic ingredients?

Foods	Cosmetics
Emulsifiers	Oils
Stabilisers and thickeners	Emollients
Fibre (soluble and insoluble)	Surfactants
Sweeteners	Conditioning polymers
Flavours	UV absorbers
Colours	Skin tanning agents
Preservatives (antioxidants and antimicrobials)	Skin lighteners
Vitamins	Fragrances
Spice extracts	Exfoliating agents
Oils and lipids	Antiwrinkle agents
Protein hydrolysates	Antioxidants
Other micronutrients	Antimicrobials
Etc., etc.	Etc., etc.

its volume of sales. Note that High fructose syrups, which is the biggest enzyme-produced product in volume terms, also generates sales in excess of the sales of the worlds best selling drug, the Statin Lipitor, from Pfizer. Also note that the value of these ingredients when formulated into final products for the consumer will be at least ten-fold higher. In addition to the data presented in Fig. 3, other high volume commodity products include acetic acid (mainly produced as vinegar), ethanol (as wine, beer etc.), maltose (in the form of brewing sugars) etc. As well as being important products in their own right some commodity bioproducts are also useful as the raw materials for the manufacture of more speciality products. Low production volume speciality products include: cyclodextrins (3–500 tpa), L-malic (400 tpa), L-carnitine (150 tpa), L-alanine 500 tpa, L-cysteine 1500 tpa and L-threonine 400 tpa. Other food ingredients manufactured by fermentation include riboflavin, glucono-δ-lactone, β-carotene, erythritol and soy sauce.

A key conclusion from Table 3 is that the biggest profits can be made from products with very big volumes of production, which requires a large market share to be won, for instance by strong advertising or price discounting versus competitors, or from speciality products that can be sold at much greater profit margins, which compensates for their much smaller volume of sales. For instance the 8 M tpa production of high fructose corn syrups selling at $0.8/kg, with a 5% profit margin, generates $320 M pa; whereas sales of 5000 tpa of flavour nucleotides at $35/kg, and a profit margin of 25%, generates just $44 M pa profits. Another conclusion is that many of these successful products depend on a microbial strain with a rather unusual metabolic feature. For instance L-glutamic acid production depends on the use of a strain of *C. glutamicum* that lacks α-ketodehydrogenase, and citric acid manufacture depends on *A. niger* strains that have inactivated aconitase. Figure 1 shows the world wide

Table 3 Production volumes and sales of food ingredient commodity products made by bioprocesses

Product	Production volume (tpa×10³)	Price ($/kg dry solids)	Sales ($M pa)
Glucose syrup	≥20,000	0.6	12,000
High fructose syrup	8,000	0.8	6,400
Lipitor	–	–	6,000
L-Glutamic acid	640	1.4	910
Citric acid	510	2.75	1,400
L-Lysine	150	3.0	450
L-Lactic acid	60	2.5	150
Ascorbic acid	53	8.0	425
Gluconic acid	46	2.0	93
Xanthan gum	33	12.4	408
Niacin	15	8.3	125
L-Phenylalanine	11	18.0	198
Flavour nucleotides	10	35.0	350
L-Aspartic acid	8.8	4.9	43

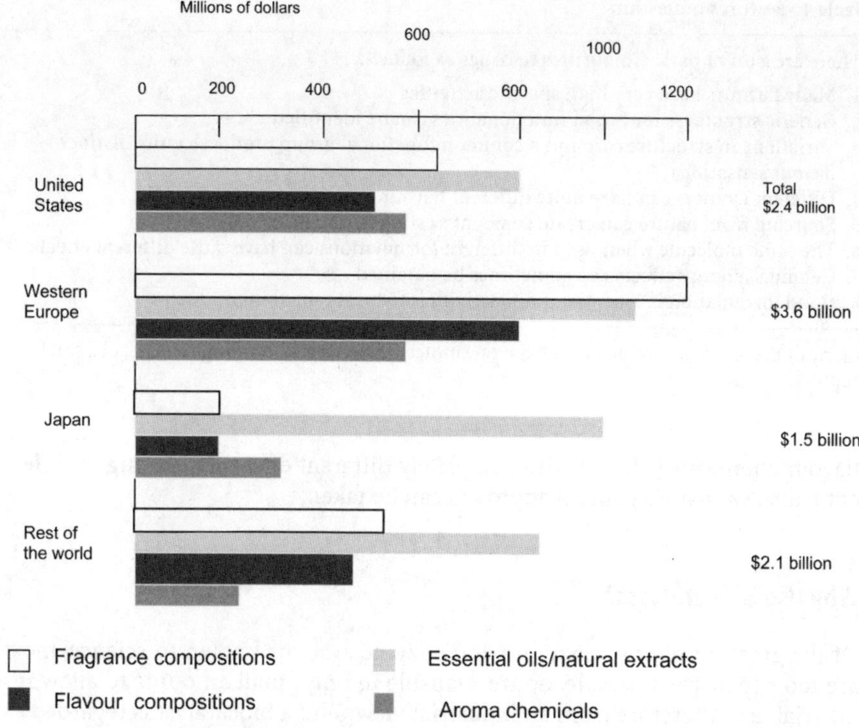

Fig. 1 Economic values of food, beverage and cosmetic ingredients: world consumption of flavour and aroma products [3]

sales of flavour and fragrance chemicals. It is of interest to note the quite different pattern of sales/ consumption in Japan (Fig. 1).

3
A Comparison with Pharmaceuticals

Some aspects of food and cosmetic ingredients are quite analogous to pharmaceuticals. For instance Table 4 shows that many of the key functional characteristics of flavours are due to their acting by binding to taste receptors, over a range of avidities of ca. 10^{16} fold. This is the same mechanism of action as for many drugs. For instance the low odour thresholds of many flavours demonstrates that they have high specific activities, generic structures for some particular types of flavour qualities can be identified, and variations in structure around a common functional group, for instance an aliphatic or aromatic aldehyde functionality, produces quite different flavour sensations, with some of these aldehydes, such as benzaldehyde and vanillin made using bioprocesses. Another complicating factor is due to the biological origin of most food and cosmetic ingredients. For instance changes in enzyme levels following harvesting of fruits can have dramatic effects on the concentrations of the required

Table 4 Features of flavours

There are a lot of basic similarities to drugs as follows:

1. Many flavours have very high specific activities
2. Generic structures for broad functionalities can be identified
3. Variations in structures around a common functional group produces quite distinct flavour sensations
4. Different isomers can have quite different flavour properties
5. Sourcing from nature can create some interesting variabilities
6. The same molecule when used in different formulations can have quite different effects
7. Genuine synergy effects can sometimes be obtained
8. Good formulation is essential if full activity is to be obtained in products

Many of these effects are due to the flavour molecules acting by binding to taste and smell receptors.

flavour chemicals [4], and with completely different effects occurring in different fruits, so that no general approach can be taken.

4
Why Use Biocatalysts?

Of the great number of enzymes and enzyme systems known to science, most are too expensive, unstable, or are available in only small amounts to allow industrial use. Therefore from an industrial viewpoint a biocatalyst is regarded as an enzyme or microorganism that is so cheap, active and easy to use that it

Type of biocatalysts used	Product made		
	Commodity	**Speciality**	**Intermediate**
Enzyme	High fructose syrup	Cyclodextrins	Isomaltulose (Isomalt^{TM})
	5'-GMP/IMP	Aspartame^{TM}	Glucose syrup (high fructose syrups)
		Enzyme modified guar gum	
Microbial	L-lysine	L-hydroxyproline	L-sorbose (ascorbic acid)
	Xanthan gum	Trehalose	D-pantoic acid (D-pantothenic acid)
	Citric acid	L-carnitine	L-phenylalanine (Aspartame^{TM})

Fig. 2 Both enzymes and microbial strains can be used to manufacture commodity and speciality ingredients and intermediates (*end products given in brackets*)

makes a cost-effective manufacturing process possible. Biocatalysts are used in two significantly different ways. One mode of use is in biotransformation processes, in which a chemically defined product is made. Or biocatalysts are used in bioprocessing in which a chemically and physically complex material is modified so as to improve flavour, solubility, viscosity or some combination of useful properties. Biocatalysts are now successfully used as processing aids in a wide range of different industries, both for high value speciality biochemicals, intermediates, and high tonnage commodities (Fig. 2) and also for chemically undefined products such as plant hydrolysates. Biocatalysts can be used in a number of different types of bioprocesses, including de novo production by fermentation such as for citric acid; precursor fermentations as for δ- and γ-decalactone flavour chemicals, bioconversions such as for sorbitol into L-sorbose in ascorbic acid synthesis, use of permeabilised non-viable cells such for malic acid, or isolated enzymes such as lipases, proteases and amylases for a whole variety of dairy, bakery, meat and alcoholic and soft drink ingredients and end products (Table 5). A good example is the use of specially selected peptidases

Table 5 Various bioroutes for manufacture of food ingredient

Methods	Complexity	Examples
'De-novo' productions by fermentation	Multi-step	Citric acid, MSG
Precursor fermentation	Multi-step	γ and δ decalactones, methylketones
Bioconversion by non-growing cells(ascorbic acid)	Single step	D-Sorbitol to L-sorbose
	Multi-step, use of different strains in sequence	L-Alanine
	Multi-step, use of different strains in combination	L-Lysine
	Multi-step, use one strain with multiple enzyme activities	D-Trehalose (2 enzymes), L-Cysteine (3 enzymes)
Isolated 'single-use' enzymes	Single step	Phosphodiesterases (5′-GMP)
	Low water reactions	Cocoa butter equivalent oil
	Synthetic reactions	Thermolysin (Aspartame)
	Plant derived biocatalyst	Locust bean gum equivalent
	Multi-step	α-Amylase/glucoamylase and pullulanase (glucose syrups)
	Combined use	Lipases and esterases for enzyme modified cheese flavours
Immobilised enzymes	Single step	Glucose isomerase (high fructose syrup)
Immobilised cells	Single step	Isomaltulose
	Combined use	L-Alanine
	Multi-step	No example known to author

Table 6 Why use enzymes?

- To enhance yields and quality in extraction processes, e.g. glycosidases for fruit juices
- To improve the processing characteristics of raw materials, e.g. β-glucanases and β-amylase in brewing and baking
- To improve nutritional quality, e.g. the addition of enzymes to animal feeds
- To improve the functional and physio-chemical characteristics of food and beverages, e.g. chymosin treatment of milk
- To remove undesirables, e.g. use of catalase on H_2O_2 treated milk
- To reduce processing costs, e.g. use of glucoamylase to saccharify starch
- To make natural versions of ingredients, e.g. natural flavour chemicals, such as δ-decalactone and vanillin
- To make ingredients not previously available as pure chemicals, e.g. L-carnitine
- To allow the production of products not possible using chemistry or extraction technologies, e.g. trehalose
- To enable significantly purer products to be made, e.g. isopropylmyristate
- Mild conditions of reaction allow labile molecules to be made, e.g. pantothenic acid
- To allow reaction of substances present at very low concentrations, e.g. urea and α-acetolactate removal from wine and beer respectively
- To enable selective reaction with substances present in complex mixtures, including the presence of different isomeric forms of the same material(s), e.g. production of L-methionine by resolution of racemic ester mixtures

for the debittering of protein hydrolysates because of their ability to selectively remove N-terminal hydrophobic amino acids from peptides that are associated with a bitter taste. It is this versatility that is an important strength of biocatalysis. Other processes include starch processing, first into glucose syrup, and then into high fructose corn sweeteners (HFCSs). Then there are bioprocesses for flavour chemicals such as methylketones and 2-methylbutyric acid and its esters, such as ethyl-2-methylbutyrate that is characteristic of apple flavours. More recently some bioprocesses have been developed for a range of healthy ingredients such as L-carnitine and ω-6 and ω-3 fatty acids etc. For these 'nutraceutical ingredients' the challenge is to demonstrate beneficial effects without the time and expense required for clinical trials. All these manufacturing bioprocesses cover conversions of carbohydrates, proteins or lipids; and even nucleic acids in the case of GMP and IMP taste enhancers.

As well as the widely cited scientific differences there are a wide range of potential industrial advantages when enzyme based systems are used to manufacture food and cosmetic ingredients (Table 6). However these different approaches must be selected carefully to solve the problems inherent in particular processes and products. As a result a good number of biocatalyst based processes are already in use to manufacture cosmetic ingredients (Table 7). These include some processes which were originally developed for food and chemicals uses, and then found useful as cosmetic ingredients.

The easiest guide to the effectiveness of a bioprocess is the concentration of product present in the process stream leaving the bioreactor. However, obviously many other factors influence the performance of a bioprocess, especially the degree of conversion of raw materials into product, as this directly affects

Table 7 Biocatalysts used to make ingredients specifically developed for cosmetic and personal-care products

Product	Raw material	Microorganism used
Isopropylmyrisate	Isopropanol and myristic acid	*Candida antarctica lipase B*
Alkylpolyglucoside	Fatty acids	*Candida antarctica lipase B*
Hyaluronic acid	Glucose	*Streptomyces zooepidermicus*
L-Malyltyrosine ethyl ester	Tyrosine ethyl ester and malic acid	*Micrococcus caseolyticus*
Kojic acid	Glucose	*Aspergillus oryzae*
D-Pantoic acid (D-pantothenate)	DL-pantolactone	*Fusarium oxysporum*
Unsaturated fatty acid esters	Isopropylmyristate and palmitate	*Rhodococcus mutant*
Ingredients cross-applied from other product areas		
Xanthan gum	Glucose	*Xanthomonas campestris*
Citric acid	Beet molasses	*Aspergillus niger*
L-Sorbose (ascorbic acid)	Sorbitol	*Acetobacter suboxydans*
L-Hydroxyproline	L-Proline	*E. coli* containing genes cloned from *Dactylosporangium strain*
Dihydroxyacetone	Glycerol	*Gluconobacter oxidans*

End-products are in brackets where the product of the reaction is an intermediate.

the raw material costs incurred, the ease (and cost) of purifying and isolating the product, the chemical and isomeric purity of the product which affects product value, and the operational stability of immobilised biocatalysts which influences processing costs (Table 8). Obviously whereas such process parameters can often be obtained from the scientific and patent literature, the most important factor, the actual production costs for a process, can only be obtained on rare occasions. Thus although the concentration of product made is perhaps the best initial guide to process efficiency other factors can be more important in determining overall process cost and product quality so that some efficient process only operates at low product intensities. For instance glucose isomerase can only economically produce a syrup in which 42% of the glucose has been converted into fructose, so that further enrichment of the fructose content requires treatment with ion-exchange resins. This additional technology took four years to introduce, following the start of commercial glucose isomerisation operations, and is still the only method used to manufacture 55% fructose syrup.

Some of the factors influencing whether to use a microorganism itself, or isolated enzymes, are shown in Table 9. Additional opportunities are to use immobilised cells or enzymes, so as to allow reuse of the biocatalyst, and/or allow continuous processing to be undertaken. Two important factors are, first, that the biosynthetic capacity of microorganisms is limited, depending on the type of metabolite produced as the product, with microbial productivities varying over a 10,000-fold range (Fig. 3) [5]. This obviously makes it difficult to

Table 8 Examples to show the processing power of biocatalysts (details from Liese et al. 2000 [12])

High percentage conversion of raw materials to products

>99% 2,2,6-trimethylcyclohexane-1,4-dione (carotinoid intermediate)
>99% for isopropylmyrstate/palmitate production
>99.5% for isomaltulose formation

High concentration of product in product stream (g/l)

800 isopropylmyristate/palmitate
ca 320 D-pantoic acid
333 L-alanine
199.5 L-aspartic acid

High yield of product from raw materials (%)

100% nicotinamide
99.5 L-carnitine
95 → 99.9% L-aspartic acid (depending on the particular process used)

High enantiomeric excess of product (%)

>99.9 Aspartame
>99.9 L-aspartic acid
>99.5 L-lysine
>99.9 L-carnitine

High chemical purity (%)

>99 L-malic acid
>99 2,2,6-trimethylcyclohexane-1,4-dione
>99.8 L-carnitine
>99 L-aspartic acid

Large scale of manufacture

Up to 250 m^3 reactors are used for lactase treatment of milk
Over 7×10^6 t of high fructose corn syrups are produced world-wide per annum

Long operational life time of immobilised biocatalyst

– 90% activity remaining after 180 days continuous operation of immobilised lactonase activity used to make D-pantoic acid
– Half-life of 243 days for L-malic acid production by immobilised cells possessing fumarase activity
– Half-life greater than 120 days for the production of L-aspartic acid using immobilised cells containing L-aspartate ammonia lyase

make vitamins or coenzymes by fermentation at the low prices required for them to become commodity ingredients. However a biotransformation approach may be more successful than fermentations, because it may be able to break the relationship between cell growth and product formation characteristic of fermentations. Advantages include eliminating the need for expensive sterile engineering, complex media, and high energy consumption because of the requirement for air compressors and agitators. This is achieved by making

Table 9   Some of the factors influencing the choice of whether to use enzymes or microorganisms as the biocatalyst

---

- Which works best (e.g. enzymes for single step hydrolysis and isomerisation reactions and microorganisms for multistep and synthetic reactions); and also as regards more specific processing requirements, for instance the biocatalysts may need to be thermostable or resistant to an organic solvent
- Ease of availability, stability etc.
- Safety, labelling and regulatory status
- Process logistics (fermentations have long down times and so are difficult to use if a continuous process is required, whereas immobilised enzymes can be easily used in a continuous process)
- Operating costs (e.g. immobilisation, to enable reuse if the biocatalyst would otherwise be an expensive cost element)
- Capital costs (sterile fermentation requires much more expensive equipment than simple hygienic food processing equipment)
- Compatibility with downstream processing

---

use of pre-formed precursor molecules and because the maintenance of cell viability during the reaction is not necessarily required.

The other important factor is that however easy or difficult the biocatalytic production of the product, the cost of product purification and isolation (downstream processing) can very often be at least 50% of the overall production costs (Fig. 4). This is important as the cost of the biocatalyst is usually less than 10% of the cost of the overall processing costs, and the total processing costs can be from 10 to 70% of the sale price of the product, depending on whether it commands a premium position in the market, or is a commodity product in hard competition with many other similar products. The practicalities of using enzymes and microbial biocatalysts are quite different. Whereas the vast majority of enzymes used are bought from enzyme supply companies, most whole cell biocatalysts are produced only by the user company, or a close

SUBSTRATE FAMILY	REACTION TYPE	TURNOVER RATE (mmol g^{-1} h^{-1})
Carbon source	Uptake, oxidation	10
Monomers/polymers	Synthesis, polymerisation	1
Amino acids	Uptake, synthesis, incorporation in protein	0.1-0.5
Bases, nucleotides	Uptake, synthesis, incorporation in RNA.	0.15
	Incorporation in DNA	0.01
Vitamins, coenzymes	Uptake, synthesis	0.001-0.005

Fig. 3   The limits to microbial productivity depend on the type of metabolite produced [5]

Product	DSP as % of Selling Price
Enzymes	50
Penicillin G	20-30
Xanthan gum	60-70

Fig. 4 Downstream processing costs for bio-products

affiliate. Therefore although the use of cells requires much more sophisticated facilities, including screening and fermentation capabilities; the process developed can be much more proprietary.

5
Established Biocatalysis Products

The earliest bioprocesses were obviously those for beer, wine, cheese, yoghurt, soy sauce etc., from which improved processes and products have been developed based on modern scientific understanding (Table 10). Following the rise of the chemical industry in the nineteenth century came other products such as ascorbic acid. This is made by a mostly chemical process, that uses just one, but vital, biotransformation step.

5.1
L-Carnitine

Building on this base, processes for some chemically defined ingredients were developed. Some examples are listed in Table 10. A good example is the manufacture of L-carnitine from butyrobetaine via 4-butyrobetaine CoA by Lonza. This process involves site-selective reaction of an unactivated carbon atom (Fig. 5). L-carnitine facilitates the transport of fatty acids across the mitochondrial membrane, and so is used as a nutritional supplement ingredient, especially in sports and health beverages. Only the L-isomer is biologically active which gives stereoselective enzyme-based processes an advantage over chemi-

Table 10 Established and well recognised bioprocessproducts include

- High fructose corn syrups
- L-Sorbose (ascorbic acid)
- Cocoa butter equivalent oil
- Isopropylpalmitate and myristate
- Aspartame
- Methylketones
- L-Aspartic acid
- L-Carnitine
  and many others

Fig. 5 New types of food ingredients – nutraceuticals

cal synthesis. The hydroxylation reaction is carried out by a carnitine-3-oxido-reductase. What is interesting is that L-carnitine has been a product for a sufficiently long period for the butyrobetaine process to be a second-generation bioprocess, replacing the original bioprocess that was based on the use of chloroacetoacetic acid ester as the precursor [6]. Furthermore Lonza have described how long it took them to develop a high yielding (300 g/l) process for butyrobetaine conversion, with further improvements becoming increasingly difficult to achieve as the concentrations of product become gradually higher.

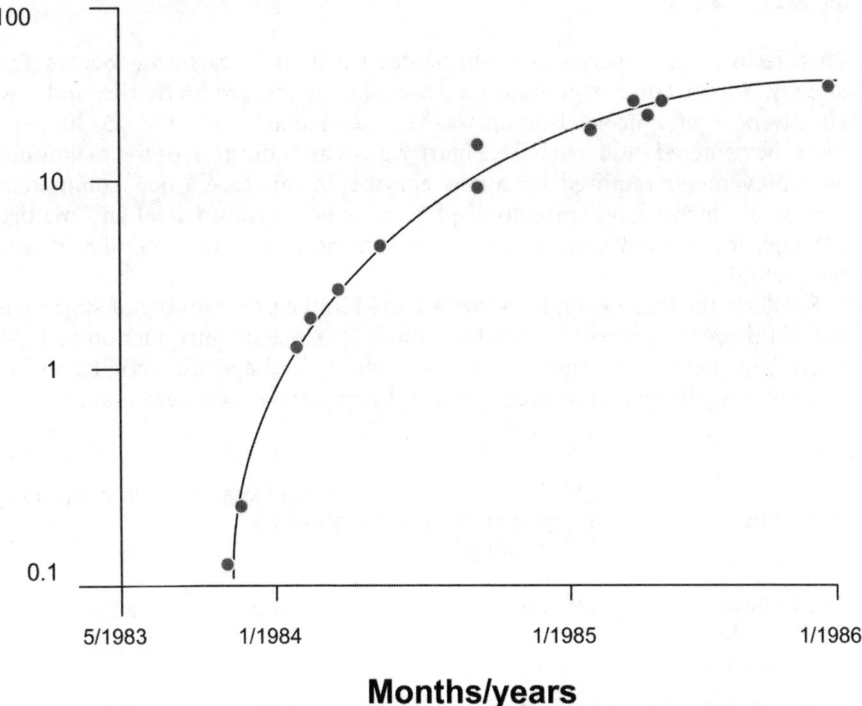

Fig. 6 An indication of the time required for research and development to optimise the concentration of a bioconversion product to a commercially viable level

This involved improving the original low yielding microbial strain, into a production strain that produces about a 2000-fold higher concentration of carnitine. This was achieved by preventing the metabolism of L-carnitine, by developing strains lacking in carnitine dehydrogenase (Fig. 6). 4-Butyrobetainyl co-enzyme A is the key intermediate. The reaction is similar to that used for fatty acid degradation, except that the oxygen atom in carnitine is derived from water, and not from divalent oxygen. Deleting the carnitine dehydrogenase from the microbe by genetic mutation increased yields by up to 30 g/l, but meant that some betaine was needed to maintain the growth and viability of the cells. Further mutation increased yields by up to 70 g/l, but was accompanied by a loss in the microbe's ability to make the butyrobetaine.

An aspect of the microorganisms physiology that greatly improves the productivity of the process is that butyrobetaine uptake is mediated by a periplasmic adenosine triphosphate-linked protein, which is closely coupled to carnitine excretion. This means that more butyrobetaine substrate is taken up by the cell for every molecule of carnitine exported. Compromise mutants of the original microbes are therefore now used for production and a 10^3-fold increase in activity has been achieved from the original mutant after two and half years of work by Lonza [7] (Fig. 6).

5.2
Glucose Isomerase

This 'technological succession effect' described for L-carnitine occurs frequently. For instance a gradual improvement in the productivities and cost effectiveness of glucose isomerases has taken place over the 25–30 years since they entered industrial use. Figure 7 gives an indication of the magnitude of improvements required for a new enzyme, in this case a new commercial version of glucose isomerase, to displace an existing commercial enzyme that although less effective, does have the advantage of customer familiarity and acceptability.

Similarly the thermostable α-amylase used in the pre-thinning of starch has evolved through at least three distinct products since its introduction in 1974. There have been improvements in heat stability and specific activity, and in particular to eliminate the need to add calcium ions, which were required as a

Achieved	'Old' GI introduced in 1974 *B. coagulans*	'New' GI from *S. murinus*	Improvement
Productivity (tonnes 42% HFCS per kg GI)	0.25-0.3	15-20	x 60
HFCS cost ($per tonne)	5	0.15	x 30

Fig. 7 Improved commercial glucose isomerase product

cofactor by the original commercial enzyme, but which inactivate glucose isomerase used later on in the corn refinery, thus requiring an ion exchange step to remove the calcium. The next improvement to be commercialised is likely to be an enzyme modified by protein engineering, that can operate at lower temperatures, and that does not have to be inactivated prior to the next (saccharification) stage in starch processing.

5.3
L-Aspartic Acid and Derived Products

This 'technological succession' effect can be wider in scope and impact. For instance, the original batch process for L-aspartic acid manufacture using immobilised cells introduced first in 1965, [8] gave way to an immobilised cell process in 1972, [9] and then to a number of much more highly active immobilised enzyme processes (Table 11). As a result L-aspartic acid has the largest production volume of the range of amino acids produced using biotransformation

Table 11 Operating characteristics of the succession of bioprocesses for making L-aspartic acid employed over the last 45 years

Suspended whole cells used batchwise (Mitsubishi 1958)

Operated at pH 9–10 and 54 °C to inactivate unwanted fumarase. The *Brevibacterium flavium* aspartase is retained in the reactor by ultrafiltration membrane, and is used in repeated batch mode. Yield is 99.99% from 1.3 M substrate

Immobilised whole cells operated continuously
i) (Tanabe Seiyaku 1973)

Uses *E. coli* aspartase in the form of whole cells immobilised in carrageenan, and operated at pH 8.5 and 37 °C. The immobilised cells have a half-life of 120 days in 1 m³ plug flow reactors and product is made in a yield of greater than 95%, and then recovered by isoelectric precipitation and then filtration

ii) (Kyowa Hakko 1974)

E. coli aspartase is used immobilised on a weak basic anion exchange resin and operated at 37 °C with a residence time of 0.75 h. The immobilised enzyme has a half-life of 18 days and converts 2 M fumarate into aspartate with a yield of 99%

iii) (BioCatalytics 1990s)

E. coli aspartase used immobilised to silica in 75L plug flow reactors with a residence time of 0.07 h. The immobilised enzyme has a half-life of 6 months at pH 8.5 and 27–37 °C. 1.5 M fumarate is converted into aspartic acid in 99.9% yield, which is recovered by isoelectric precipitation at pH 2.8

Recombinant enzyme immobilised on surface of support (Nippon Shokubai 2000)

Recombinant *E. coli* aspartase immobilised as a surface film on particles. 20% ammonium fumarate is reacted in a flow through bioreactor at 20 °C and with a residence time of 0.1 h achieving a conversion of 99.4%. The L-aspartic acid is then crystallised and the mother liquor recycled

Table 12 Amino acids manufactured using bioconversions [10]

Product	Scale of production world wide (tpa)
Enzyme processes	
L-Aspartate	7000
L-Cysteine	1500 (also by extraction)
L-Alanine	500 (also by chemical synthesis)
Precursor addition processes	
L-Threonine	4000 (also by fermentation)
L-Tryptophan	500 (also by fermentation)
L-Isoleucine	400

processes (Table 12) [10]. This immobilised cell process was a very important advance as it proved to the world that this technology could be successfully used on an industrial scale. The major impact of using a continuous immobilised cell catalyst, rather than operating a batch process, was to reduce the contribution of the catalyst to the total operating cost of the process by about tenfold [11]. In addition this process to make aspartic acid also stimulated a number of other related bioprocesses, such as the use of L-aspartic acid as the starting material to make L-alanine, the use of L-aspartic acid as a component of Aspartame, and then to make other products such as D-aspartic acid (Fig. 8), which is useful as an intermediate in Amoxycillin antibiotic synthesis. This process uses DL-aspartic acid, as the raw material. The aspartate decarboxylase not only produces L-alanine, but also unreacted D-aspartic acid which can be isolated following crystallisation of the L-alanine. This process is very efficient, operating with a 99% conversion and a 86% recovered yield of product. As a result the manufacturing process produces 317 g D-aspartate/l/day and 170 g L-alanine/l/day [12].

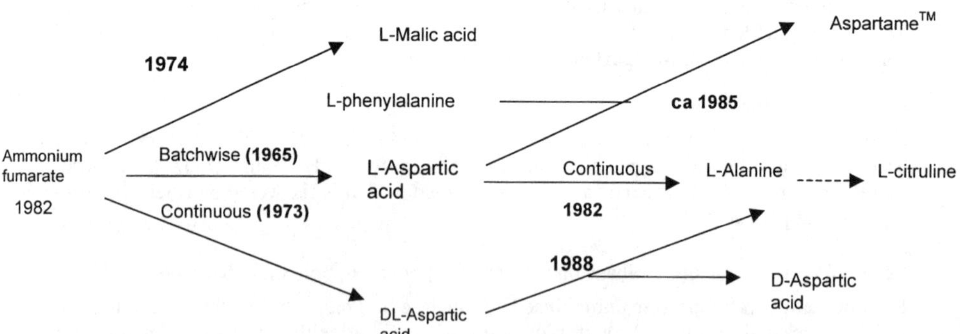

Fig. 8 Product innovation over 25 years based on L-aspartic acid and related products. Dates indicate the approximate years in which the various products were first manufactured

Other amino acids manufactured using biotransformations include L-methionine, by resolution of racemic methionine ester using immobilised aminoacylase, and L-lysine using two microbial biocatalysts together, one possessing a lactamase, and the other a α-amino-ε-caprolactam racemase that allows nearly quantitative yields to be obtained from cheap racemic precursor. The processing costs of making the L-methionine are minimised, and the product stream quality can be maintained relatively constant by regenerating the immobilised aminoacylase columns, once their activity has decayed appreciably, by desorbing the inactive enzyme from the support and adsorbing back on fresh active enzyme [13].

5.4
L-Cysteine

L-Cysteine is another amino acid produced by a biotransformation process. It is produced on a 1000 tpa scale, and used as a food additive and in hair treatment products. Originally it was obtained by chemical extract from waste hair, but this proved unsatisfactory due to environmental problems, excessive energy costs and poor supplies of hair. Ajinomoto now operate a process to produce L-cysteine from chemically synthesised D,L-2-amino-Δ2-thiazoline-4-carboxylate (L-ATC) using three different enzymes (Fig. 9) [14, 15]. These enzymes are responsible for converting the L-ATC into *S*-carbamoyl L-cysteine (L-ATC hydrolase) and then an *S*-carbamoyl-L-cysteine hydrolase forms L-cysteine. These two enzymes are augmented by ATC racemase that concerts the D-ATC isomer into L-ATC so as to ensure a stoichiometric conversion into L-cysteine product. A strain of *Pseudomonas thiazolinophilum* is used with the enzymes required being induced by the addition of D,L-ATC to the growth medium. A mutant strain has been developed in which L-cysteine degradation by cysteine desulfhydrase does not take place. Degradation can also be prevented by addition of hydroxylamine. But for production the mutant strain lacking the hydratase is used. During the reaction the L-cysteine is oxidised to L-cystine due to the vigorous aeration, which precipitates as crystals in 98% yield, and with a productivity of over 30 g/l.

5.5
Aspartame

The use of the *B. stearothermophilus* protease Thermolysin to manufacture the dipeptide ester high intensity sweetener Aspartame from protected aspartic acid and phenylalanine methyl ester is another industrially successful synthetic enzyme process [16]. Thermolysin is essential for this process, as it is the only enzyme to have the required combination of selectivity for just the α-carboxyl of aspartic acid, stereoselectivity for just the L-phenylalanine methyl ester precursor, and with no tendency to hydrolyse this methyl ester. The commercial success of Aspartame has created a derived demand for L-phenylalanine, which is now manufactured by fermentation using over-producing cells on a scale of 8–10,000 tpa. Fermentation is used in preference to a number of bioconversion

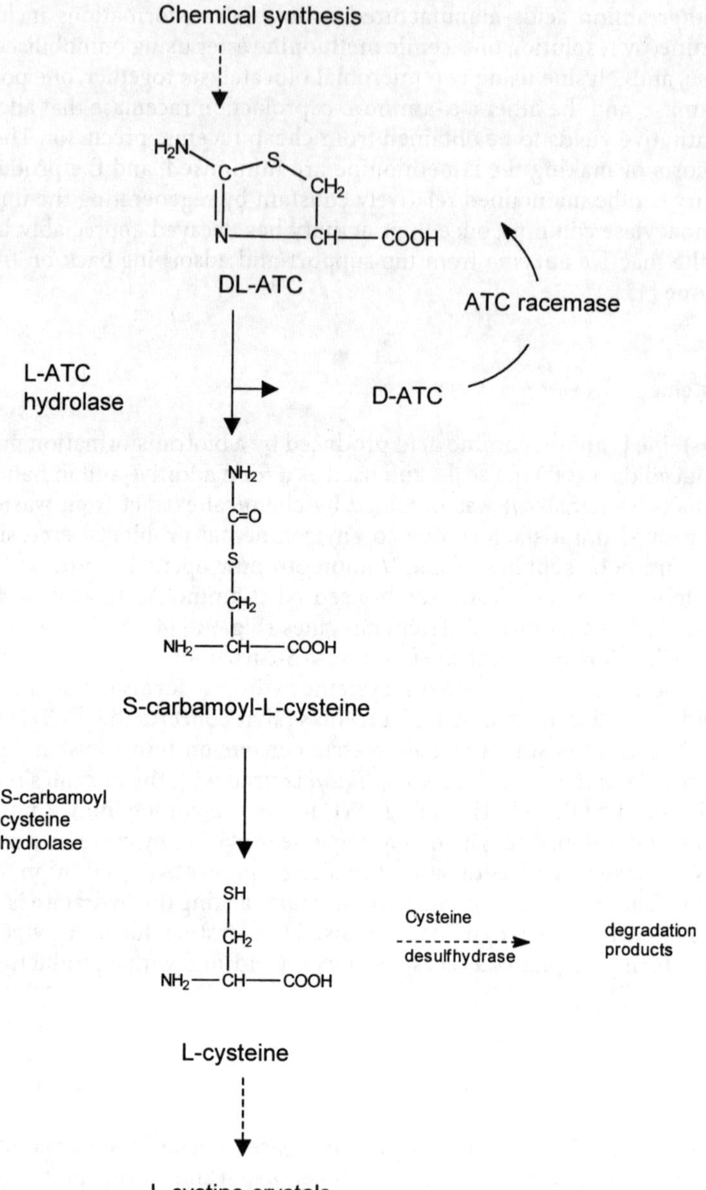

Fig. 9 Chemical systhesis of L-cystine crystals (Ajinomoto)

processes which have been developed to an advanced stage but proved less economic than the fermentation.

An important competitive force for this enzyme process is from long established chemical processes that are fully depreciated and highly optimised after many years of operation and investment. For instance the enzyme process for making Aspartame competes with the earlier chemical process operated by Nutrasweet/Monsanto. A further factor is that Regulatory Authorities state that products containing Aspartame must be labelled as "containing a source of phenylalanine" because of safety concerns for phenylketonuriacs (EC Directive 96/21/EC).

5.6
Nucleotide Savoury Taste Enhancers

Another example of how different production technologies are in competition is provided by nucleotide taste enhancers. About half of 5′-nucleotide production is by the enzyme hydrolysis of yeast RNA into nucleotide phosphates, including 5′-guanosine monophosphate, using phosphodiesterase; followed by isolation of the 5′-adenosine monophosphate and 5′-guanosine monophosphate, and then selective deamination of the 5′-AMP into 5′-inosine monophosphate using adenyl deamidase. This is the only example of the large scale use of a deamidase enzyme (Figs. 10 and 11) [17]. Competition exists because 5′-IMP

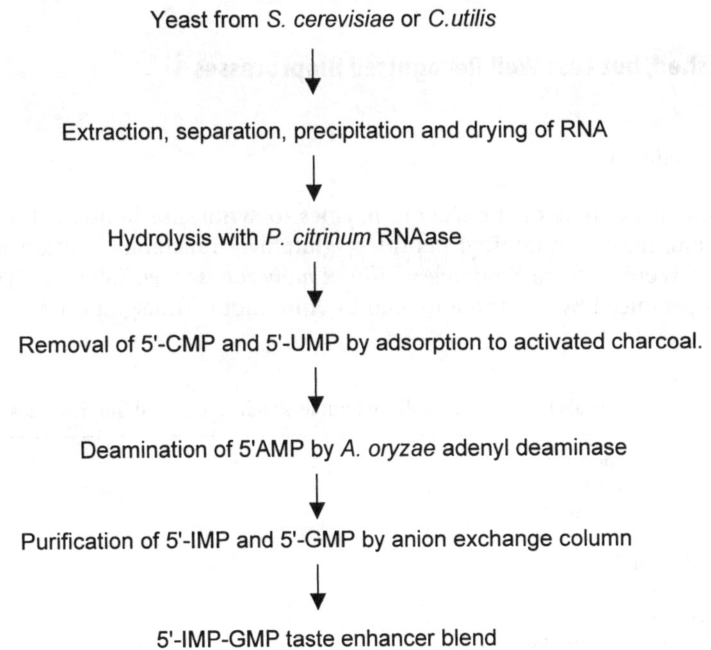

Fig. 10 Bioprocess to make 5′-purine nucleotide taste-enhancers from food grade yeast

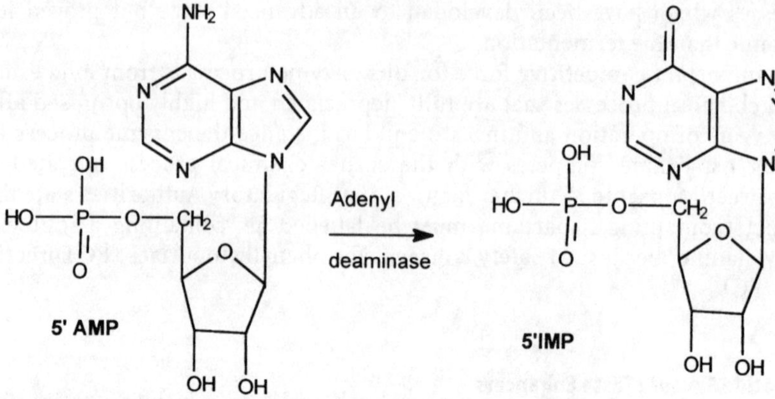

Fig. 11 Oxidative deamination of 5′-adenine monophosphate into 5′-inosine monophosphate taste enhancer

is also manufactured both by fermentation, and by a fermentation to produce inosine at 30–35 g l^{-1} followed by chemical phosphorylation. Similarly guanosine can be produced by fermentation, and then the 5′ GMP produced by chemical phosphorylation; or another approach is to make the intermediate 5-amino-4-imidazole carboxamide ribose by fermentation, followed by its chemical conversion into 5′-GMP.

6
Established, but Less Well Recognised Bioprocesses

6.1
Transglutaminase

An excellent example of the use of enzymes to synthesise bonds is the use of transglutaminase (glutaminyl-peptide-γ-glutamyl transferase, obtained after lengthy screening, from *Streptoverticillium mobaraense* (see Table 13). This enzyme is produced by Amano and sold by Ajinomoto. Transglutaminase cross-

Table 13 Some materials for which established but less well recognised bioprocesses exist

- Ethyl-2-methylbutyrate
- δ and γ-decalactones
- Fructo-oligosaccharides
- Cyclodextrins
- 2-Phenylethanol
- Isomaltulose
- Flavour nucleotides
- Enzyme modified guar gum
- Human milk equivalent triglycerides

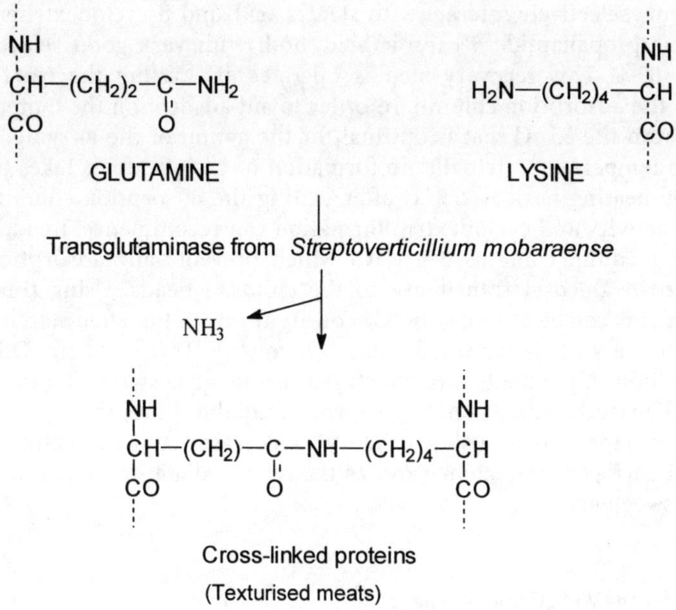

$$NH$$
$$CH-(CH_2)_2-C-NH_2 \qquad H_2N-(CH_2)_4-CH$$
$$CO \qquad O \qquad CO$$
$$\text{GLUTAMINE} \qquad \text{LYSINE}$$

Transglutaminase from *Streptoverticillium mobaraense*

NH_3

$$NH \qquad NH$$
$$CH-(CH_2)-C-NH-(CH_2)_4-CH$$
$$CO \qquad O \qquad CO$$

Cross-linked proteins

(Texturised meats)

Fig. 12 Transglutaminase (Ajinomoto)

links proteins by means of an acyltransferase reaction between the γ-carbox-amide groups of peptide bound glutamine and primary amines, such as the ε-amino groups of lysine residues in adjacent peptide chains (Fig. 12) [18]. Transglutaminase occurs naturally and is involved in several traditional processes, such as the formation of gels in surimi fish products. The microbial transglutaminase product has found a number of uses, such as for reforming mechanically recovered meat, improved meat hydration, and in sausage manufacture. It is particularly useful for protein processing because unlike previously described transglutaminase its activity does not require the addition of calcium and it has a smaller molecular weight and so presumably can more easily penetrate into meat. However it is a relatively unstable enzyme, that restricts its applications.

6.2
Cyclodextrins

The key problem in developing an effective manufacturing process for cy-clodextrins, that are useful as encapsulating agents for flavours and other active molecules, is not about how best to use the cyclodextrin glucosyltransferase, that makes a mixture of α-, β- and γ-cyclodextrins from starch, but rather how to extract pure individual cyclodextrins from the reaction [19]. Isolation is essential because cyclodextrins easily inhibit the cyclodextrin glucosyltransferase. Isolation is carried out by the selective adsorption of the α and β cyclodextrins onto chitosan beads modified with appropriate ligands. For instance, α-cyclodextrins selectively interact with stearic acid, and β-cyclodextrins with cyclohexane propanamide-*n*-caproic acid, both with very good selectivities of almost 100%. This recovery step is operated by cycling the reaction fluid through the adsorption column. In order to aid adsorption the temperature is cooled from the 55 °C that is optimal for the action of the enzyme, to 30 °C, at which temperature virtually no formation of cyclodextrins takes place, followed by heating back to 55 °C after exiting the adsorption column so that enzyme activity and cyclodextrin formation can recommence. In addition the reaction medium contains 3% NaCl which prevents any adsorption of the cyclodextrin glucosyl transferase to the chitosan beads. Using this process cyclodextrins can be obtained in 95% purity from 8% liquefied starch in yields of 5% for the γ-, 11% for the β-, and 22% for the δ-cyclodextrin. Other commercial food ingredients manufactured using glycosyltransferases include fructo-oligosaccharides, that are a form of soluble fibre, that is particularly popular in Japan, made using fructosyltransferases, and isomaltulose, useful as a non-cariogenic sweetener and as the intermediate in the manufacture of Isomalt sweetener.

6.3
Enzymes for the Wet Milling of Wheat

For several decades the largest volume use of enzymes has been of α- and β-amylases and glucoamylases to manufacture glucose and maltose syrups from wet milled maize on a 20 billion tonnes per annum scale. Major uses of the glucose are for HFCS manufacture and fermentation into ethanol for fuel and solvent uses and maltose syrups for brewing. Wet milling of wheat has always been much more difficult because of the presence of lysolecithin that is a good emulsifier, and also of arabinoxylans and arabinogalactans that have high water binding capacity, and so cause filtration problems. Wet milling is now possible by first hydrolysing these materials with phospholysolecithinase and hemicellulases; provided these enzymes are free of any amylases or proteases that could degrade the wheat starch or gluten respectively [20].

6.4
L-Phenylalanine

L-Phenylalanine is important as a raw material for the synthesis of the high intensity sweetener Aspartame. Coca-Cola have made efforts to backwards integrate by developing a process to make L-phenylalanine by resolving D,L-phenylalanine isopropyl ester by the selective hydrolysis of just the L-isopropyl ester, using the protease subtilisin, which is produced by *Bacillus licheniformis* [21]. The unconverted D-isopropyl phenylalanine is continuously extracted from the pH 7.5 aqueous phase in which the hydrolysis takes place, into a supported liquid membrane consisting of 33% *N,N*-diethyldodecanamide in dodecane that is immobilised in a microporous membrane. From the liquid membrane it passes into a pH 3.5 aqueous phase, from which it is racemised by refluxing in anhydrous toluene using an immobilised salicaldehyde catalyst, and then recycled.

The selectivity of the extraction across the liquid membrane is because the carboxylic acid group of the L-phenylalanine is charged at pH 7.5, and so is not extracted into the acid aqueous phase; whereas the unreacted D-phenylalanine ester is uncharged, and so will partition across the membrane into the acid aqueous phase. As a result L-phenylalanine is obtained in a 73% yield from the L-ester present in the racemic starting material, and with an enantiomeric excess value of 95%.

6.5
Zeaxanthin

2,2,6-Trimethylcyclohexane-1,4-dione, important as an intermediate in the chemical synthesis of carotinoids such as zeaxanthin, is produced using the reducing activity of baker's yeast in 86% yield and 99% purity at a rate of 2.8 g/l reactor volume/day. Oxoisophorone precursor is supplied in fed-batch manner together with sucrose. The product is toxic to the cells that make it, and so the process is operated at a sub-optimal temperature of 20 °C to encourage the precipitation of the product, which is recovered as crystals, and also to prevent the reduction of the 4-keto group [22]. Zeaxanthin as well as providing colour to foods, has, together with another carotinoid, lutein, been associated in clinical studies with a reduced incidence of acute macular degeneration. For instance spinach contains high concentrations of zeaxanthin [23].

6.6
Isomalt

Isomaltulose is used as a reduced sweetness, non-cariogenic and non-browning sugar, and especially as the intermediate to make Isomalt. This is a product of Sudzucker on a 45,000 tpa scale. Isomalt is a non-cariogenic, reduced calorie, non-hygroscopic, crystalline bulk sweetener used in boiled sweets, chewing gum, and tablets etc. Isomaltulose is formed from concentrated sucrose solutions using immobilised *Protaminobacter rubrum* cells in high yield. It is then crystallised, and then hydrogenated to form isomalt [24]. This process is distinguished by the use of high concentrations of sucrose of about 350 g/l and by the long operational lifetime of the immobilised cell activity, with a half-life of around a year, by the production of trehalulose as a side-product, and by the ease with which the isomaltulose can be recovered in high yield and purity by crystallisation, ready for hydrogenation into Isomalt.

6.7
Ethylglycoside Ester Surfactants

Ethylglycoside ester surfactants are biodegradable with low foaming characteristics, and are beginning to be used in cosmetic products. The esters are formed using *C. antarctica* B lipase to achieve regioselective reaction with just the primary hydroxyl group of the sugar. Reaction is at elevated temperature and at reduced pressure so as to remove the water formed. This allows the reaction to proceed without any need for organic solvents since the C8–C18 fatty acids raw materials dissolve in the ethyl, or isopropyl glucoside.

6.8
Locust Bean Gum Equivalent Polysaccharide

Plants often possess enzymes that are not readily available from microorganisms, even from Basidiomycetes strains that seem to have metabolic processes that are most similar to those of plants. An example is the commercialised process for converting guar galactomannan into a form that is functionally equivalent to that of locust bean gum (LBG). This uses an enzyme obtained from a plant that is a well known crop. The scope for obtaining useful enzymes from plant sources is illustrated by the statistic that whereas some 300,000 to 500,000 plant species have been identified only about 3000 have been brought into agricultural cultivation. LBG is used as a viscosifier and gelling agent in food, feed and personal care products. LBG has more valuable functional properties than guar gum, but has become very variable in quality and supply. Formation from guar gum requires partial removal of the galactose side chains from the polymannan 'backbone', from the 38–40% galactose content of guar gum to the 22–24% characteristic of LBG. But debranching of the guar must be done without any depolymerisation (hydrolysis of the polymannan backbone), that would reduce its viscosifying and gelling properties. A suitable α-galactosidase was obtained by cloning its gene from locust bean seed aleurone cells into yeast for ease of production, and also to eliminate the β-mannanase that can hydrolyse the polysaccharide and reduce its molecular weight and gelling properties. Another interesting technical feature of this process is that a superior process intensity is possible because the enzyme can work effectively on very concentrated guar gum, of about 30% (w/v), that is sufficiently concentrated to form a dough-like consistency [25, 26] (Fig. 13). The action of an enzyme on similar very viscous substrates is already well established on a very large scale for the pre-thinning of the maize starch by thermostable α-amylase, so as to liquefy the starch prior to further hydrolysis by glucoamylase into glucose syrup. The chief driver for the commercialisation of this process was that whereas guar is a raw material of reliable supply from the Indian subcontinent; LBG, which is obtained from Mediterranean areas, suffers from big variations in supply and price and a decline in production with difficulties in increasing production since locust bean trees take many years to mature. This allowed the new product to be introduced onto the market at a time of shortages and high prices for LBG, in the same way that high fructose corn syrups were first successfully launched during a period of sugar shortage in the 1970s.

Another process that uses low water, dough-like processing conditions is for the important aroma chemicals α- and β-ionones. These are formed by the co-oxidation of β-carotene with fatty acids by lipoxygenase obtained from soy flour. Reaction is enhanced by the concentrated consistency of the reactants, by enhancing oxygen supply, and by the in situ generation of fatty acids from vegetable oil by lipase [27]. Note that lipoxygenase can also be used in combination

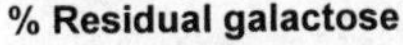

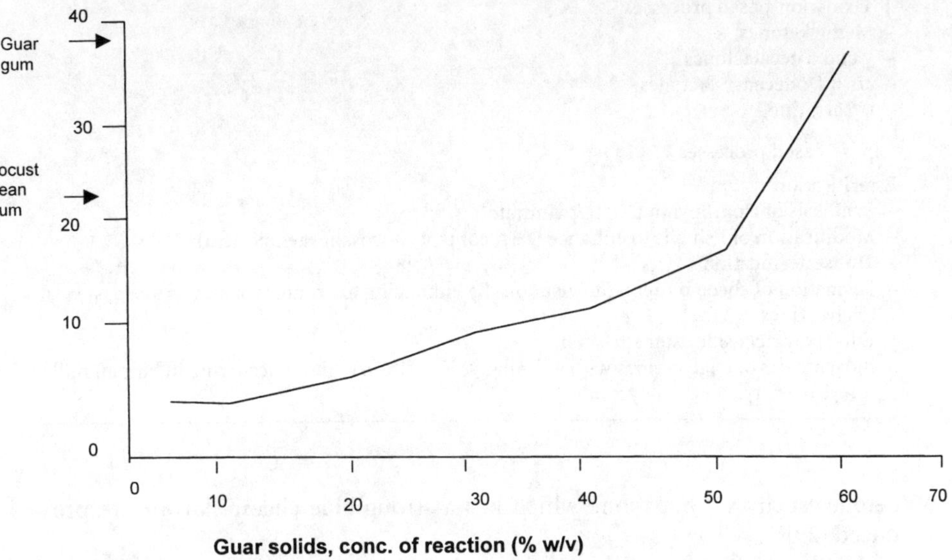

Fig. 13 Use of α-galactosidase to convert guar gum into a locust bean gum equivalent by partial removal of the galactose side chains (reproduced with permission of the publishers of biocatalysis, see http://www.tandta.co.uk)

with plant hydroperoxide lyases to make the 'green'/fresh tasting flavour chemicals *cis*-3-hexenol and *trans*-2-hexenol. These flavours are characteristic of fresh 'green' vegetables and fruits. Actually *cis*-3-hexenal is the immediate product of the reaction, but is unstable and is very rapidly converted into the corresponding alcohols that are stable molecules, but which have significantly lower flavour intensities.

6.9
Generic Areas of Biocatalysis

Sufficient industrial processes have been developed to allow the identification of some generic areas (Table 14). These are important as they allow synergies in research, process and product development, manufacturing, and raw materials purchasing etc. across a whole range of processes and products. One generic area is for processes that depend on microorganisms that carry out β-oxidation or some variant of this pathway. These include methylketone formation (Fig. 14) [28], δ- and γ-decalactone production, *cis*-6-dodecen-γ-lactone formation (Fig. 15), and L-carnitine production. Some metabolic differences occur. For instance whereas the last enzyme involved in β-oxidation is a thioesterase; by contrast in the microbial strains, such as *Penicillium rockfortii*, that make methylketones, it is replaced by a decarboxylase. So when fatty acids, such as octanoic acid present in coconut oil, are metabolised, methyl-

Table 14 Examples of generic bioprocessing approaches

'β-Oxidation' based processes
- Methylketones
- γ and δ decalactones
- *cis*-6-Dodecen-γ-lactones
- L-Carnitine

'Lipase' based processes

Esterification
- Synthesis of isopropylmyristate/palmitate
- Modification of fish oils to enhance DHA content (docosahexaenoic acid)
- Transesterification
- Formation of cocoa butter equivalent oils by enhancing the content of stearyl, oleyl, stearyl triglycerides
- Ethyl polyglucoside ester surfactant
- Biosynthesis of triglycerides with a similar composition to those occurring in human milk. (oleyl, palmityl, oleyl triglycerides)

---

ketones such as 2-hepanone which has a strong blue cheese flavour are produced [29].

Another generic technology is lipase based processes. These perfectly illustrate that, as well as hydrolytic and other degradative reaction, biocatalysts can perform synthetic reactions on a manufacturing scale. In the synthesis of the cosmetic emolients isopropylmyristate or isopropylpalmitate using lipases, two methods are used to encourage product formation, despite the formation of water by the esterification reaction, that, if allowed to accumulate, would encourages the back reaction (ester hydrolysis). Pervaporation or azeotropic distillation can be used to remove the water, with isopropanol fed continuously into the reactor to replace that lost by distillation. As a result ester products are made that are purer than that formed by the earlier chemical process.

More recently Kao have launched a diglyceride oil in Japan. This is made enzymically, and is claimed to have healthier properties as it reduces the build-up of fat in the human body. Another new development is the use of lipases to incorporate omega-fatty acids such as linolenic acid and DHA and EPA, into glycerides to improve the nutritional quality of food materials, such as edible spreads.

6.10
DHA Containing Glycerides

Lipases are also used in the synthesis of glycerides of DHA, as a means of purifying this fatty acid (Fig. 16). Ability to make commercial quantities of purified ω-3 fatty acid docosahexaenoic acid (C22:6 DHA) is a good example of the power of biocatalysis. DHA is likely to be important in neonatal development, and is found in high concentrations in human brain, retina and heart. DHA is available from mother's milk, but is not present in cows milk. DHA also occurs in fish oils, but together with eicosapentaenoic acid (C20:5, EPA) which inter-

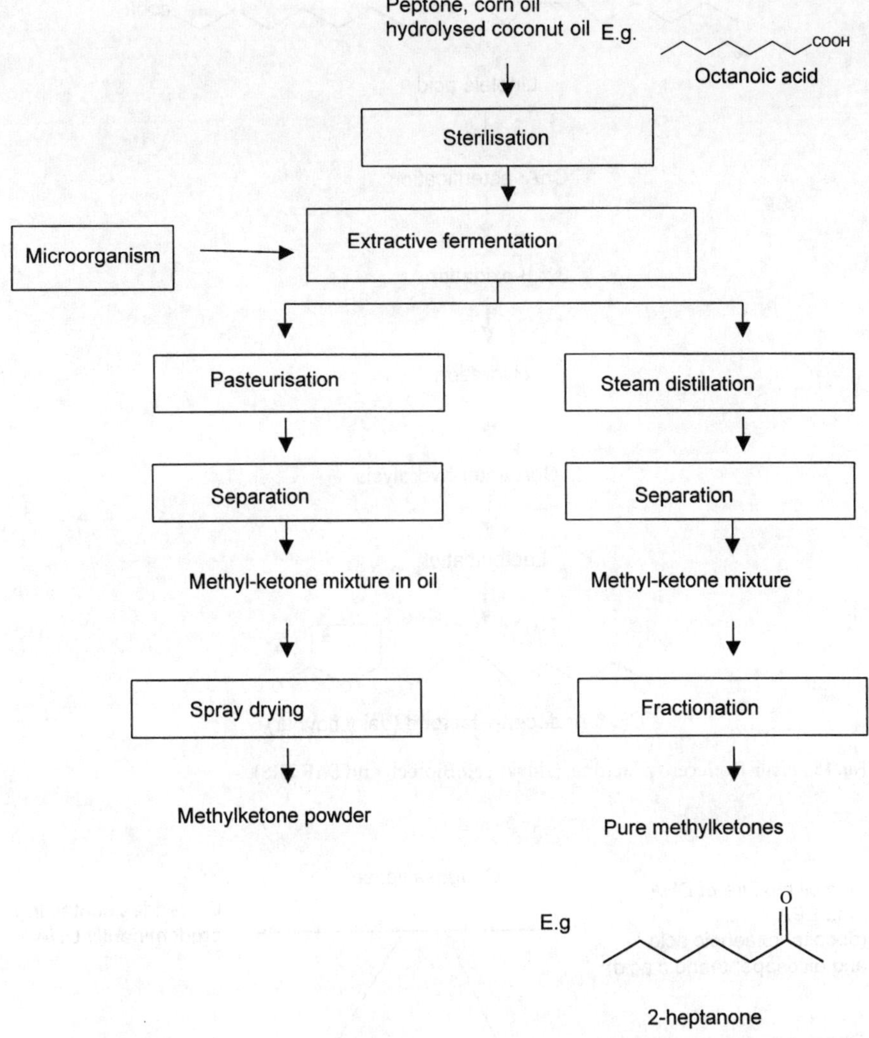

Fig. 14 Methylketone (blue cheese note) production. Reproduced by permission of the Royal Society of Chemistry

feres with DHAs beneficial effects, and so needs to be removed. This problem has been overcome using *Candida rugosa* lipase to make glycerides containing high proportions of DHA from fish oils. This process is possible because the enzyme used is more active towards DHA than EPA, so that DHA is incorporated preferentially. This enzyme based process use in competition with the use a marine algae, *Crythecodinium cohnii*, that has been discovered to produce high levels of DHA enriched oil. Rather than purify the DHA and formulate it the algae can be fed to hens to incorporate the DHA into eggs for human consumption [30–32].

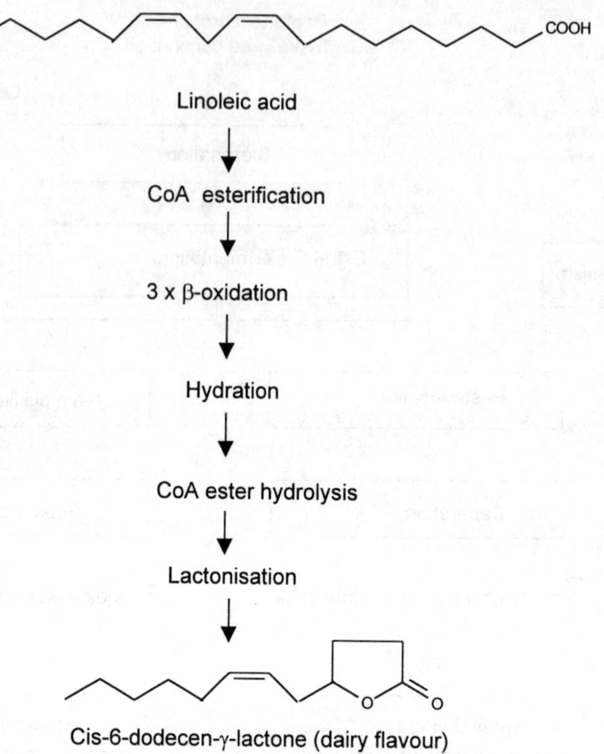

Fig. 15 *cis*-6-Dodecen-γ-lactone. (Advanced Biotech and SAF ISIS)

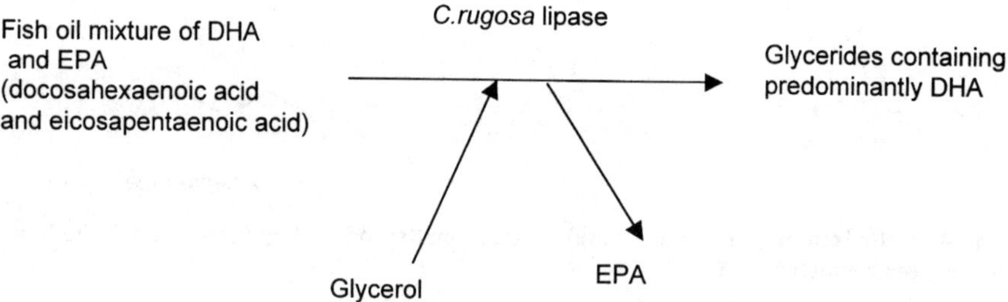

Fig. 16 Bioprocess to prepare fish oil glycerides enriched in nutritionally desirable DHA

6.11
Cocoa Butter Equivalent Oils

Lipases can also be used to change the composition of glycerides by transesterification (interesterification) technology. For instance immobilised 1,3-specific lipases are used to make cocoa butter equivalent oils by companies. The attraction is the relatively high price of cocoa butter, varying from ca $3000–$8500/t depending on market conditions. The required properties are achieved by exchanging some of the palmitoyl substituents present in the 1 and 3 (or 1') positions of the triglycerides with stearic acid. This gives about 80% of the glycerides the structure 1-stearoyl-2-oleyl-3-stearoyl-glycerol, which has the same melting point (of about 37 °C) and composition, as cocoa butter oil, so that it melts in the mouth [33, 34]. This reaction is carried out using an organic solvent, to provide a low water environment, to ensure that the lipase carries out the desired interesterification reaction, rather than hydrolysis. However, the organic solvent has to be saturated with water to provide a sufficiently high water activity to maintain enzyme what activity. The immobilised enzyme has a productivity of about 10 t product/kg enzyme over three half lives of use. This process is remarkable because it was developed to an advanced stage, with patent applications filed as early as 1976, well in advance of basic research on the use of enzymes in organic solvents [35]. This important product is a good example that often 'step wise' advances are often made by technologists working directly in response to market needs; and then with the scientific background and rational for they have done, worked out only afterwards by researchers in Universities.

7
Emerging New Bioprocesses

7.1
D-Trehalose

This novel sugar has excellent preserving properties, protecting proteins from inactivation and denaturation on freezing (see Table 15). The key to the process is the use of a specially selected microbial strain of *Arthrobacter ramosus* that possesses both maltooligosyltrehalose synthetase and maltooligosyltrehalase activities [36]. It is also of note that even after the trehalose has been synthesised a six step down-stream processing regime is then required to make the pure crystalline trehalose product (Fig. 17).

Table 15 Some materials for which new and emerging new bioprocesses exist

- Trehalose
- Hydoxyproline
- *cis*-6-Dodecen-γ-lactone
- Allicin
- Theanine
- Unsaturated fatty acid esters
- D-Pantoic acid
- Arabinoxylan
- Vanillin
- Biomelanin

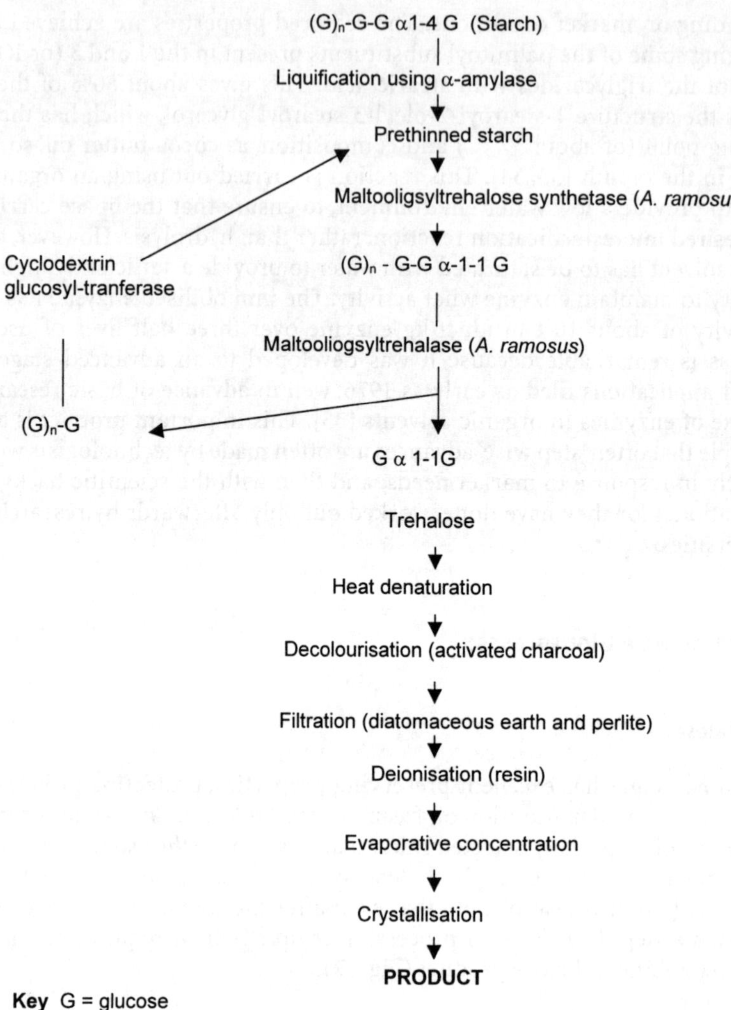

Fig. 17 Trehalose manufacture (Hayashibara)

7.2
trans-4-Hydroxyproline

trans-4-Hydroxyproline has good water binding properties and so is useful as a moisturising ingredient in skin creams, as well as being a useful pharmaceutical ingredient. Manufacture by the acid hydrolysis of animal collagen requires a complex process and generates a lot of wastes. Therefore a microbial approach was sought. The hydroxylation of proline was achieved using the proline 4-hydroxylase activity (2-ketoglutarate-dependant dioxygenase) of a *Dactylosporanigium* strain. This was isolated from a screening programme based on etamycin producing Actinomycetes. These were selected

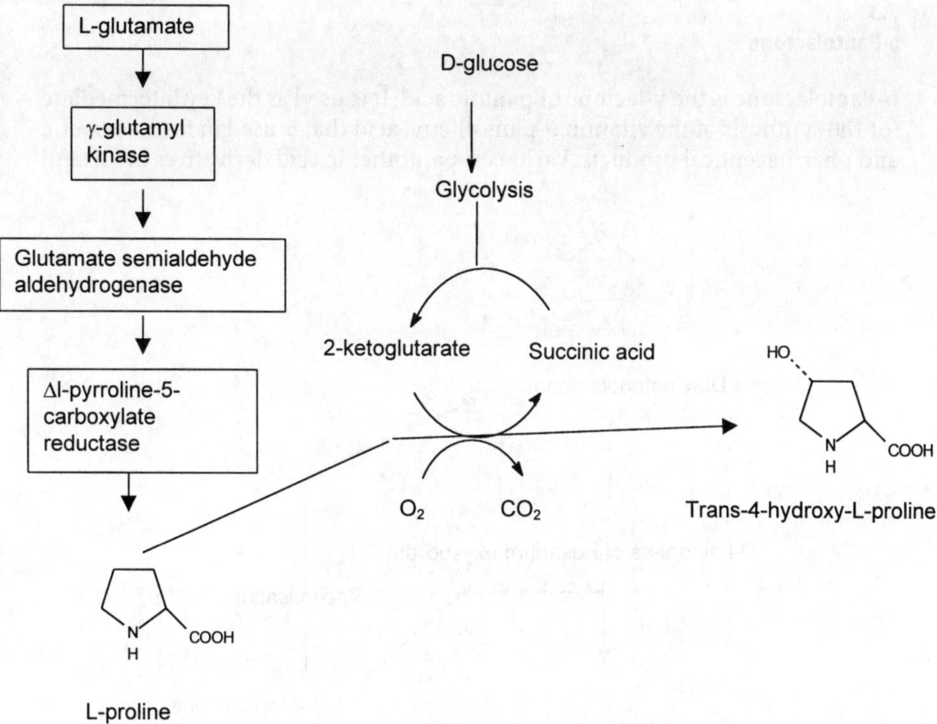

Fig. 18 *Trans*-4-Hydroxy-L-proline (Kyowo Hakko)

because etamycin contains 4-hydroxyproline as part of its structure (Fig. 18) [37].

The enzyme acts with good regioselectivity and stereoselectivity, but needs a supply of 2-ketoglutarate that requires regeneration. Also microbial production of proline is limited by feedback inhibition of γ-glutamyl kinase. Production of *trans*-4-hydroxyproline from proline, without the supply of glucose, can be achieved using a recombinant *E. coli*. This *E. coli* strain contains a plasmid carrying the proline hydroxylase gene; and also a gene for a feedback resistant γ-glutamyl kinase, so as to allow the *E. coli* to over-produce proline as the precursor for the proline hydroxylase. Cloning also required modification of N-terminus codons to match the coden usage of *E. coli*, and the promoter was doubly expressed to achieve over-expression of the enzyme. As a result the transformed *E. coli* has a 1400-fold higher 4-hydroxylase activity than the original strain.

7.3
D-Pantolactone

D-Pantolactone is the γ-lactone of pantoic acid. It is used as the key intermediate for the synthesis of the vitamin D-pantothenic acid that is used in feed, cosmetic and pharmaceutical products. Various D-pantothenic acid derivatives are useful

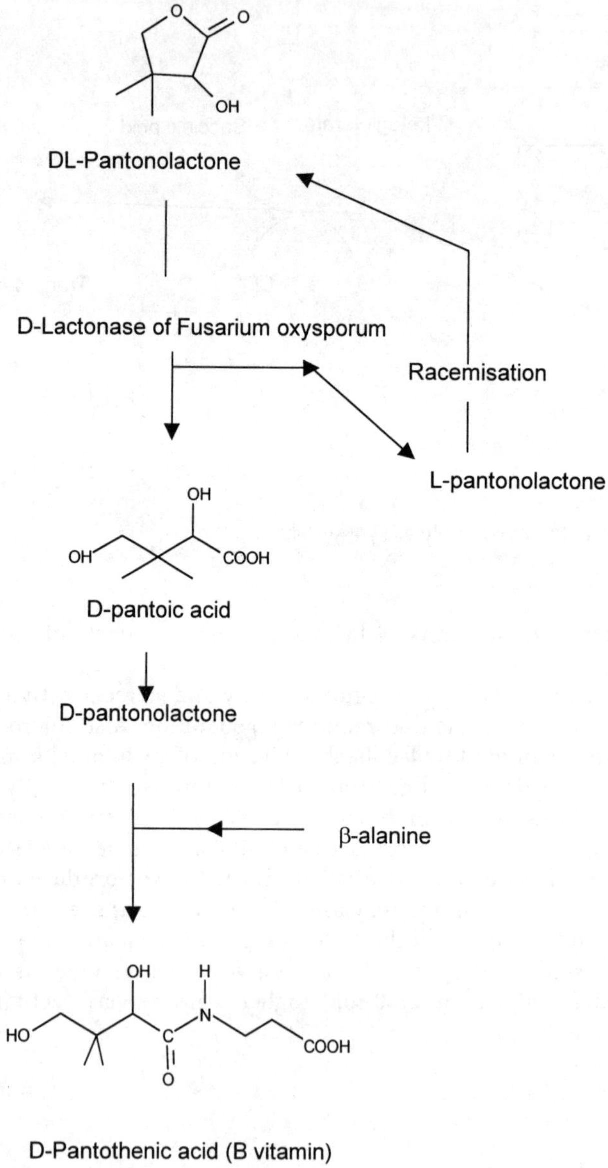

Fig. 19 D-Pantothenic acid (Fuji)

such as panthenyl alcohol, pantheine and Coenzyme A. Chemical synthesis is difficult. A lactone hydrolase based process has now entered production using an enzyme found to be present in *Fusarium oxysporum* that specifically hydrolyses D-pantolactone into D-pantoic acid. The cells are used immobilised in alginate, and remaining L-pantolactone easily recovered and racemised for recycling [38] (Fig. 19). D-Pantoic acid is obtained in a 90–95% yield and with an ee of 90–97% from 350 g/l racemic lactone. The D-pantoic acid is then converted back into D-pantonolactone, and reacted with β-alanine to form D-pantothenic acid. A big advantage of the bioprocess over chemical production is the reduced number of process steps involved.

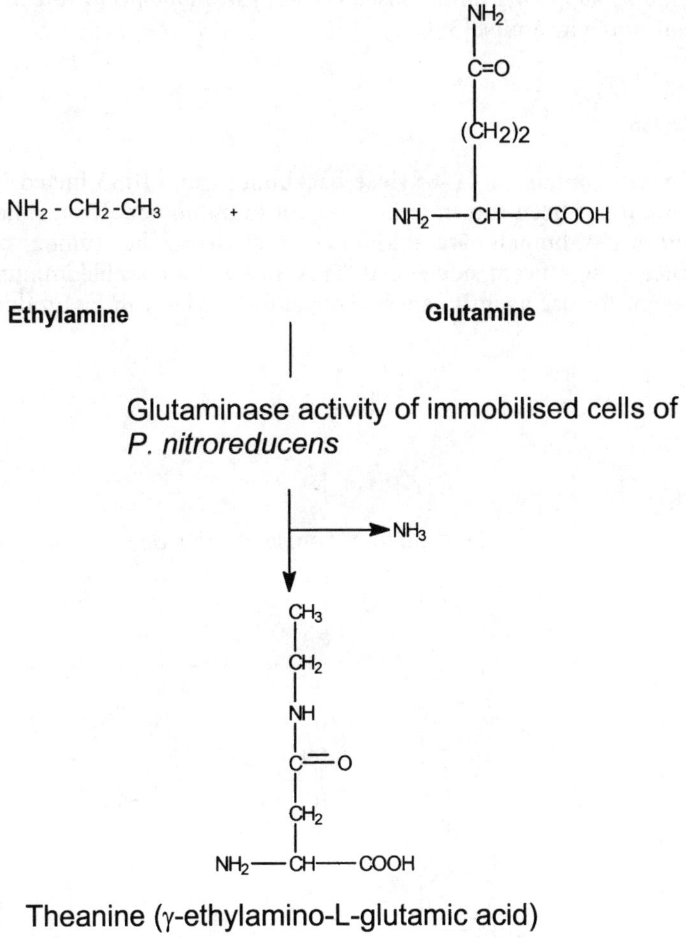

Fig. 20 Bioprocess to make Theanine (Taiyo Kagaku)

7.4
L-Theanine

An exciting nutraceutical opportunity is the non-protein amino acid L-theanine (γ-ethylamino-L-glutamic acid) (Fig. 20). Theanine is found naturally in tea. It is synthesised in the tea plant roots, and is translocated to the leaves where it is metabolised into catechins. Theanine is a multifunctional chemical with some taste properties. More importantly it counters the effect of coffee, reduces the blood pressure of hypertensive rats, and in humans has been shown to stimulate the α-brain waves that are characteristic of relaxed behaviour. Rather than just extract theanine from tea, it can be produced in good yield for at least a 50-day period by supplying immobilised cells of *Pseudomonas nitroreducens* with glutamine and ethylamine [39].

7.5
Arabinoxylan

Arabinoxylan contains a β1-4-xylose backbone with a β1–3 linked arabinose side chain. It has been shown to be a potent immunomodulator, enhancing in vitro and in vivo human natural killer cells activity against tumour cells, with an absence of significant side effects. This makes it a possible immuno-therapeutic agent for use against cancer. This arabinoxylan can be produced from

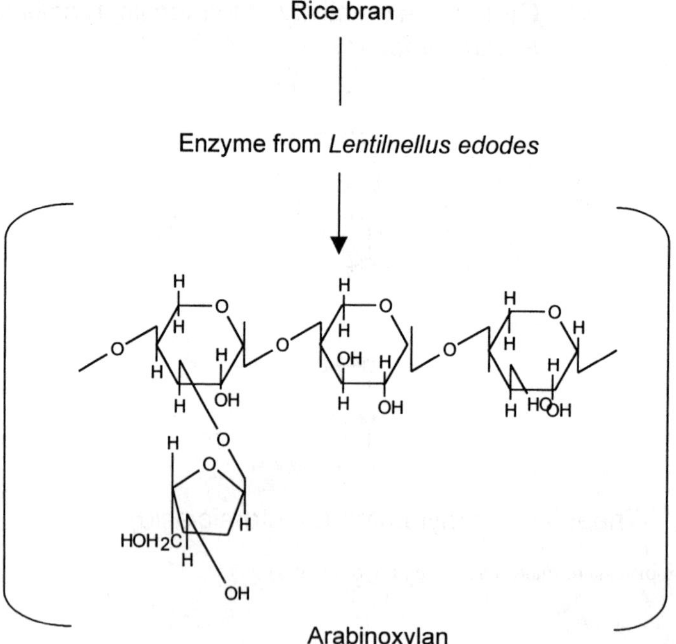

Fig. 21 Arabinoxylan (Daiwa Health-Sha)

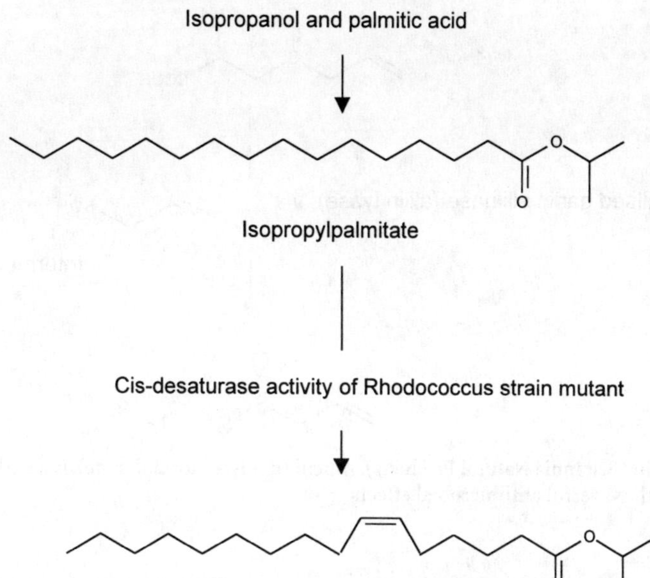

Fig. 22 Unsaturated fatty esters (Kao Corp)

rice bran by treatment with an enzyme extract from Shiitake mushroom (Fig. 21) [40].

7.6
Unsaturated Fatty Acid Esters

The Kao Corporation in Japan has developed a process for making unsaturated derivatives of cosmetic emollient esters such as isopropylmyristate and isopropylpalmitate. This involves the use of a mutant microbial strain possessing mid-chain desaturase activity and which requires the supply of thiamine and magnesium (Fig. 22) [41]. The reaction starts as an oil in water emulsion, but this inverts to a water in oil emulsion as isopropylpalmitate is added. The product and residual substrate are removed using a hydrophobic hollow fibre filtration step, and the product separated by forming a complex with urea, followed by silica gel chromatography.

7.7
Allicin

This is a natural antimicrobial present in fresh garlic. For instance, allicin has minimum inhibitory concentrations of 8 ppm and 32 ppm against *Listeria monocytogenes* and *E. coli* 0157 respectively. It can be produced using an immobilised enzyme process (Fig. 23) [42]. The major problem overcome is the

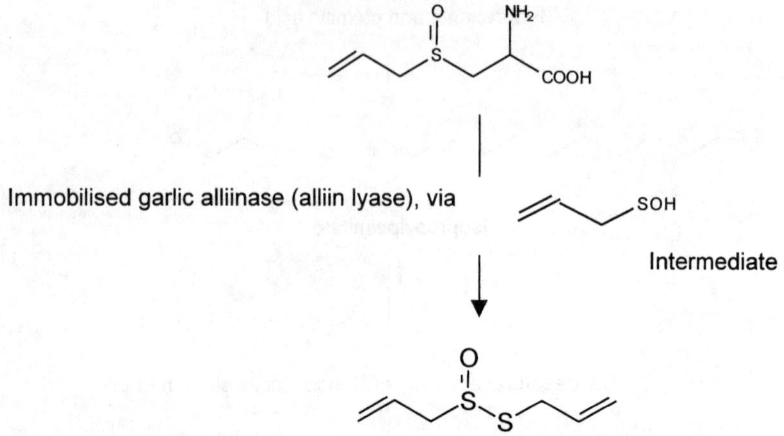

Fig. 23 Allicin (Britannia Natural Products). Allicin (diallyl-thiosulphinate), is a garlic derived molecule with powerful antimicrobial effects

chemical instability of the various metabolites derived from the allicin raw material present in the garlic flesh.

8
Bioprocessing Approach

As well as the use of biocatalysts to biotransform pure chemicals supplied as raw materials, they can also be used to process chemically complex plant or animal extracts (Fig. 24) in which the value of the product depends a lot on its chemical and physical complexity, such as in beer, wine, cheese, yoghurt, soy sauce etc., which gives quality flavours and mouthfeel. Whereas the 'biotransformations' approach is commonly used for both food and pharmaceutical products, the bioprocessing approach is only really used for food products. As a consequence the activity of the biocatalysts used, assessed in terms of the amount of product formed per unit reactor volume per unit time is usually much lower than in biotransformation processes in which the product is simply just one defined chemical. Another difference is that rather than the development of an entirely new biotransformation process, the bioprocessing approach usually requires careful integration into an existing, sometimes traditional process, and with preservation of the traditional identity of the product. For instance proteases are used on a large scale to make protein hydrolysates, useful for instance in savoury flavourings. The main problem is to prevent the formation of bitter tastes, due to the presence of peptides containing terminal hydrophobic amino acids. This is now possible by using special peptidases. Other examples include the use of α-amylase, proteases, β-glucanase and pullulanase during brewing. They are added to ground malted cereal during the mashing stage. In addition protease, amyloglucosidase and β-glucanase can be added to brewed beer to aid in its conditioning. When used in bioprocessing the

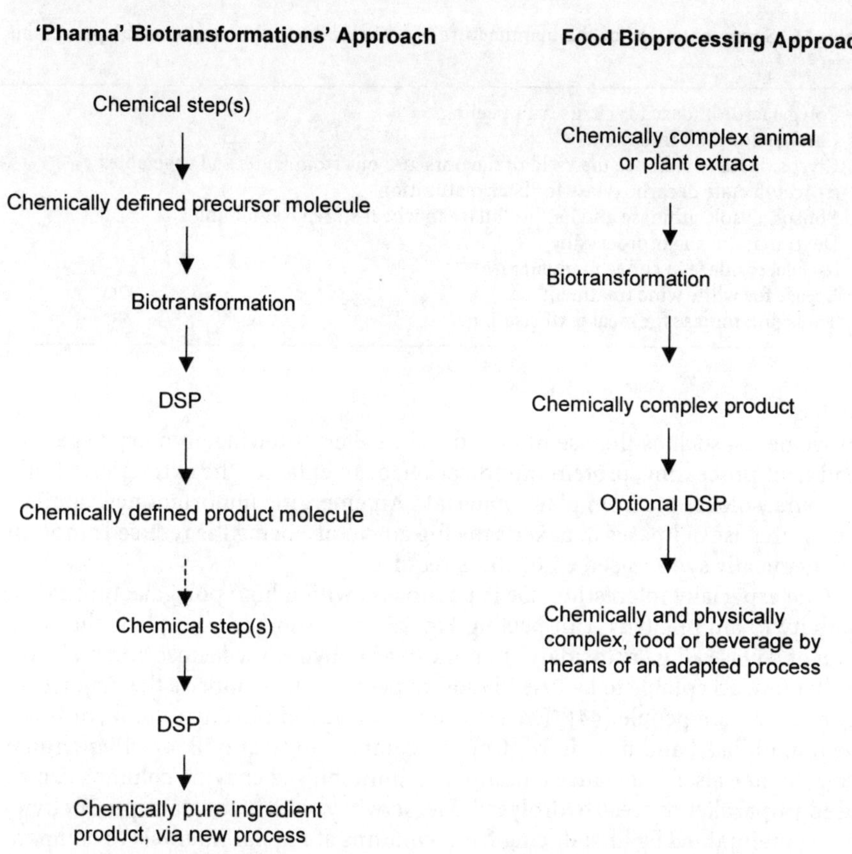

Fig. 24 Biotransformations – two approaches are used

enzymes used usually have to have become inactivated by the end of the reaction, or to be easily inactivated by heating or such-like, or to be food-grade enzymes that have been proved to be safe to be consumed when still present in the product.

In some cases the bioprocessing approach involves the removal of undesirable materials, such as α-acetolactate and urea from beer and wine respectively (see later). These uses depend on the selective reaction properties of enzymes, combined with their ability to react with their substrates even when it is present at only very low concentrations (that is the enzyme has to have a low Km value).

An important aspect is that only enzymes approved for food use can be used. This is because the enzyme(s) used can still be present in the final product consumed by customers and so just like any other ingredient of a food they have to be proved safe by showing no adverse effects in a series of safety tests. Examples of food products made using bioprocessing are given in Table 16. These are having a major impact both in achieving desirable processing im-

Table 16 Some materials for the manufacture of which industrial examples of bioprocessing exist

- Polygalacturonidase for citrus fruit peeling
- Catalase for milk treatment
- Glycosidases to enhance the yield of flavours and oils from fruits and vegetables
- α-Acetolactate decarboxylase for beer maturation
- Phospholysolecithinase and hemicellulase in wheat starch wet milling
- Dextranase in sugar processing
- α-Galactosidase in coffee manufacture
- Urease for white wine treatment
- Transglutaminase for meat texturisation.

provements, such as the use of enzymes in bakery, brewing, in dairy, vegetable and fruit processing, protein modification, or to enhance the extraction of oils, flavours, colours etc from plant materials. An emerging important new application is the use of lipases in bakery, the big advantage being the reduced amounts of chemically synthesised emulsifiers required.

One especially interesting use is pectinases with a high polygalacturonidase activity to aid in citrus fruit peeling Fig. 25 [43]. Another example is the addition of lactase (β-galactosidase) to milk so as to hydrolyse lactose and make the milk more acceptable to lactose intolerant people, who comprise the majority of non-Caucasian people [44]. The lactase can be added directly to milk, or it can be immobilised and used in continuous columns to treat milk or other similar dairy materials, for instance a number of immobilised enzyme columns can be used in parallel to treat hydrolysed cheese whey, with a constant productivity being maintained by introducing fresh columns at regular intervals to compensate for the gradual decay in activity of the immobilised enzyme, which has an operating half-life of 50 days [45]. The scale of the opportunity for using enzymes in the dairy industry is illustrated by the estimation that 145 M tpa of whey is generated world-wide, containing some 6.5 M t of lactose, of which only 10% is sold.

A recently developed processing approach is the use of glycosidases to enhance the extraction from plant materials of aroma terpenes such as linalool and geranol. This is by the hydrolysis of the α-L-arabinoside, α-L-rhamnoside, β-D-xyloside and other bonds that link the terpenes to plant materials. DSM

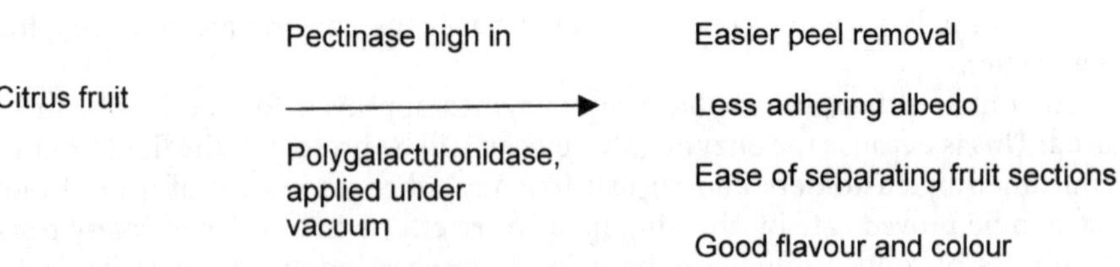

Fig. 25 Citrus fruit treatment (Biocatalysts and others)

Food Specialities in particular have developed a range of enzymes specifically to treat fruits. These are invariably complex mixtures of enzymes with different activities and specificities. This is because the physico-chemical structure of each type of fruit is different and so offers different technical challenges. Also the enzyme used will change depending on whether the extract required is to be clear, cloudy or pureed; dilute or concentrated; and also whether the fruit is fresh or stored, as lower yields of juice are obtained using stored fruits and higher enzyme doses are required to compensate. Usually a mixture of glycosidases of different specificities and that have good activities at low pHs is required. An additional benefit is that acceptable yields of flavour from some hitherto waste materials can be achieved [46]. This approach has also been useful to increase the yield of the flavour chemical 2,5-dimethyl-4-hydroxy-3(2H) furanone (Furaneol) from pineapples, to release rhamnose from flavanone glycoside precursors, such as naringin, which is useful as a precursor for the synthesis of Furaneol [47], and to release C13 norisoprenoid aroma chemicals such as ionones from a variety of plant sources [48].

Some of the ways in which enzymes and microbial biocatalysts have been used in dairy processing and cheese manufacture are shown in Fig. 26. Whereas the traditional craft use of microorganisms to improve cheese flavour have been in use for very many years, the use of microbial and cloned rennets (chymosin) for coagulating milk, and the use of blends of lipases and proteases for manufacturing intense cheese flavours (enzyme modified cheese flavours) are more recent developments. Chymosin is especially interesting as a microbial enzyme had to be found to replace the enzyme traditionally extracted from animal sources, that very specifically hydrolyses just the peptide bond between phenylalanine 105 and methionine 106 of milk kappa-casein. Also cloned chymosin was the first cloned protein to be approved as safe for general use. Enzyme modified cheese flavours are interesting as flavour intensities much higher than those normally present in traditional cheeses are needed. This requires the selective hydrolysis of fatty acids to give flavour molecules, and also the hydrolysis of proteins to create the peptides that give the required mouthfeel.

8.1
α-Acetolactate Decarboxylase

Another recently introduced example is the use of α-acetolactate decarboxylase to convert α-acetolactate into acetoin in lager beers so as to reduce maturation times (Fig. 27) [49]. This is because α-acetolactate, produced by the yeast during brewing, is normally decarboxylated into diacetyl, which gives the freshly brewed beer an unpleasant buttery taste. This diacetyl is only slowly reduced to acetoin, which has a much lower taste; by yeast alcohol dehydrogenase during the traditional maturation of beers, especially lagers. Hence addition of acetolactate dehydrogenase to freshly brewed beer greatly reduces its maturation time, and therefore significantly reduces brewing costs, despite the extra cost of the enzyme.

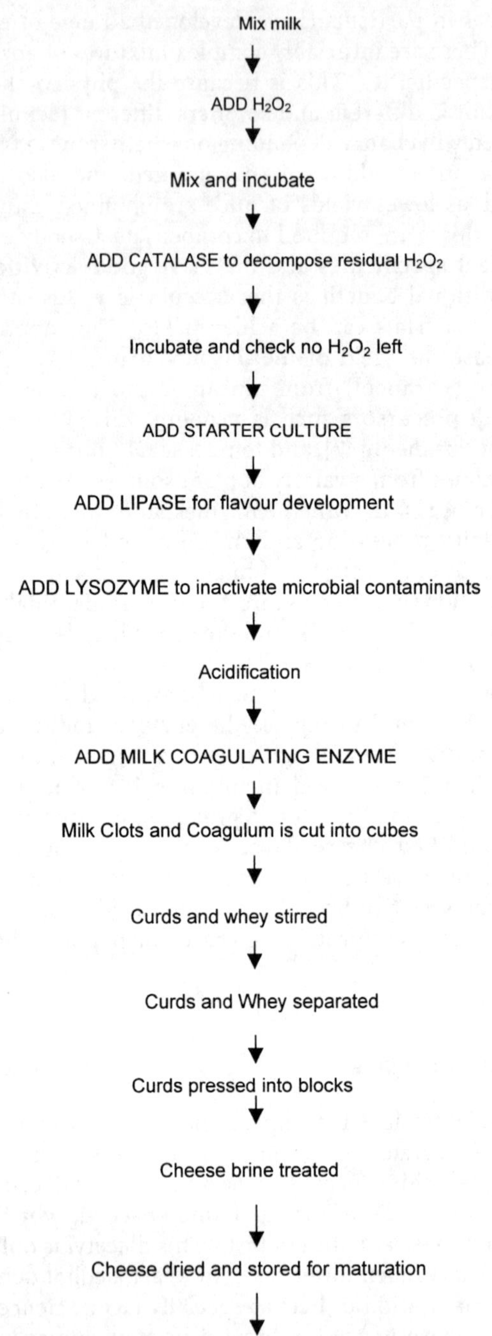

Fig. 26   Idealised use of enzymes and microorganisms in cheese making (note that not all of these operations will be used together)

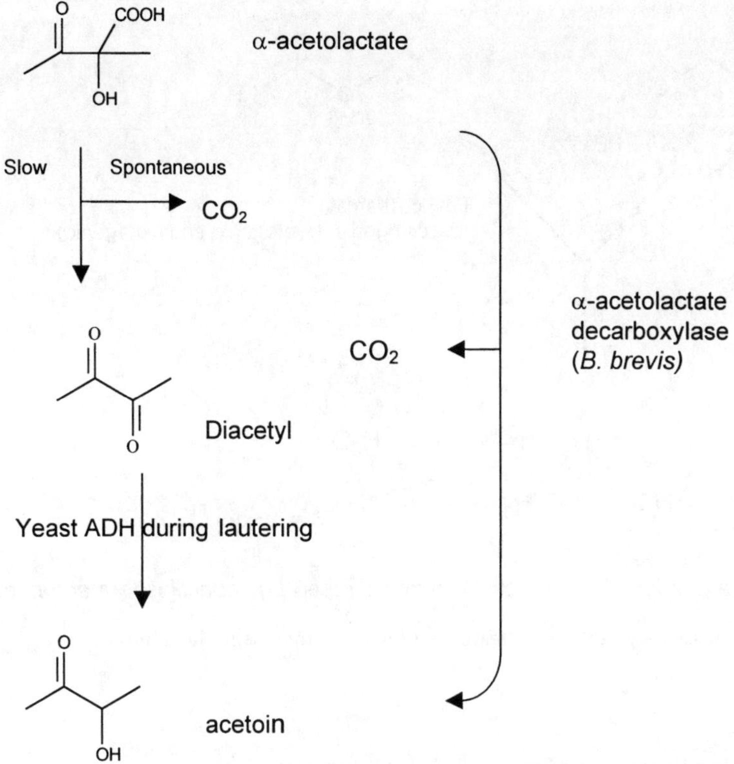

Fig. 27 α-Acetolactate decarboxylase for beer maturation (Novozyme)

8.2
Urease

Another application of biocatalysts in beverages involves the use of urease to prevent the formation of ethylcarbamate in white wines. Low concentrations of urea occur in wine and some other fermented beverages and upon storage slowly reacts with ethanol to form ethylcarbamate that is carcinognic, teratogenic and mutagenic. This problem can be prevented by treating the beverage with urease, that can reduce the urea concentration to just a few ppm, even at low temperature and low pH, and despite the urea only being present at low concentrations (Fig. 28) [50]. For cost-effective industrial use the urease is immobilised onto porous chitosan particles and these are used in a continuous process, with the enzyme having an operational life time of over 150 days. The only problem with this process is that urease is inhibited by tannins, and so this method is confined to the processing of wines with low tannin contents, especially white wines.

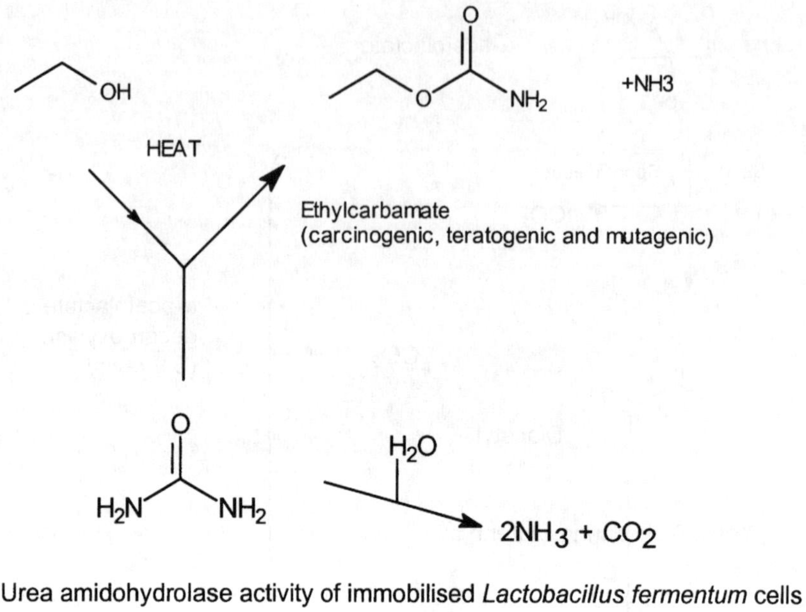
OH
O
O
NH2
+NH3
HEAT
Ethylcarbamate
(carcinogenic, teratogenic and mutagenic)
O
H2N
NH2
H2O
2NH3 + CO2
Urea amidohydrolase activity of immobilised *Lactobacillus fermentum* cells

Fig. 28 Urease for white wine treatment (Toyo Joza and Asahi Chem Ind)

9 Limiting Factors in Bioprocess Development

Each aspect of bioprocess and bioproduct development is key, but certain aspects appear particularly vital, such that they may very well limit future developments. These include biocatalyst identification and development. There are three main approaches. The simplest is to find new uses for existing commercial enzymes and biocatalysts, then new forms of known enzymes can be searched for, with, for instance different operating conditions and/or substrate specificities. In addition there is the more difficult search for entirely new enzymes that will carry-out entirely new reactions.

9.1 Biocatalyst Identification and Development

Various strategies can be pursued. Databases are a good way to identify enzymes and microbial strains that may be suitable biocatalysts [51]. Another is to try to use commercially available biocatalysts, that can be bought cheaply and in standardised forms, particularly if the enzymes have broad substrate specificities, in the hope that they can react with 'new' substrate molecules. These are obviously preferred, especially for low value products that cannot justify the cost of extensive R&D unless very large volumes of sales can be made. But this approach is usually not possible because of the limited range of enzyme activities available from suppliers. It may be possible to devise a new way to use an existing commercial enzyme. For instance Le Moult et al. [52] found that the widely used *C. antarctica* lipase B will also catalyse the Baeyer-Villiger reaction, reacting with 2-hexylcyclopentanone to produce the corresponding racemic lactones in 73% yield. This reaction takes place via the formation of the peracid, through hydroperoxide attack on the acylated enzyme. This approach provides a potential new approach to the synthesis of flavour and fragrance lactones, currently made by intramolecular chemical esterification.

Another alternative approach is to identify a whole cell biocatalyst already used in pharmaceutical or fine chemical manufacture and test whether it has the required substrate specificity. For instance whole cell nitrile hydratase biocatalysts were originally developed to produce acrylamide monomer, and then later cross- applied by Lonza to the manufacture of nicotinamide [53]. This is remarkably successful both in terms of the activity of the biocatalyst, and the yield and purity of the product obtained.

9.2
Microbial Screening

Microorganisms are the main sources of biocatalysts, although a few enzymes are extracted from plant sources. The initial step is usually to test pure strains available from culture collections. If this is not successful then the isolation of strains with particularly useful enzymes and/or metabolic activities by selective screening of environmental samples is vital, so called 'small game hunting' [54]. This depends on the abundant diversity of microbial strains present in soil and other natural sources [55], especially as most naturally occurring microbial strains have not yet been isolated, yet alone characterised and tested as possible biocatalysts [56]. The screening approach is perfectly illustrated by a current advertisement by Kyowa Hakko: "At this very moment the cure for cancer may be rotting in a ditch somewhere … the answers are out there".

Although very time consuming if manual methods are used, and expensive if high-throughput equipment is used, screening is invaluable because absolutely new strains, often with novel activities and properties, or with higher activities can be obtained. Such new strains are usually patentable. Success is more likely if very specific and selective screening techniques can be used. An example is the isolation of superior producer strains of the carotinoid pigment astaxanthin by means of their resistance to diphenylamine [57]. Methods for detecting microbial strains with the required properties, more selectively and/or more rapidly, are obviously of great value in screening. An example of a new and novel approach is the use of mid-infra red diffuse-reflectance absorbance spectroscopy for the detecting metabolite over-production [58]. Success may be more likely if areas of high biodiversity, such as rain forests, and new environmental niches can be searched, such as deep-sea hydrothermal vents, and the resulting enzymes exploited by cloning for easy and lower cost production [59].

If a microbial strain is the biocatalyst the microbial strain ought to have easy and reproducible growth characteristics on cheap undefined media, and the user must have some fermentation capability, otherwise the strain must be available as dried cells for ease of storage and transportation. In some cases the biocatalyst must achieve a high, reproducible and cost-effective performance for month after month of operation on a large scale. For example on an industrial scale glucose isomerase columns are normally operated continuously for 10–12 months until replaced by new more active enzyme. This requires general and specific engineering criteria to be met. For instance immobilised enzyme particles have to be sufficiently strong to resist compaction when used in packed bed reactors. Therefore the columns containing immobilised enzyme must be treated as pressure vessels because of the big hydraulic forces involved, so that appropriate safety and insurance provisions must be made.

9.3
Vanillin

Natural vanillin, useful as a flavour characteristic of vanilla, provides an excellent example of the value of microbial screening. Early on there was the discovery that strains of *Proteus vulgaris* could deaminate methoxytyrosine into the intermediate 3-methyoxy-4-hydroxy phenylpyruvic acid, which can then be converted into vanillin [60]. Then workers at the Institute of Food Research found a strain of *Pseudomonas fluorescens* that uses ferulic as a precursor, and possesses a previously undescribed enzyme called hydroxycinnomoyl–CoA hydratase/lyase, that carries out two distinct reactions. First a hydration of the side chain of ferulic acid takes place, and then an retroaldol cleavage to produce vanillin and liberate free acetyl CoA [60]. INRA working with Pernod-Ricard used *Pycnoporous cinnabarinus* to convert ferulic acid into vanillin. Only yields of 560 mg l^{-1} after seven days were reported, even when the fermentation was optimised by the addition of cellobiose to minimise the further metabolism of vanillin, which acts either as a carbon source and/or as an inducer of cellobiose quinone oxidoreductase, which is a known inhibitor of vanillic acid decarboxylation [62]. Similarly Zylepsis has developed three routes to making vanillin, directly from ferulic acid, via vanillic acid as an intermediate, and also from ferulic acid glycoside, the form of ferulate that exists in common plant sources, through to vanillic acid, and then conversion to vanillin using a *Micromucor isabellensis* strain that has good carboxylic acid reductase activity [63] (Fig. 29). This reaction is of potentially wider utility because of the more general requirement for 'limited oxidation' of acids to aldehydes without further reaction. Haarmann and Reimer have used a novel strain of *Amycolatopsis* to produce vanillin from ferulic acid in high molar yields of 80%, and at a concentration of 11 g l^{-1} [64], and have shown that this strain uses a similar pathway to that described for *P. fluorescens* [65]. In addition they have also found microbial strains that will convert eugenol into ferulic acid, prior to its bioconversion into vanillin (Fig. 29). Eugenol is of value because it is a relatively cheap and natural starting material, and its use is of scientific interest because eugenol has powerful antimicrobial properties, and so any strains that metabolise eugenol are exceptional in also being resistant to it.

Another use for ferulic acid has been found, by means of a microbial biotransformation using *Paenobacillus polymyxa* into vinylguaiacol that has good antioxidant and antimicrobial activities (Fig. 30) [66].

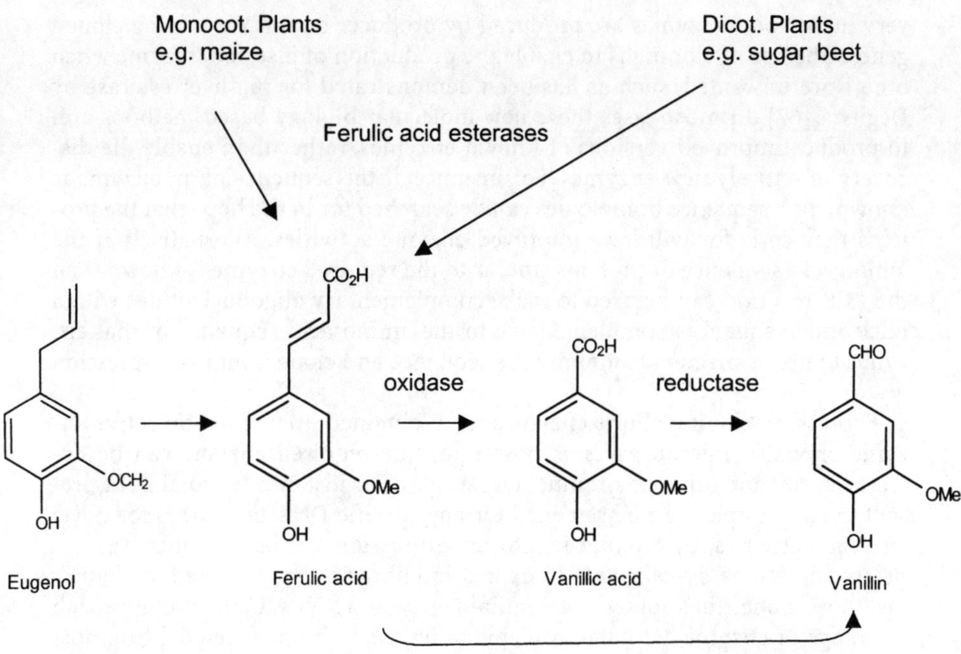

Fig. 29 Microbial routes to vanillin

Fig. 30 Conversion of ferulic acid into vinylguaiacol by Paenobacillus polymyxa

10
Molecular Biology Approaches

New approaches are emerging to supplement more traditional screening methods. If they prove to have wide application then very many new opportunities will open-up. These methods have mostly found applications first for pharmaceutical uses, but the big enzyme producing companies are using them, and they of course supply enzymes for food and cosmetic uses. Consequentially very many food enzymes are produced by producer strains containing cloned genes. One use of cloning is to enable the production of just one isozyme when others are unwanted, such as has been demonstrated for pig liver esterase by Degussa [67]. In most cases these new molecular biology based methods aim to produce improved versions of known enzymes, rather than enable the discovery of entirely new enzymes. For instance, if the sequence of an enzyme is known, then sequence homologies can be searched for in the hope that the proteins they code for will have improved enzyme activities. Alternatively if the amino acid sequence of proteins similar to the required enzyme is known then the PCR reaction can be used to make complementary oligonucleotides with a nucleotide sequences complementary to the amino acid sequence of that enzyme, to use as primers to amplify the sequence and clone it into an expression vector.

Expression cloning allows enzymes to be obtained even when no active enzyme or viable microorganisms possessing the required enzyme can be detected in soil and other environmental samples. For instance the total DNA present in soil samples can be screened for any specific DNA that codes for novel enzyme activities, by cloning it into an expression vector, and then the expressed activities, so-called cDNA expression libraries, are screened to identify just those clones that possess the required enzyme activity. Using this approach the range of enzyme activities available to industry can be extended from just those present in microorganisms that can be easily isolated in a viable form and grown on a large scale. Novozyme have used this approach to search for new carbohydrases, such as pectin methyl esterase and rhamnogalacturonidase [68], and DSM have used a similar expression cloning approach to develop an entirely new enzyme for fruit treatment. This is endo-xylogalacturonan hydrolase. This new enzyme is likely to be useful in improving yields of juice, because up to now only enzymes active against the pectin and cellulose components of the fruit structure have been available, and none that will degrade the xyloglucan component [69].

Further genetic techniques are now being developed to improve the activities of existing enzymes. This can be done at an enzyme level by shuffling the DNA coding for an enzyme by means of multiple cycles of random mutagenesis and recombination. These generate a range of variants, which are then subject to selective screening to detect genes that produce enzymes that are more active under the required conditions or that have superior enzyme activities or properties, such as thermostability or pH optima, by the technique of 'directed evolution' [70, 71] or 'gene shuffling', that involves the random interchange of gene sequences coding for domains in enzymes within a group of related genes. This allows the incorporation of multiple changes into just one enzyme. For instance the activity (Kcat/Km) of aspartase has been increased [72], and the substrate specificity of β-glucuronidase for the primary hydroxyl group of its glucuronic acid substrate has been modified [73].

The major effort still lies in the biochemical screening of the library to identify the very, very few with the required properties which still requires a relatively traditional time consuming approach. Even then a positive strain may not be an effective biocatalyst because the shuffling has had a detrimental effect, in which case a fresh round of shuffling will be required to correct the deficit, e.g. by shuffling with the original wild type enzyme.

Another method the 'Protein engineering' approach, has been worked on for over ten years now, and is has given some success. For instance, Novozymes have used protein engineering to develop a new thermostable α-amylase that does not require the addition of calcium for activity. This is a great advantage in starch processing because until the introduction of this enzyme in 1998 the calcium had to be removed from the pre-thinned starch because it inhibits the glucose isomerase used later in the starch processing.

Using whole cell biocatalysts the technique of 'pathway engineering' (or 'metabolic engineering') can be used to improve more complex biotransformations. This could involve the introduction of genes for new enzyme activities from exogenous sources that will modify or completely change the type of product made by the cloned cells. Improvements could include reducing product inhibition, or could even allow the creating of whole new pathways using genes from different sources. However a challenging problem is ensuring efficient expression and regulation of the new pathways. The great potential of this approach is shown by the new BASF process for making ascorbic acid. Another example of how this new approach could be useful is the bioproduction of carotinoids as food and cosmetic ingredients. β-Carotene is currently manufactured using *Blakeslea trispora*. It is used as a food colour and as a pro-vitamin, and is increasingly valued for its antioxidant properties. Whereas the β-carotene produced chemically is entirely the *trans*-β-caroteine isomer, natural β-carotene contains a higher proportion of *cis*-β-carotene which is of value as it is only this *cis*-isomer of β-caroteine that appears to have tumour suppressant activity [74]. β-carotene is also useful as it is an intermediate in the production of another carotinoid, lycopene, that is found in tomatoes, and is now associated with healthy antioxidant properties. Lycopene production can be achieved by carrying out the fermentation at a slightly alkaline pH so as to prevent the cyclisation of lycopene into β-carotene [75]. Furthermore genes have been introduced into the food grade yeast *Candida utilis* to create a novel pathway, not just to make lycopene and β-carotene, but also astaxanthin, which is the carotinoid present in salmon and lobsters [76]. Moreover pathway engineering is now being used to create a whole range of novel carotinoids that may have new and useful properties [77]. Currently these molecular biology methods require quite expensive R&D, so that they tend to be only used for those few opportunities where a major economic benefit can be obtained. However it is to be expected that costs will fall with time so that they can then find wider uses.

11
Product Inhibition

One common problem in biocatalysis that has not yet been solved properly is product inhibition. As the reaction proceeds and product begin to accumulate, it can have a number of adverse effects on the process, affecting the rates of reaction, the degree of conversion of raw materials achieved, the concentration of reactants that can be reached, and the ease and cost of downstream processing. Therefore product inhibition effects often greatly compromise the commercial potential of biocatalysis processes. Product inhibition is a characteristic not just of the particular enzyme(s) involved, but also of the permeability properties of the cell membrane and intracellular membranes, such as of the peroxisome that contains the enzymes involved in fatty acid metabolism involved in methyl ketone and flavour lactone formation. Effects may be by end-product inhibition of a key enzyme, or toxicity of the product to the producing cells. In other cases the equilibrium point of the enzyme catalysed reaction is not far enough towards complete conversion of substrate into product. This is the case for the 55% fructose syrup required for cola beverages, because the equilibrium position of the glucose isomerase reaction is only partially towards fructose. Therefore only syrup containing 42% fructose can be produced, despite the partial relief of product inhibition afforded by the plug flow kinetics of immobilised biocatalyst systems. To produce syrups containing a higher concentration of fructose an additional separations step, such as chromatographic separation, has to be used.

12
Downstream Processing – from Bioreactor to Product

Once the biocatalyst has made the desired molecule considerable time, effort and costs must be spent on down-stream processing (DSP) to make the product in a purified, and concentrated or dry form, suitable for sale. The purity that must be achieved to meet the specification for the final product is just as important a determinant of the extent of downstream processing required as is the concentration and purity of the reactant stream emerging from the bioreactor. The importance of DSP is shown by the process flow diagram for trehalose (Table 17) in which the majority of the unit operations and therefore the majority of processing costs are for DSP. This process also shows that upstream processing, the preparation of the precursors in suitable form for bioprocessing, can also be a considerable challenge. Therefore to minimise the cost of processing it is often more desirable to minimise the number of process steps than to optimise the performance of the biocatalyst. This is despite the biocatalyst being very much the lynch-pin in the process, with a strategic value far in excess of its actual contribution to processing costs. This is because, without it, absolutely no product could be made. DSP costs can easily exceed the costs of the bioproduction step, particularly when a rigorous final product specification is required. The unit operations involved in DSP which include chromatography, solvent extraction, evaporative concentration and crystallisation are frequently expensive. To avoid such expensive steps it is extremely useful if the product can be easily precipitated, as in the industrial processes to make Aspartame and L-cysteine, as this greatly simplifies product purification and isolation. Product purification and isolation cost can be greatly reduced by using concentrated raw materials, and by achieving high conversions of raw material into product, that is by operating as intense a process as is possible [78] (Fig. 31). Similarly, over the whole process yields of product at each step are maximised, and the number of steps used are minimised [79].

Table 17 Survey of down stream processing operations employed in some commerical processes

	Glucose iso-merase	Aspar-tame	As-partic acid	Malic acid	L-Lysine	L-Car-nitine	Iso-mal-tulose	Cyclo-dextrins	IPM/ IPP	Meth-ionine	Pantoic acid	L-Ala-nine	Δ8–9 fatty acid esters	L-Phenyl alanine
Filtration	×	×	×	×	×	×	×						×	×
Concentration	×			×			×							
Adsorption								×						
Crystallisation				×	×	×	×	×		×	×	×		×
Distillation									×					
Pervaporation									×					
Extraction											×			
Racemisation		×			×					×	×			×
Adduct formation		×											×	
Centrifugation														
Precipitation		×	×											
Chromatography	× for 55% HFCS												×	
Removal of pro-tecting group		×												
Ion-exchange						×								
Hydrogenation							(×) for isomalt							
Lactonisation											×			

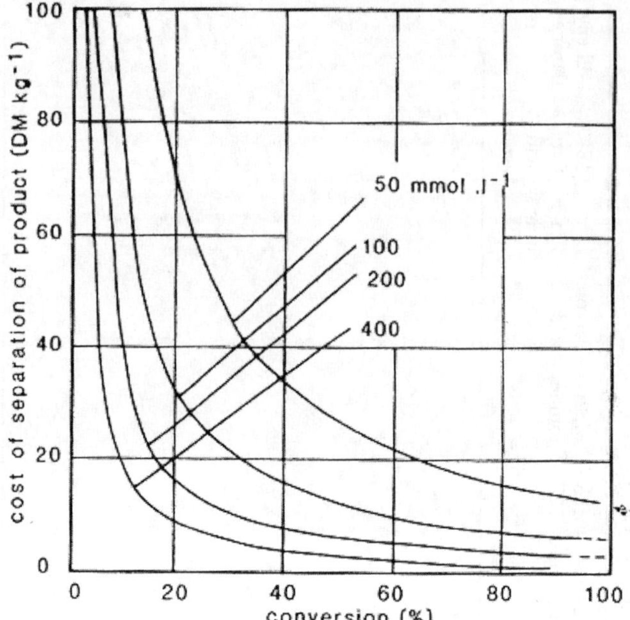

Fig. 31 Cost of separation per unit weight of product as a function of conversion at different starting substrate concentrations. From *Wandrey and Flaschel* [78]

13
In Situ Product Recovery (ISPR)

The variety of DSP operations used in commercial biocatalysts processes is surveyed in Table 18. Ideally the first step in DSP would be selective isolation of the product directly from the bioreactor, such as by partitioning across selective membranes, a technique called in situ product recovery [80]. An ISPR approach is doubly valuable as often inhibition or toxicity effects on the biocatalyst by the product can be relieved, allowing greater rates of reaction and making continuous production possible. ISPR can also reduce by-product formation, achieve favourable shifts in reaction equilibria, decrease the number of process steps required, and reduce product isolation and purification costs. These advantages are important because low product concentrations is the number one reason why bioprocesses fail to become commercial successes.

ISPR is already proven for some speciality chemical and pharmaceutical uses, such as the process to make L-tertiary leucine from trimethylpyrivate using leucine dehydrogenase, and then formate dehydrogenase to regenerate cofactor bound to polyethylene glycol [81]. A similar membrane based process has been developed by Sepracor to make *p*-hydroxyphenylglycine using enzyme impregnated in a membrane, with the substrate supplied from one side of the membrane, and product removed from the other.

Table 18 Advances in biocatalysis and bioprocessing pioneered by manufacturing processes for food products

Advance	Example
Reaction on water-insoluble substrate	Thermostable α-amylase for prethinning maize starch
Thermostable enzymes	
Low water enzyme reactions	Lipase interesterification process to make cocoa butter equivalent
Solvent-less reactions	Isopropylmyristate/palmitate
Synthetic reactions	Use of thermolysin to synthesise the high intensity sweetener Aspartame
Continuous processes	Vinegar, milk souring
Use of sequential enzyme reactions	Production of L-aspartic acid and L-alanine from ammonium fumarate
Coupled enzyme reactions	L-Lysine production using lactamase and caprolactam hydroylase
Plant derived enzymes	Cloned locust bean α-galactosidase to make partially debranched guar gum
Fungal spores as biocatalysts	Methylketone manufacture (blue cheese flavour)
Algae as biocatalysts	Docosahexaneoic acid production
Cloned protein	Chymosin of milk clotting
Biopolymer cross-linking	Transglutaminase
Biopolymerised product	MelaneZe (Biomelanin)
Aldehyde product	AromaZe (Vanillin)

This ISPR approach is now being applied to food ingredients such as 2-phenylethanol, which is used for its rose like aroma, useful for many fruit flavours [82]. Another approach to preventing product inhibition is possible if the product is volatile, in which case gas-phase processing may be feasible. An example is the conversion of ethanol into acetaldehyde, also useful as a natural flavour, that gives a fresh impression. This reaction is carried out by the alcohol oxidase of *Pichia pastoris* [83].

14
Process Integration and Optimisation

Process integration is very important if economically viable processes are to be created. This is particularly important because bioprocesses for foods and cosmetic ingredients can involve not just single step reactions, but also multi-step reactions, and combined reaction step reaction; involving whole cells, isolation enzymes, immobilised cells and immobilised enzymes. The particular form of biocatalyst used is selected on a bespoke basis (Table 14).

Important goals include good process reproducibility and reliability, to ensure excellent product quality and to minimise downtime during manufacture [84]. A good example is the long established industrial process for ascorbic acid (vitamin C). Several positive features combine to make this a very cost-effective process. The bioconversion step makes the key intermediate L-sorbose in a concentration of 250 g/l on a 70 m^3 scale. The precursor used, sorbitol, has significant sales in its own right. Also the 2-keto-L-gulonate precursor of ascorbic acid is easily recovered as its acetone adduct. Furthermore hydrogen for the reduction of glucose earlier in the process, and chlorine required for the oxidation of the diacetone adduct of L-sorbose later on in the process, are both generated easily by the electrolysis of sodium chloride [85].

Undoubtedly one of the very best examples of a fully integrated bioprocess is corn (maize) refining. The range of products made from a single raw material includes glucose syrups, and the high fructose syrups produced from them, starch is a viable product in its own right, as are corn oil, gluten, maize fibre and corn steep liquor (Fig. 32). In addition there is the wide range of fermentation products produced using glucose syrups as the carbon source and corn steep liquor as a source of trace nutrients. These fermentations include processes to

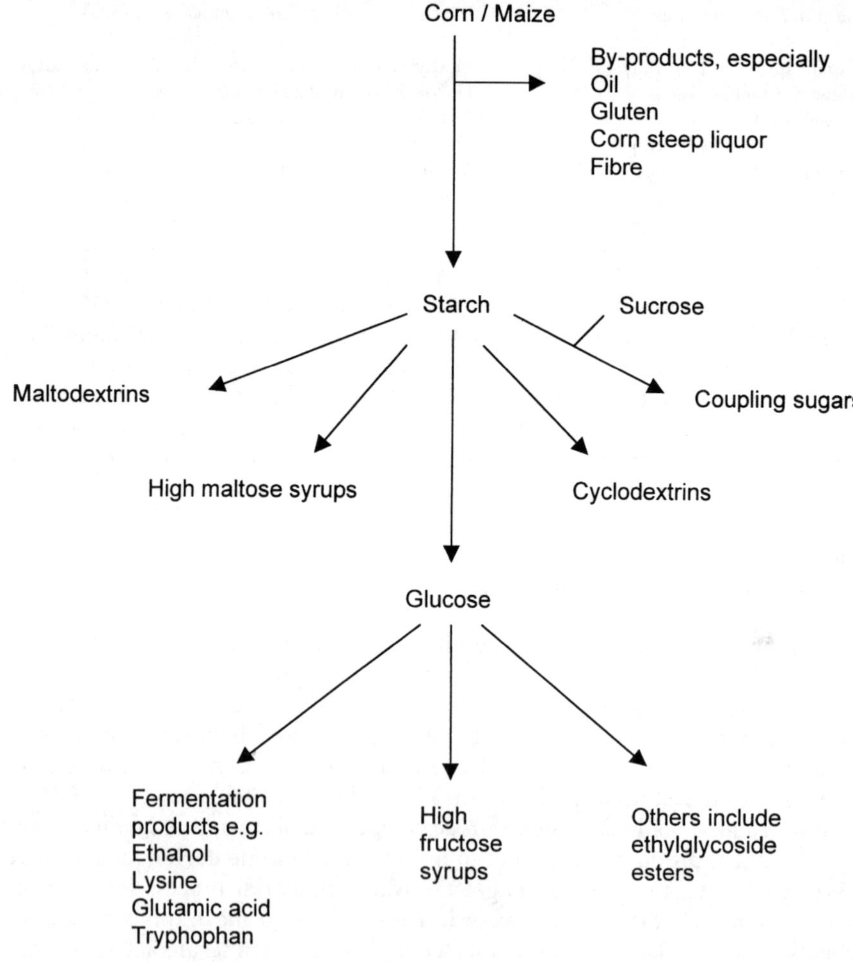

Fig. 32   The range of products produced by bioprocessing of maize

make the α-amylase and glucoamylase needed to hydrolyse the starch, and also fermentations to make ethanol, xanthan gum, lysine, threonine, glutamic acid, lactic acid, and other products. The use of the by-products from corn refining is carried even further. For instance ADM use the waste hot water and CO_2 to grow cucumbers, lettuces and herbs; and even to farm Tilapia fish (ADM Annual Report 2001).

For the glucose isomerase step the complexity of the interrelationships between all the process variables and enzyme activity and stability has been thoroughly studied. This shows not just the expected effects of pH and temperature, but also marked effects on enzyme performance of substrate purity, colour, contact time, syrup concentration (viscosity), enzyme particle size, and the concentrations of magnesium and calcium ions and oxygen present (Fig. 33) [86]. Even temperature has opposing effects, enzyme activity increasing with temperature; but enzyme stability decreasing with temperature so that for long term operation a compromise operating temperature of around 62 °C is used. This is also sufficiently high to help control microbial contamination, so that long term reactor productivity is maximised. As a result high productivities have been achieved. Current commercial glucose isomerase products have productivities of 1.4–6.0 column bed-volumes of syrup converted per hour. This equates to about 250–1200 kg/m³/h of fructose formed.

In carrying out such optimisation studies it is important to appreciate that operational parameters such as the pH optimum and the temperature optimum of enzymes are not absolutely fixed and invariant characteristics of particular enzymes. For instance it has been reported that the temperature optimum of

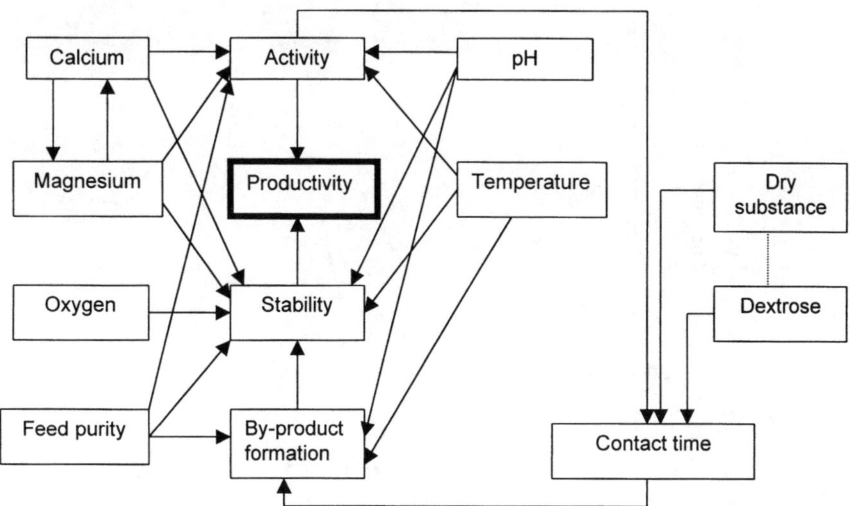

Fig. 33 Interaction logic diagram for the operation of an immobilised glucose isomerase reactor. The diagram shows the influence of individual process parameters on the activity and stability of the enzyme in a qualitative way and hence indicates the overall effect on the productivity of the enzyme

pullulanase, a microbial enzyme used to hydrolyse the α1–6 glycosidic 'branch' bonds of starch, changes from around 45 °C when the enzymies is used at pH 4, to an optimum of about 55 °C when used at pH 5. (Fig. 34) Similarly the thermostable α-amylase produced by *B. licheniformis*, and used to pre-thin starch at the beginning of the starch refining process, has a pH optimum of around 6.0 when used at 37 °C, but when used at its normal operating temperature of 95 °C it has a pH optimum of about 6.75 [87]. Because of the subtleties of these interactions there is considerable scope for fine-optimising many bioprocesses. The optimum conditions for an enzyme reaction can also vary with the precise type of substrate present, especially when peptides or oligosaccharides are involved. Then the degree of polymerisation of the substrate can change during the course of a hydrolytic reaction, so that the actual substrate available for the enzyme becomes gradually smaller as the reaction proceeds.

Use of several enzymes together also requires special optimisation skills. An example is the use of proteases, peptidases, lipases and esterases to make Enzyme Modified Cheese Flavours.

A more recent good example of process integration has been provided for L-carnitine [84]. This stresses the importance of the correct choice of reactor, raw materials, and also of minimising the number of unit operations. In particular this study shows that both operating and capital costs are minimised if inter-

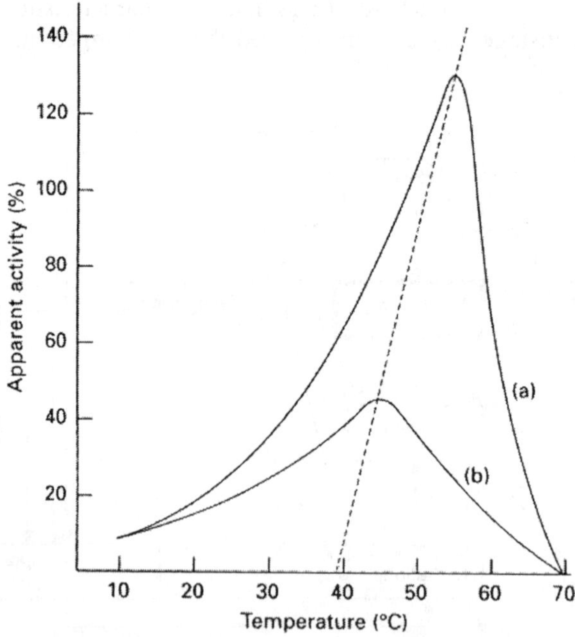

Fig. 34 Interaction of the variables on the activity of an enzyme. The figure shows the combined effect of pH and temperature on the apparent activity of *Klebsiella aerogenes* pullulanase (E.C 3.2.1.41). Curve (a) shows activity as a function of temperature at pH 5.0 and curve (b) at pH 4

mediates do not have to be isolated, especially in pure and concentrated forms, as this requires additional processing steps, and therefore adds costs to the product.

Another good example of process integration is the Tanake Seiyaku processes for making L-malic, L-aspartic acids and L-alanine from fumaric acid [88] (Fig. 8). One advantage is that the L-malic and L-aspartic acids are both made from ammonium fumarate using immobilised catalysts. Originally the immobilised cell process for aspartic acid was operated as a batch reaction, with aspartase catalysed conversion of fumaric acid into L-aspartic acid as the first step, followed by the above decarboxylation reaction to produced the L-alanine; that is as a two step conversion. This was because the two reactions have quite different pH optima, of pH 8.5 and 6.2 respectively. This process was then improved by using plug flow column reactors, operated in sequence. Now a single reactor containing co-immobilised cells with separate aspartase and aspartate decarboxylase activities is operated. A major problem in developing this process was the large quantities of CO_2 produced. This is because CO_2 is generated as a stoichiometric by-product of the decarboxylase of the L-aspartic acid into L-alanine. This CO_2 production is equivalent to some 50 l of CO_2 for each litre of fluid processed, as it contains ca 300 g/l ($\cong$2.5 mol/l) aspartic acid. The large quantities of CO_2 generated disrupt the plug-flow fluid dynamics of the immobilised cell column, and also lowers the pH, away from the optimum for the decarboxylase at pH 6.2. Therefore the manufacturing process uses a pressurised column to keep the CO_2 in solution. The stability of the enzyme is unaffected by continuous operation at the 10 bar excess pressure used, and the enzyme is additionally stabilised by the addition of pyruvate and pyridoxial phosphate.

A very recent example, that has commenced manufacture only in 2001, is the biopolymer MelaneZe. This is a multi-functional cosmetic ingredient, with good antioxidant activity and aqueous solubility properties, an attractive bronze colour, and also some UVA and UVB absorbing and DNA protecting

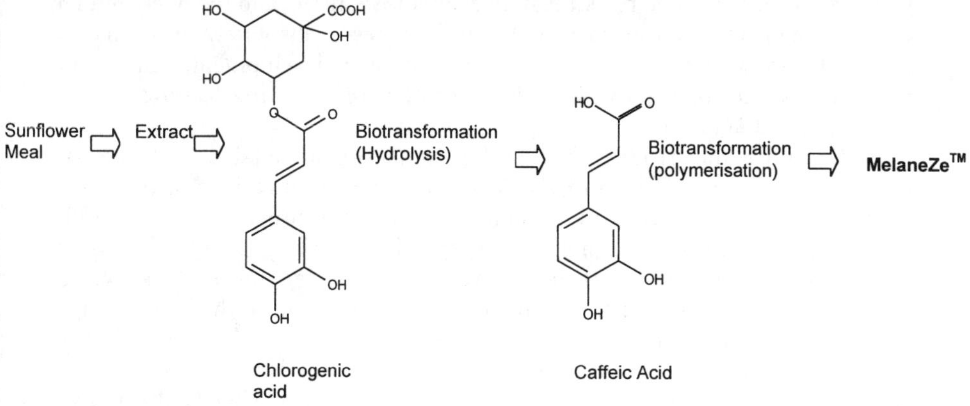

Fig. 35 MelaneZe

properties [89]. Its manufacture by Zylepsis requires a careful integration of all the steps from the extraction of caffeic acid precursor from enzyme treated sunflower seed meal, through to the maintenance of a good colour specification, molecular weight, and antioxidant activity in the final product. This is believed to be the first biopolymerisation process to enter production (Fig. 35).

15
New Bioproduct Development

Although successful new science is often an important and necessary factor for success, it is not necessarily the most important factor involved. It appears that very often advances are first made by technologists working in response to market needs, and then only later are the full scientific details elucidated, often by academic researchers. Then this mechanistic, analytical and other data is an excellent starting point for the next generation of technologists to improve the original process. For instance although glucose isomerase had been used on a massive scale since ca 1974 it was only in 1984 that the actual initial product of the enzyme reaction was shown to be α-fructofuranose, which then quickly isomerises to a stable mixture of β-fructofuranose and β-fructopyranose [90].

Other important factors for success include patenting and trade marking, raw materials selection and sourcing, product formulation and applications testing, marketing and consumer acceptability trials, and safety testing and regulatory approval including Kosher approval. In particular a market justification for the proposed new product has to be developed. This is more difficult to do for new products than if the proposal is for an improved way to make an existing ingredient, for which data on the performance of the existing ingredient products, existing market sales volumes and revenues, growth rates and sometimes even profit margins, may be available as a guide and benchmark. Such financial analysis is very important, especially to establish the likely demand for the proposed new product, and the prices that could be charged, which will determine the amounts and phasing of investments required, and the likely return on investment that can be expected. This analysis is absolutely necessary to justify the initial investments in research, product development, safety and regulatory approval, and the establishment of manufacturing facilities and a marketing and sales force.

Furthermore the product has to be regulatory acceptable and safe in each country it is sold in and may require other certifications such as for Kosher/Halel, gluten-free, and 'organic'. Similarly the process must be both safe to operate and environmentally acceptable, especially as regards wastes emissions. Gaining the necessary approvals can be lengthy and expensive and may rate-limit sales of the product. It is only when all of these steps have been completed can we make additions to the growing list of commercially successful bioprocesses and bioproducts. While the complexities of new product development can make it expensive, the time and costs involved can also provide a big barrier to competitors, disincentivising them to enter the area.

Furthermore the complexities involved can also be a rich sources of opportunities. These can arise, for instance; from changes in raw material supplies or regulatory requirements, the expiry of competitor patents, or changes in consumer preferences.

16
Use of Bioproducts as Starting Materials for New Products

Another way in which new product development occurs is by the use of bioproducts, with uses in their own rights, as the starting materials for the synthesis of other products, provided these starting materials can be made cheaply enough. Examples include commodity products such as glucose syrups and glutamic acid, and also speciality biochemicals, including L-aspartic acid and isomaltulose.

17
Process Development

Ideally any process should involve a minimum of operations, exclude any that require expensive capital equipment and those that require expensive capital equipment and those that have high operating costs, such as requiring the concentration of dilute solutions of product by evaporation, achieve high process intensities by using high concentrations of raw material and achieving high degrees of conversion into products, and operate at high through puts, which means a low residence time in immobilised cell or enzyme reactors. Capital expenditure is particularly important, since unlike consumable cost such as for energy and raw materials, capital expenditure has to be made well in advance of any product being made for sale, so that appreciable interest or other costs have to be borne, and because of the additional costs of installation and trialing of new equipment. Therefore whereever possible existing proven equipment should be used, especially if its capital cost has become significantly depreciated.

Simple process logistics factors and the physico-chemical characteristics of the product are also important. For instance in processing the ease of pumping and filtering, and resistance to colour development and microbial contamination are very important. In particular the ideal process needs to be robust and reproducible. Process robustness means that process operating parameters do not have to be very exactly controlled for the process to work well; so the small fluctuations and deviations in pH, temperature, reactant concentrations etc. that invariably occur do not have dramatic effects on productivity, and that the process is very reproducible with only small run-to-run variations and little out-of-specification product made that needs to be reworked or disposed of. Unfortunately many bioprocesses are far from being robust and reproducible as expected from any new technology for which little experience of large-scale use has been built up.

17.1
Product Specifications

Every product has a quality specification, that must be delivered by the process, otherwise the product cannot be sold. In that case the out-of-specification material has be reprocessed or disposed of; both of which are expensive options. Quality features can include texture/mouthfeel, flavour and colour, microbial counts and oxidative stability. For instance maintenance of good microbial quality will depend on the quality of raw materials, requiring good supply chain management; efficient processing, especially good analysis of the process to identify the critical control points, the moisture content, pH and other factors, use of antimicrobials and antimicrobial packaging.

17.2
Process and Product Improvement

Once the process is established, programmes of improvements are set in place. The aim is finite reductions in production costs and improvements in product quality to be achieved each year. These could be achieved by the use of new, cheaper or better raw materials, reduced energy consumption or labour costs, and reduced waste disposal etc. These are often supported by R&D efforts and aided by the progressive depreciation of capital equipment. A major benefit is if productivity can be improved so that the marginal cost of production is reduced.

17.3
Environmental Aspects

Biotransformations has often been cited as an environmentally friendly processing approach. Certainly because they operate at ambient or close to ambient temperatures bioprocesses appear to be less energy intensive than chemical processes operating at elevated temperatures and pressures. Two particular advantages of bioprocesses are their use of renewable raw materials and the production of biodegradable products. One consequence of the use of natural raw materials is their variable quality, unless a defined monoculture crop such as maize is used, or else these variations have to be accommodated, for instance by blending. Use of maize as the raw materials for glucose syrup production, and also for the manufacture of high fructose syrups, ethanol and the range of other glucose derived products such as xanthan gum, monosodium glutamate, lysine etc. is a good example. But a detailed economic analysis disputes whether fermentation processes, such as for polyhydroxalkanoate polymers, such as poly-3-hydroxybutanoate, really are more energy efficient than comparable chemical processes [91]. A good example of a biodegradable product is the ethylglycoside ester surfactants made by enzyme mediated esterification. A future opportunity could be more biodegradable fragrance materials. This is because current chemically synthesised musks are poorly biodegraded. Also because of their high volumes of usage, high partitioning coefficients and high chemical stabilities, they have accumulated in the environment; not just in river water and fish, but also in human fat and milk [92]. Therefore a bioprocess to make macrocyclic musks could offer good environmental benefits and also perhaps safety advantages.

18
Discussion

This chapter is not just a simple review of the science underpinning successful bioprocesses, but rather an exploration of how the many technical and commercial requirements for the creation of new and successful bio-products interrelate to create successful new products. It also demonstrates the rapid growth in the industrial use of bioprocesses, their extensive use to manufacture a wide range of different food and cosmetic ingredients, and the significant and increasing economic values they generate.

The many examples provided prove that the biosynthesis of chemically complex and pure chiral molecules is not the sole preserve of the pharmaceutical industry, that cheap, abundant and renewable raw materials really can be utilised on a manufacturing scale, and that bioprocesses can both improve the production of existing products by making them cheaper, or by making them in a purer form, or from new raw material sources. In addition, entirely new chemical entity products with new functional properties can often be created by exploiting the novel catalytic properties of biocatalysis. These examples also show how applications of biocatalysis to food and personal care products can be used both in competition and in combination with other methods of production such as agriculture, chemistry, and fermentation.

An interesting comparison can be made with the pharmaceutical use of biocatalysis. Pharmaceutical use is mostly confined to the biotransformation of relatively small molecule substrates, and to the production of pure products. By contrast many food and personal care processes have also been developed with two important differences, to make chemically complex products, and by utilising the hydrolysis or modification of biopolymers such as starch, proteins, nucleic acids, pectins and gums. This has involved the extension of biocatalyst use to incompletely water soluble raw materials, such as some polysaccharides and cheese; and also to treat chemically and physically complex raw materials to produce improved products such as meat, yoghurt, wine and beer etc. For these products their chemical composition is not necessarily the prime indication of quality, but rather the taste, mouthfeel, aroma, or some combination of such organoleptic characteristics.

A particular important conclusion is that the first manufacturing use of very many technical advances in enzyme technology has actually occurred for the production of food ingredients; rather than for pharmaceutical and healthcare products as is generally assumed. These include the use of thermostable and cloned enzymes, the carrying out of enzyme reactions in low water reaction media, and synthetic and coupled enzyme reactions etc. (Table 18).

The large number of well-established and emerging new bioprocesses and bioproducts described in this chapter demonstrate the large-scale use of an im-

Table 19 Biocatalysis reactions for food and cosmetic ingredients carried out on a large scale

Hydrolysis

α-Amylases, glucoamylases and pullulanases for starch hydrolysis
β-Amylase to make high maltose syrups for brewing
β-Glucanase also for brewing
Glucomannanase to hydrolyse gums in coffee
Glycosidase mixtures to release bound aroma chemicals and colour etc from plant materials
Dextranase to treat liquid sugar
Lactase to hydrolyse lactose in milk
Polygalacturonidase for citrus fruit peeling
α-Galactosidase for raffinose hydrolysis in beet sugars
α-Galactosidase for the conversion of guar gum into a locust bean gum equivalent
Hemicellulase for wheat wet milling
Maltooligotrehalose trehalohydrolase in trehalose manufacture
Pectin and pectate lyases
Chymosin for milk clotting
Proteases and lipases used in enzyme modified cheese manufacture
Amino-peptidase for debittering of protein hydrolysates
Phosphodiesterase for 5′-GMP production from yeast RNA
Urea amidohydrolase (urease) for white wine treatment
Aminothiazoline carboxylate and carbamoyl-cysteine hydrolases for L-cysteine manufacture
Lipases for triglyceride hydrolysis

Esterase

Aminoacylase to resolve D,L-methionine
Chlorogenic acid hydrolysis to release caffeic acid
Lactamase used in L-lysine manufacture
Pectin methylesterase
D-Pantonolactonase used to prepare D-pantoic acid for D-pantothenic acid production

Isomerisation

Glucose to fructose for HFCSs using glucose isomerase
Sucrose to isomaltulose for Isomalt production using isomaltulose synthase
Maltooligosyltrehalose synthetase for trehalose manufacture

Transesterification

Cocoa butter substitute manufacture
Human milk glyceride preparation
Deamidation
Conversion of 5′-AMP into the taste enhancer 5′-IMP using adenyl deamidase

Decarboxylation

L-Aspartic acid decarboxylase for L-alanine manufacture
α-Acetolactate decarboxylase (for beer maturing)

Oxidations

Sorbitol into L-sorbose for ascorbic acid production
Carnitine-3-oxido-reductase to make L-carnitine
Proline 4-hydroxylase (2-ketoglutarate dependant dioxygenase)
Glutathione -DHAA oxidoreductase in bread making
Methylbutanol oxidation into methybutanoic acid flavour chemicals
Lipoxygenase for β-ionone
Use of peroxidase to manufacture MelaneZe
Catalase to remove residual hydrogen peroxide from milk
Dihydroxyacetone formation from glycerol

Table 19 (continued)

β-Oxidation and variants

Fatty acids conversion into methylketone cheese flavours
Hydroxy fatty acid into δ and γ-decalactone flavour chemicals
Ferulic acid into vanillin

Glycoside synthesis

Starch into cyclodextrins using cyclodextrin glucosyltransferase
Formation of BioEcola oligosaccharide
Fructo-oligosaccharides

Lyases

Aspartic acid production from fumaric acid
Hydroperoxide lyases for flavour aldehydes
Alliinase use in allicin production

Esterification

Ethyl-2-methyl butyrate flavour production
Decosahexaneoic acid (DHA) glyceride synthesis
Isopropylmyristate and palmitate synthesis

Peptide bond formation

Thermolysin for Aspartame manufacture
trans-Glutaminase for cross linking proteins (Acyltransferase)
Malyltyrosine synthesis using peptidase

Hydratase

Nicotinamide manufacture using nitrile hydratase
L-Malic acid acidulant from fumaric acid

Desaturase

Fatty acid desaturase used to make unsaturated fatty acid esters

Racemases

D-Aminothiazoline carboxylate racemase in L-cysteine manufacture
Caprolactam racemase used for L-lysine manufacture

---

pressive range of raw materials and also of different types of enzymes, not just simple hydrolases (Table 19).These include fatty acids and fatty acid esters, sugars, phytophenols, and especially amino acids, amino acid derivatives and proteins as the starting materials. The wide range of enzyme systems used include 'hydroxylases', β-oxidation, hydratase, desaturase, and especially enzymes making or breaking carbon-nitrogen bonds, such as glutaminase, aminotransferase and amidohydrolase enzymes. The only class of enzymes that has not yet fully contributed to the biocatalysts used by industry appears to be the ligases. Allinase (Fig. 23), and one of the enzymes involved in cysteine production (Fig. 9) are especially interesting as rare examples of the use of carbon-sulphur lyases.

From these examples a 'league table' for the activities of bioprocesses to make food and cosmetic ingredients can be made (Table 20), with details of some bio-

Table 20 Bioreactor league table

Product	Reactor productivity (g//l/h)
L-Aspartic acid (BioCatalytics process)	3000
L-Aspartic acid (Nippon Shakubai)	2000
Glucose Isomerase (various operators)	250–500[a]
Nitrile hydratase	80
L-Malic acid (Tanabe Seiyaku)	50
Isopropylmyristate (Unichema)	42
Isomaltulose (Sudzucker)	40x
L-Methionine (Degussa Huls)	25
Penicillin amidase	18.5–33
L-Pantoic acid (Fuji)	15
L-Phenylalanine (Coca Cola)	14
L-Alanine (Tanabe Seiyaku)	7
L-Carnitine (Lonza)	5.4
Glucoamylase (various)	5.3
L-Lysine (Toray)	4
L-Tryptophane (Amino)	3.1

[a] Values calculated for immobilised enzymes. Pharmaceutical processes (in italics) are provided as benchmarks.

processes for pharmaceutical products also given for comparison. The real situation is more subtle as biocatalyst performance also depends on other factors. For instance the key requirement for the enzyme synthesis of Aspartame is very good reaction selectivity. Similarly glucose isomerase has only a quite low activity, but this is more than compensated for by its very long operational stability. Furthermore rather than the activity of the biocatalyst step, it may be the process intensity achieved, or the chemical or isomeric purity of the product made that is the most important contribution to the value of the product. Also, the activity of the biocatalyst, and thus its 'cost-in-use', is just one of the factors that contribute to production costs. These other factors include the costs of raw materials, capital equipment costs (which may be interest repayments on the purchase price), and DSP costs.

These successful manufacturing processes for food and cosmetic ingredients are additionally impressive as they employ almost the full range of bioprocessing technologies, from precursor fermentations to the use of immobilised enzymes (Table 5). The one exception is that no manufacturing process employing a multi-enzyme immobilised cell process has been identified. Nevertheless despite the successes achieved to-date there is not yet a fully developed 'Toolkit' to allow an easy start to new projects. For instance there is still not yet a generally applicable large-scale method for immobilising enzymes. The processes listed in Table 5 range from processes that focus just on the production of a particular chemical such as Aspartame, to those that take a refinery approach, producing a range of different products, such as glucose, maltose and fructose syrups from maize. Notable exceptions from the list of industrial bioprocesses are any processes using plant cells as the biocatalyst, presumably be-

Strengths

> Shorter Regulatory process (except for food additives)
> Larger market sizes in some cases
> Upstream and Downstream processing is not required in some cases

Weaknesses

> Products are less highly valued
> Genetic engineering is mostly viewed as negative by consumers
> New ingredient acceptance and uptake by customers can be slow
> Bioprocesses can be expensive to research, develop and operate

Opportunities

> Nutraceuticals and cosmaceuticals
> Ingredients derived from foreign foods
> Genetic engineering (once consumer and regulatory acceptability is gained)

Threats

> Improved plant breeding
> Consumer 'chemophobia'.
> Reduced biodiversity

Fig. 36 S.W.O.T. Analysis of the use of biocatalysts to make food and cosmetic ingredients (as compared to the use of biocatalysts to make pharmaceuticals)

cause of their slow growth rates and susceptibility to contamination; or any process using immobilised cells possessing multiple enzyme reactions (Table 5), probably because of the difficulties in maintaining cells in a viable state long enough to obtain an economically acceptable performance.

Some of the general advantages and disadvantages of using biotransformation and bioprocessing approaches are shown in Fig. 36, and eight ways in which competitive advantages can be achieved are identified in Table 21. Having the ability to achieve sufficiently low cost production so as to be able to sell products

Table 21 Competitive advantages

May be gained from the use of new, or more effective, bioprocesses by achieving one or more of:
- Lower cost product
- Improved quality product
- Lower wastes and/or better by-product credits
- Use of cheaper and/or more available raw materials
- IP position created
- Safer product produced
- Improved labelling/marketing opportunity
- New or increased range of end-use applications

Table 22 Costs of bioprocesses

- Naturally derived raw materials are often more expensive and more variable in quality when compared with petrochemical feed stocks, except for the few that are the products of massive mono-cultures, such as maize
- Bioprocess reactions are usually slow, with long batch reaction times, or high residence time if continuous processes are used
- As regards operating costs biocatalysts tend to have significantly lower volumetric activities and shorter operational stabilities than chemical catalysts (Fig. 37)
- Bioprocess streams are generally dilute, impure and microbiologically unstable
- Downstream processing costs, for the concentration, purification and isolation of products, are high because of the low concentration and purity, and labile nature of many bioprocess product streams; and also because of the increasing standards of purity required for biochemical products
- Bioprocesses often use new technology that requires new equipment. So if manufactured in-house savings from the use of depreciated equipment may be low. Alternatively if toll or contract manufacturers are used they will charge premium prices because of the larger amount of development effort they will have to make and the higher risks of failed production runs
- Bioprocesses are often not 'robust', as they require precise control of many process parameters, which can lead to a greater frequency of failed batches or reduced productivities
- R&D costs can be high and time scales lengthy, so that only very substantial products and markets can justify the investments required

at competitive prices is crucially important. This is because however pure and active the ingredient, and however novel the science used; very few will buy the product unless it is cheap enough. Furthermore the more efficient the process, the cheaper the raw materials, and the lower the capital and energy costs; then the greater the profit margins (and the more money that can be reinvested back into R&D for the next generation of products).

Several factors contribute to the relatively high cost of using enzymes, as compared to chemical catalysts (Table 22) [93]. Therefore when they are able to carry out the desired reaction, chemical catalysts currently tend to be significantly cheaper than biocatalysts (Fig. 37). So biocatalysts will only be used if they offer an advantage. Such advantages include improved chemo-, regio- or

Catalyst	Role	Useful life (yr)	% product price
Fe/A1$_2$O$_3$/CaO/K$_2$O	Ammonia synthesis	5-10	0.7
Ag (unsupported)	Formaldehyde synthesis	0.3-1	0.6
Immobilised Glucose isomerase	Formation high fructose syrups	0.7	3
Amyloglucosidase	Saccharification of starch	0.005	1
Detergent proteases	Hydrolysis of stains	0.0001	1

Fig. 37 A comparison of commercial chemical and biocatalyst [93]

enantioselectivity, which is useful if there is a need to make very specific molecules; or reaction under mild conditions, useful if any of the reactants are labile; or a need for reduced by-product formation, or a natural image to aid in environmental compliance and marketing etc.

It is particularly interesting to note that as well as these more general trends, very many of these manufacturing processes have special features, particular to an individual process, that have greatly helped them to achieve success (Table 22). Therefore it would appear that attempts to develop generic bioprocess approaches can be expected to have only limited success, as each process and product has to be substantially researched and developed on an individual basis. One of the reasons for the absence of a truly generic approach may be because bioprocesses are operated by a very wide range of different

Table 23 Special features of biocatalysts that have helped them to become successful

Product/biocatalyst	Bioreaction effects
Use of α-amylase in the manufacture of glucose and high fructose syrups	Thermostable α-amylase reacts with insoluble starch. At >100°C pre-thinning it suffficiently to allow further liquidication, saccharification and then isomerisation.
Aspartame manufacture using thermolysin	The protease thermolysin reacts with just the α-carboxyl of the aspartic acid precursor, so that no blocking of the β-carboxyl is necessary. Aspartame product precipitates with excess phenylalanine methyl esters to make recovery necessary
Monosodium glutamate (taste enhancer) manufacture by *C. glutamicam*	Use of a fermentation medium deficient in biotin increases cell permeability to MSG, which greatly increases the productivity of the process.
High fructose corn syrup by glucose isomerase	The enzyme works without forming a phosphorylated intermediate, and is stable for months in continuous use
Enzyme modified cheeses	Quite different enzymes (proteases, lipases etc) all work effectively under the same conditions (viscous cheese)
L-malic acid production by immobilised cells	Addition of bile salts surpressed undesirable side-reaction and increased activity.
Isopropylmyristrate/palmitate	Solvent-less processing is possible (the isopropanol and acid act as the solvents).
γ-decalactone	The ricinoleic acid precursor is readily available, as it is ca 85% of the fatty acids in castor oil triglycerides.
Isomaltulose using immobilised cells	Stability is increased by high substrate concentrations, which also facilitate easy product recovery by crystallisation
Nicotinamide	The process operates using precursor that is suffficiently concentrated to cause the nicotinamide product to precipitate out during the reaction in an easy to recover form.
L-cysteine	Three enzymes required are all possessed by a single microblal strain.
Urease treatment of wine	Both products of the reaction are gasses.

companies, with quite different strengths (and weaknesses), priorities, and requirements; and with these companies ranging from raw material refiners, to chemical ingredient suppliers, to consumer product manufacturers; for both food and cosmetic industries (Table 24).

Another factor that makes a general approach to bioprocess development difficult is that for any particular process under development, all the many technical and commercial aspects of the overall process and product need to be satisfied, with the biocatalyst being just one of the factors required to make a successful bioproduct, The successful process very often has the best combination and balance of these features, rather than just using the most exciting science, the lowest capital costs, or the easiest IP or regulatory position etc. Therefore because of the complexity of the process and product development pathways, and the variety of quite different factors involved, it can be argued that the most defining aspect of biotechnology is not any one particular scientific or engineering activity, however multidisciplinary, but rather the management skills needed to successfully integrate all of the technical, legal, commercial and other requirements necessary to develop new bioproducts successfully to market. This absolutely vital requirement for success is well illustrated by new product selection procedures, which are a critically important step in the innovation process, as answers are required to a whole range of important questions, all of which need to be answered if an efficient process and cost effective product are to be developed.

Consequentially there does not appear to be any 'standard model' for success, especially as there seems to be little relationship between the type and form of biocatalyst used, the types of products made and their uses etc. [94] (especially Table 4.9). The common factor is that for each successful product, when all of the factors are combined, then the result must have a combination of a process that is sufficiently productive and cost-effective, and a product that achieves

Table 24 Examples of the use of bioprocesses for food and cosmetic ingredients throughout the supply chain

Type of company		Products	Operating companies
Raw materials refiners	Food	High fructose corn syrups	ADM (USA) and Staley (USA) etc
		Isomaltulose (Palatinit)	Mitsui Sugar (Jpn) and Sudzucker (Ger)
	Cosmetic	Isopropylmyristate/ palmitate	Unichema (NL)
Chemical ingredient manufacturers	Food	L-Carnitine	Lonza (Switz)
		Aspartame	Holland Sweetener (NL)
	Cosmetic	Hydroxyproline	Kyowa Hakko (Jpn)
Consumer product manufacturers	Food	L-Phenylalanine	Coca Cola (USA)
		Diacylglycerides	Kao (Jpn)
	Cosmetic	Δ9-Unsaturated fatty acid esters	Kao (Jpn)

Table 25 Conclusions – I hope I have shown

- The great variety and rapid growth of biocatalyst based processes successful on a manufacturing scale, their increasing economic importance, and the diverse range of companies involved
- The use of bioprocesses to make both entirely new ingredients, as well as improved forms of existing ingredients
- The use of biotransformations to make not just pure chemical products, but also chemically complex, and in some cases chemically undefined products
- The very wide range of enzyme types and forms of biocatalysts employed
- The special features of individual process that have made them successful
- Especially the value of new biocatalyst searches and generation, and process integration and DSP
- The importance of proving the functional efficacy of the ingredient produced, and its performance when formulated into end-products suitable for use by consumers
- Development of sufficiently low production costs is vital to allow products to be sold at competitive prices and good profit margins
- The importance of setting good market targets and fully considering all the economic, safety, regulatory, IP, and other factors necessary for success
- Very many of the technical advances made in biocatalysis have actually been made while developing processes for food ingredients, rather than for drugs
- Most successful bioprocesses and products have actually arisen as a response to a defined market need or customer problem, rather than by academic research that has been transferred to industry; and have required an adaptive heuristic approach during their R&D and market introduction
- Japanese companies can still perhaps be regarded as the leading innovators
- The essence of biotechnology is actually the effective management of the necessary technical, commercial, legal factors necessary for success, rather than ability, however great, in just one or a few of these areas of expertise

good volumes of sales at sufficiently high prices to generate good profits and so provide a good return on R&D costs.

In conclusion I hope to have demonstrated the current technical and commercial value of biocatalysts for the manufacture of food and cosmetic ingredients, the variety of enzyme types and forms of biocatalyst used, their advantages and disadvantages, and also their future potential (Table 25).

19
References

1. Cheetham PSJ (1998) J Biotechnol 66:3–10
2. Cheetham PSJ (1999) Enzymes for flavour productionIn: Flickinger MC, Drew SW (eds) Encyclopaedia of bioprocess technology. Wiley, p 1004
3. Somogyi LP (1996) Chem Ind (March) 170
4. Evenhuis B (1992) Agro-Food Ind High Tech Sept/Oct 19
5. Witholt B, Lageveen RG, Kok H (1985) Biotrans In: Tramper J et al. (eds) Organic synthesis. Elsevier, Amsterdam, p 239
6. Zimmerman TP, Robins KT, Werlen KT, Hoeks FW (1997) In: Collins AW et al. (eds) Chirality in industry. Wiley, New York, p287
7. Hoeks FWJMM (1991) Chimica 45:81
8. Terasawa M, Yukawa H, Takayama Y (1985) Proc Biochem 20:124
9. Sato T, Tetsuya T (1993) In: Tanaka A et al. (eds) Industrial applications of biocatalysts. Dekker, New York, p 15

10. Kumagai H (1999) Advs Biochem Eng Biotechnol 69:68
11. Chibata I, Tosa T (1977) Advs Appl Microbiol 22:1
12. Liese A, Seelbach K, Wandrey C (2000) Industrial biotransformations. Wiley-VCH, Weinheim
13. Chibata I, Tosa T, Sato T, Mori T (1976) In: Mosbach K (ed) Methods Enzymol 44:746
14. Sano K, Yokezeki K, Tamura F, Yasuda N, Noda I, Mitsugi K (1977) Appl Environ Microbiol 34:806
15. Yokozeki K, Majima E, Izawa K, Kubota K (1987) Agric Biol Chem 51:963
16. Oyama K (1992) In: Collins AW et al. (eds) Chirality in industry. Wiley, New York, p 237
17. Nagodawithana TW (1994) In: Gabelman A (ed) Bioprocess production of flavour, fragrance and colour ingredients. Wiley Interscience, New York, p 135
18. Ando H, Adachi M, Umeda A, Matsuura M, Nonaka R, Uchio H, Tanaka H, Motaki M (1989) Agric Biol Chem 49:2283
19. Okabe M, Tsuchiyama Y, Okamoto R (1993) In: Tanaka A et al. (eds) Industrial applications of immobilised biocatalysts. Dekker, New York, p 109
20. Bentley IS, Williams EC (1996) In: Godfrey T, West S (eds) Industrial enzymology, 2nd edn. Macmillan Press, London, Basingstoke, p 341
21. Ricks EE, Estrada-Valdes MC, McClean TL, Lacobucci GA (1992) Biotechnol Prog 8:197
22. Leuenbeger HGW, Boguth W, Widner E, Zell R (1976) Helv Chem Acta 59:1832
23. Hankinson S (1992) BMJ 305:335
24. Cheetham PSJ (1987) Production of isomaltulose using immobilised microbial cells. In: Mosbach K (ed) Methods Enzymol 136:432
25. McCleary BV, Critchley P, Bulpin PV (1984) Eur Pat Applic 0 121 960 A2
26. Cheetham PSJ, Underwood DR (1993) Biocatalysis 7:237
27. Ropert F, Dumont B, Belin JM (1995) Abs Bioflavour'95 Dijon (France) ed INRA
28. Takken HI, Groesbeck NM, Roos R (1992) In: Patterson RLS (ed) Biotransformation of flavours. Royal Society of Chemistry, p 186
29. Humphrey A (1993) US Patent 5,185,252
30. Moore SR, McNeil GP (1996) J Am Oil Chem Soc 73:1403
31. McNeil GP, Ackman RG, Moore SR (1996) J Am Oil Chem Soc 73:1409
32. Kyle DJ (1996) Lipid Technol 8:107
33. Coleman MH, Macrae AR (1980) US Patent 1,577,933
34. Macrae AR, Brench AW (1983) Eur Pat Applic 0 069 599
35. Laane C, Boeren S, Hilhorst R, Veeger C (1987) Biocatalysis in organic media. Elsevier, Amsterdam
36. Ohguchi M, Kubota N, Wada T, Yoshinaga K, Uritani M, Yagisawa M, Ohishi K, Yamagishi M, Ohta T, Ishkawa K (1997) J Ferment Bioeng 94:358
37. Mori H, Shibasaki Y, Yano K, Ozaki A (1997) J Bacteriol 179:5677
38. Kataoka M, Shimizu K, Sakamoto H, Yamada H, Shimizu S (1995) Appl Microbiol Biotechnol 43:974
39. Chu D-C, Kobayashi K, Juneja LR, Yamamoto T (1997) In: Yamamoto T et al. (eds) Chemical applications of green tea. CRC Press, Boca Raton, New York, p 127
40. Ghoneum M (1998) Int J Immunother 14:89
41. Takeuchi K, Koike K, Ito S (1990) J Biotechnol 14:179
42. Hirota Y, Kohori JK, Kawakara Y (1989) Ew Pat Applic 307154
43. Berry RE, Baker RA, Breummer JH (1988) In: Goren R, Mendel K (eds) Proceedings of the 6th International Citrus Congress, Tel Aviv. Margraf Scientific Books, Weikersheim, p 1711
44. Honda Y, Kako M, Abiko K, Sogo Y (1993) In: Tanaka A et al. (eds) Industrial applications of immobilised biocatalysts. Dekker, New York, p 109
45. Pitcher WH (1978) In: Brown GG et al. (eds) Enzyme engineering, vol 4. Plenum Press, p 67
46. Winterhalter P, Schreier P (1994) Flav Frag J 9:281
47. Cheetham PSJ, Quail MA (1991) US Patent 5,007,206
48. Winterhalter P, Skouroumounis GK (1997) In: Scheper T (ed) Adv Biochem Eng/Biotechnol. Springer, Berlin Heidelberg New York, 55:73
49. Pedersen S, Lang NK, Nissen AM (1985) Ann New York Acad Sci 750:376

50. Matsumoto K (1993) In: Tanaka A et al. (ed) Industrial applications of immobilized biocataalysts. Dekker, New York, p 255
51. Wubbolts MG, Bucke C, Bielecki S (2000) In: Straathof AJJ, Adlercreutz P (eds) Applied biocatalysis, 2nd edn, p 153
52. Le Moult SC, Richardson PF, Roberts SM (1995) JCS Perkin Trans 1:89
53. Petersen M, Kiener A (1999) Green Chem 4:99
54. Cheetham PSJ (1987) Enz Microb Technol 9:194
55. Bull AT, Goodfellow M, Slater HJ (1992) Ann Rev Microbiol 46:219
56. Rondon MR, Goodman RM, Handelsman J (1999) TIBTECH 17:403
57. Chumpolkalwong N (1997) J Ferm Bioeng 83:429
58. Winson MK, Goodacre R, Timmins EM, Jones A, Asberg BK, Woodward AM, Rowland JJ, Kell DB (1997) Anal Chim Acta 348:273
59. Madigan MT, Marrs BL (1997) Sci Amer 276:82
60. Casey J, Dobb R (1992) Enzyme Microb Technol 14:739
61. Gasson MJ, Kitamura Y, McLauchan WR, Narbad A, Parr J, Parsons EL, Payne J, Rhodes MJC, Walton NJ (1988) J Biol Chem 273:4163
62. Lesage-Meessen L, Haon M, Deattre M, Thibault JF, Ceccaldi BC, Asther M (1997) Appl Microbiol Biotechnol 47:393
63. Cheetham PSJ, Gradley ML, Sime JT (2000) Pat Applic PCT/GB00/00654
64. Rabbenhorst J (1996) Appl Microbiol Biotechnol 46:470
65. Overhage J, Priefert H, Steinbuchel A (1999) Appl Envir Microbiol 65:4837
66. Gradley ML, Willetts AJ, Sime JT (2000) Pat Applic PCT/GB/00488
67. Lange S, Musidlowska A, Schmidt-Dannert C, Schmitt J, Bornscheuer UT (2001) Chem Biochem 2:576–582
68. Dalboge H, Lang I (1998) TIBTECH 16:265
69. Herweijer M (2002) Presented at Biocatalysis in Food and Drink Industries, University of Westminster, London
70. Giver L, Gershenson A, Freskgard PO, Arnold FH (1998) Proc Natl Acad Sci USA 95:12,809
71. Joern J, Meinhold P, Arnold FH (2002) J Mol Biol 316:641
72. Wang L, Kong X, Zhang N, Wang H, Zhang J (2000) BioChim Biophys Res Commun 276:346
73. Matsumura I, Ellington AD (2001) J Mol Biol 305:331
74. Margalith PZ (1992) Pigment microbiology. Chapman and Hall, London
75. Ninet L, Renaut J (1979) In: Peppler HJ, Perlman D (eds) Microbial technology 2nd edn. Academic press, New York, p 529
76. Miura Y, Kondo K, Saito T, Shimada H, Fraser PD, Misawa N (1988) Appl Environ Microb 64:1226
77. Schmidt-Dannert C, Umeno D, Arnold FH (2000) Nat Biotechnol 18:750
78. Wandrey C, Flasichel E (1979) Adv Biochem Eng 12:147
79. Fish NM, Lilley MD (1984) Bio/technology 2:623
80. Lye GJ, Woodley JM (1999) TIBTECH 17:395
81. Kragl U (1996) In: Godfrey T, West S (eds) Industrial enzymology, 2nd edn. Macmillan Press, London, Basingstoke
82. Fabre CE, Blanc PJ, Marty A, Goma G, Souchen I, Voilley A (1996) Perfumer and Flavourist 21:27
83. Duff SJB, Murray WB (1990) Process Biochem: April 40–41
84. Hoeks FWJMM, Muhle J, Boehlen L, Psenicka I (1996) BioChem Eng J 61:53
85. Roberts SM, Turner NJ, Willetts AJ, Turner MK (1995) Introduction to biocatalysis. Cambridge University Press
86. Poulsen PB, Zitton L (1976) Methods Enzymol (Mosbach K (ed)) 44:809
87. Fullbrook PD (1996) In: Godfrey T, West S (eds) Industrial enzymology, 2nd edn, p 504
88. Furui M, Yamashita K (1983) J Ferment Technol 61:587
89. Banister NE, Cheetham PSJ, (2001) US Patent 6,303,106BI
90. Makkee M, Kieboom APG, van Bekkum H (1984) J R Netherlands Chem Soc 103:361
91. Gerngross TU (1999) Nat Biotechnol 17:541
92. Ford R (1999) Soap Perfumery Cosmetics Feb 32

93. Cheetham PSJ (1988) Food Biotechnol 2:117
94. Cheetham PSJ (2000) In: Straathof AJJ, Adlercreutz P (eds) Applied biocatalysis, 2nd edn. Harwood Academic Publisher, Amsterdam, pp 93
95. Lilly MD (1997) J Biotech 59:11

Received: July 2002

Supplement

During the printing process of this volume several new developments occurred in the bioproduction of ingredients for food and cosmetics. The most important ones are summarized in the following text.

Ceramides

Ceramides are complex lipids that are responsible for much of the water barrier properties of human skin, preventing the skin from drying out, and in cosmetic terms giving the skin a smooth 'young' appearance. The ceramide content of human skin is reduced with age and in some dermatological conditions such as psoriasis. A number of different ceramide structures are present in human skin, varying in the type of fatty acids esterified to the C18 sphingosine or phytosphingosine base, or in the case of ceramide 1, also containing a long chain amide linked omega-hydroxy fatty acid.

Ceramides are manufactured by a process based on the use of a very highly evolved yeast, *Pichia cifferrii*, to produce the key precursor tetraacetyl phytospingosine, via desaturation of palmitic acid precursor, and then coupling with serine; followed by its extraction, deacetylation; and then acylation, such as with stearic acid [1a]. Spingosines are also useful as cosmetic ingredients in their own right as they have some antimicrobial and anti-inflammatory properties; and an ester salicyloyl phytosphingosine is marketed as a skin clarifying and exfoliating agent as part of a range of sphingosine and ceramide products by Goldschmidt.

Isopropylmyristate and Palmitate

A number of other opportunities for lipases to produce lipid products show promise, but despite considerable technical advantages in some cases, few have yet proved sufficiently good to justify commercialisation. These include the use of lipases for fat hydrolysis (lipolysis), to selectively produce glycerides containing unsaturated fatty acids, to separate conjugated linoleic acid isomers, to make monoglycerides that are useful as emulsifiers or surfactants, or diglycerides that are effective for retarding fat crystallisation, with the possibility of both 1,2- and 1,3-diglyceride products.

For instance *Candida rugosa* lipases are non-specific and so can act on a range of fats; and are not inhibited by high concentrations of free fatty acids, and so can achieve high degrees of lipolysis of concentrated triglycerides. *Fusarium oxysporum* lipases or *Geotrichum candidum* B lipase are selective for unsaturated fatty acids, and so are capable of producing low saturated fatty acid oils by selective hydrolysis. 1,2-diglycerides can be produced by selective hydrolysis, involving kinetic resolution, using 1,3-specific lipases; except that if high degrees of conversion of triglyceride raw materials is required excessive monoglyceride is formed. 1(3)-monoglycerides have been made in over 95% purity using the lipase patatin, which is almost completely specific for the hydrolysis of monoglycerides. However this lipase is not very heat stable, which makes it difficult to esterify long chain saturated fatty acids. Nevertheless large-scale manufacture has been pioneered for several products.

Diglycerides

New developments in fatty acid technology include the process developed by the Kao Corporation for making 1,3-diglycerides using specially selected lipases [2a]. These diglycerides have been shown to be beneficial, because they do not accumulate as adipose deposits in the human body.

Vitamins

The use of lipases for the large scale preparation of the vitamins pantothenic acid, retinol (vitamin A), and α-tocopherol (vitamin E) have been developed by Roche Vitamins [3a]. The processes using enzymes to make retinylacetate and tocopherylactate were both more effective than the comparable chemical process.

In the synthesis of retinylacetate the key intermediate is a monoester. This could be produced very selectively (>97%), and in good conversion (>99%) using immobilised Chirazyme L-2 lipase and vinyl acetate as the donar, whereas chemistry produces a mixture of mono and diacetylated intermediates, resulting in lower yields of the retinylacetate following dehydration/isomerisation of the intermediate. For larger scale use the enzyme stability had to be improved by purifying the feedstock. This was done by using a pre-column containing tetrasodium EDTA, and addition of hydroquinone antioxidant and triethylamine base. Then the immobilised enzyme remained stable for at least 100 days operation. The scaled-up process used 30% w/v substrate, and acetone as a cosolvent with the vinylacetate. Using this process 1.6 Kg of product could be produced/day with substrate solution supplied at a rate of 0.6 Kg/h.

Alpha-tocopherol acetate, which is the major application form of vitamin E is produced by a chemical process from isophytol and trimethylhydroquinone. An alternative route starts with a cheap and accessible trimethylhydroquinone diacetate, but requires selective deacetylation. This was achieved using the thermostable *Thermomyces lanoginosus* lipase which can be immobilised onto Accurel MP1001. Water-saturated tert-butyl methyl ether was the most suitable solvent, and at 55 °C no unwanted mono ester regio-isomer was formed, or any hydrolysis to trimethylhydroquinone took place.

Tagetose

D-Tagetose, is beginning to be used as a non-calorific, non-laxative bulking agent in foods. It was originally made chemically by Biospherics Inc., but a high yielding process, suitable for manufacturing has been developed by the Tong Yang Confectionery Co. [4a] based on an earlier observation that D-galactose can be isomerised into D-tagetose by L-arabinose isomerase [5a]. An immobilised L-arabinose isomerase from *E.coli* is used accumulating about 100 g product/L in about 48h and with little loss of activity on repeated use. Although the immobilised enzyme has only 38% of the activity of the free enzyme, it productivity was 4 fold better.

Octadecenedioic Acid

Octadecenedioic acid is a natural derived cosmetic ingredient made as a mixture of $C18_1$, $C18_2$ dicarboxylic acids sold by Uniqema. It is similar to the dioic acid azelaic acid (C9) that has long been known to be efficacious in the treatment of acne and dandruff. Octadecenedioic acid has good skin penetration properties in oil in water formulations and antidandruff, deodorancy and some skin lightening activities; together with low skin irritancy and eye irritancy potential.

Octadecenedioic acid is made using a mutant yeast, in which the yeasts ability to further metabolise by β-oxidation of dioic acids in the peroxisomes has been eliminated. Thus dioic acid over production takes place in the microsomes from alkanes or fatty acids, provided that small amounts of glucose are supplied to satisfy the maintenance requirements of the microorganisms, since the mutant cells are unable to derive energy from the metabolism of acetyl CoA produced by the beta-oxidation of the dioic acids as occurs in the wild-type strain [6a].

Downstream Processing – from Bioreactor to Product

The ideal DSP is if the product can be easily separated and isolated in a pure, solid form in one or a very small number of steps. This can be achieved if the product precipitates easily.

For instance muconic acid can be precipitated from the reaction liquor supply by acidification or cooling. The industrial use of this approach has been pioneered by the pharmaceutical industry. Thus amoxicillin is easily recovered because it is insoluble under the reaction conditions and so precipitates as it is formed. By contrast recovery of cephalexin as an insoluble crystalline product requires the addition of a complexing agent, β-napthol. In this process this approach is used again, with unreacted phenyglcineamide being precipitated as a Schiffs base with benzaldehyde.

This approach also proved useful for food ingredients since Aspartame is recovered as an insoluble product because it complexes with excess of phenylalanine methyl ester precursor. Similarly in the manufacture of nicotinamide, and of β-malic acid (as its calcium salt), such concentrated solution of precursors can be used that the products precipitate out at the end of the reactions.

New Developments

While this review deals with just science and technology that have already reached a large manufacturing scale future progress depends on new research involving new enzyme activities and making new products with novel functionalities.

One excellent example is the use of the enol CoA hydratase of *Galactomyces reessii* to make 3-hydroxy-3-methylbutyric acid (HMBA) by Rosazza and co-workers [7a]. HMBA is a recently introduced dietary supplement that is claimed to improve animal growth and health, and human strength during exercise programmes. The biotransformation uses 3-methylbutyric acid as the starting material, producing up to 38g/L of HMBA via isovaleryl CoA, 3-methylcrotonyl CoA and 3-hydroxy-3-methyl butyryl CoA intermediates, which are all part of the cells metabolic pathway to HMBA, from leucine. The *G. reessii* cells are grown on 3MBA, and used as a cell free extract, and the enzyme appears to exist as a homotetramer.

Researchers at Nestle [8a] have found a good method to make the precursor of 2-acetyl-2-thiazoline (2A2T), that has an intense roasted, popcorn, bread crust like aroma, via the bioconversion of cysteamine, ethyl-L-lactate and glucose using bakers yeast. Upon heating (cooking) the precursor liberates 2A2T. Alternatively the precursor could be made by reduction of 2A2T, also by bakers yeast, in up to 60% yield.

Another promising biotransformation is the use of the polyphenol oxidase laccase from Botyritis species to carry out the allylic oxidation of valencene into nootkatone, which is a key grapefruit flavour molecule. Valencene hydroperoxide is the intermediate, and nootkatone was produced in 25–30% yield from 5.5% valencene, which is obtained from orange oil [9a].

A quite different type of commercial use of a polyphenol oxidase has been developed by Novozymes to remove the phenolic off-tastes in wine that can arise from corks, as well as the even mouldy-tasting anisoles that can be formed from the cork phenols. Removal of the off-tastes in achieved by washing the corks with a selected polyphenoloxidase, suberase, that polymerises the phenols, so making them organoleptically inactive, and unable to convert into anisoles [10a].

References

1a. Casey J, Cheetham PSJ, Harries PC, Hyliands D, Mitchell JT, Rawlings AV (1994) PCT Application WO 94/10131
2a. Hirota Y, Kohori JK, Kawakara Y (1989) Eur. Patent Applic. EP307154
3a. Bonrath W, Karge R, Netscher T (2002) J Mol Catal B 19–20, 67
4a. Pil K, Sang-Hyun Y, Hoe-Jin R, Jin-Hwan C (2001) Biotechnol Prog 17, 208
5a. Cheetham PSJ, Wootton AN (1993) Enz Microb Technol 15, 105
6a. Wiechers JW, Groenhof FJ, Wortel VAL, Hindel NAL, Miller RM (2002) SOFW-Journal 128, 2
7a. Dhar A, Dhar K, Rosazza JPN (2002) J Ind Microbiol & Biotechnol 28, 81
8a. Rey YF, Bel-Rhlid R, Juillerat M-A (2002) J Mol Catal B 19–20, 473
9a. Huang R, Christenson PA, Labuda IM (2001) US Patent 6,200786 B1
10a. Conrad IS (2000) BioTimes 2, 6

Adv Biochem Engin/Biotechnol (2004) 86: 159–189
DOI 10.1007/b12442

Smart Biocatalysts: Design and Applications

Ipsita Roy · Shweta Sharma · M. N. Gupta[1]

[1] Chemistry Department, Indian Institute of Technology, Delhi, Hauz Khas,
New Delhi 110016, India. *E-mail:* mn_gupta@hotmail.com

Abstract Smart materials respond to chemical or physical changes in their environment in a predictable fashion. One class of such materials are smart polymers which can be used to design reversibly soluble–insoluble biocatalysts. One important advantage of such soluble polymer enzyme conjugates is in bioconversion of macromolecular or insoluble substrates. In addition, they share the advantage of reusability with conventional immobilized enzymes. Stimuli that are used to "recover" smart polymer – enzyme conjugates for reuse include changes in pH, temperature, ionic strength and addition of chemical species like calcium. In addition to these, enzymes linked to photoresponsive polymers have also been described in the literature. Both adsorption and covalent coupling have been used to create such polymer conjugates. End-group conjugation and site-specific conjugation are recently described strategies to obtain biocatalysts with better designs for solving mass transfer constraints. Some important applications of such smart biocatalysts are hydrolysis of starch, cellulose and proteins. Work has also been carried out on hydrolysis of pectins and xylans. All the above applications involve hydrolysis and are hence carried out in aqueous media. For synthetic applications such as synthesis of peptides, some photoresponsive polymers linked to proteases have recently been described.

Keywords Cellulose hydrolysis · Photoresponsive polymers · Protein hydrolysis · Smart polymers · Soluble biocatalysts · Starch hydrolysis · Xylan hydrolysis

Abbreviations

DCC	dicyclohexyl carbodiimide
LCST	lower critical solution temperature
IgG	immunoglobulin G
NIPAAm	N-isopropylacrylamide
PEG	polyethylene glycol
BAPNA	N-α-benzoyl-DL-arginine p-nitroanilide
NASI	N-acryloxy succinimide
AIBN	Azobisisobutyronitrile

1
Introduction

Some of the greatest inspirations for searching for design principles come from living organisms and their behavior. One key attribute of living systems is the swift way they respond to changes in their environment. This stimulus-sensitive response is best illustrated by examples from the cellular level. The change in enzyme activity in response to change in levels of specific metabolite level is one such important example. Such changes are fast and dramatic, without which the living systems cannot function. Thus, it is not surprising that scientists have, over the years, tried to develop materials that exhibit this stimulus-sensitivity or smartness of biological systems.

Smart materials respond to chemical or physical changes in their environment in a predictable fashion. Smartness of such materials is evaluated in terms of (i) the extent of "change", and (ii) how fast they respond to the stimuli. An example of a smart material is donor-doped barium titanate ceramics [1]. The resistivity of these ceramics rises by about six orders of magnitude when the temperature changes from 350 to 450 K. Thus, these positive temperature coefficient resistance materials allow design of a smart self-regulating heating circuit. The ultimate aim is to create smart analogs of biological organisms.

Some of the emerging applications of smart materials relevant to life sciences and biotechnology are [1–7]:

- Bioseparation
  - Affinity precipitation [8–14]
  - Aqueous two-phase affinity extractions [15]

- Thermosensitive immunospheres for antibody purification [16]
- Thermoresponsive chromatography [5]
- Cell separation [17]
- Separation/dehydrating process using thermoreactive hydrogels [3, 18]
- Immobilized biocatalysts
  - Thermoreactive hydrogels as a matrix for enzymes [19] and cells [20]
  - Reversibly soluble–insoluble biocatalysts [21]
- Biomimetic actuators [22]
- Chemical valves [23]
- Responsive polymer systems for controlled delivery of therapeutics [24]
- Immunoassays [25]
- Biomedical engineering [26]
- Applications to environmental problems [27]
- Renaturation of proteins [28]

While different smart materials have been used for the purposes listed above, relatively less work has been reported related to smart polymers which change their solubility in water and nonaqueous media as a result of changes in their microenvironment. In recent years, such polymers have been successfully used for the design of macroaffinity ligands for use in a bioseparation strategy known as affinity precipitation [8–14]. Yet another application of such polymers is in crafting smart hybrid conjugates with enzymes that are reusable, often have enhanced stability, and can be used with considerable advantage with macromolecular or insoluble substrates. Such bioconjugates form a subclass of immobilized enzymes [29]. Immobilized enzymes are conventionally prepared from insoluble supports by adsorption, covalent coupling and entrapment techniques.

1.1
Immobilization of Enzymes

Immobilized enzymes have been widely used for synthesis and bioanalysis, even at the industrial level.

At this stage, it may be pertinent to critically examine why enzymes in the immobilized form have been so popular. In fact, enzymes, when immobilized, invariably show a decrease in their biological activity. This is due to two factors:

(A) *Binding procedure*: The various ways by which immobilized enzymes are created may be listed as follows:
  (i)   Absorptive or ionic binding to carriers or matrices [30, 31].
  (ii) Covalent coupling to prefabricated matrices [32].
  (iii) Inclusion methods such as entrapment and microencapsulation [33].
  (iv) Bioaffinity immobilization [34].
  (v)   Chemical aggregates [35, 36] and crosslinked enzyme crystals [37]. These forms do not involve any carrier.

In principle, adsorption and bioaffinity immobilization are more gentle procedures as compared to others. With the advent of fusion proteins [38, 39], we may see more frequent applications of bioaffinity immobilization. Also, the intended application plays a role in deciding the choice of the binding method. In the case of chemical coupling (and, to some extent, in the case of entrapment), the enzyme may be subjected to harsh chemicals/extreme pH conditions. In any event, chemical coupling modifies surface residues. These chemical modifications, plus changes in conformation as a result of immobilization, contribute to lower activity after immobilization.

(B) *Mass transfer effects*: These consist of:
  (i) Decreased availability of enzyme molecules within the pores of the carrier.
  (ii) Steric hindrance by the carrier even if enzyme molecule is on the surface. Slow diffusion by the substrate contributes to these effects (we will come back to this later).

On the other hand, there are two compensating factors which confer advantages on immobilized forms.

(I) *Improved stability*: Unfortunately this is somewhat overemphasized. Furthermore, it is quite often not evaluated rigorously. Sadana has proposed deactivation models which have not yet become widely adopted [40]. Often, a first-order kinetics of deactivation is presumed. Also, conditions like temperature, pH and time of measurement are arbitrarily chosen and "enhanced stability" is illustrated under these chosen conditions. An immobilized enzyme never reaches the stability of some of the enzymes isolated from thermophiles, and, in addition, immobilization sometimes even leads to a decrease in stability [41, 42]. Often, evaluating experimental stability with solutions of a substrate in aqueous buffers may be misleading. A more realistic assessment is obtained by using the substrate in a medium in which it might be encountered in the actual intended application. For example, the half-life of immobilized lactase decreased from 89 d to 7 d when the substrate was changed from 5% lactose solution to acid whey [34]. Nevertheless, very often, immobilization results in limited enhancement in stabilization. Sometimes, the observed enhancement of stability may be an artifact especially when porous matrices are used. In the beginning, at limited substrate concentration, the more accessible enzyme molecules on the surface are used. As deactivation begins, these enzyme molecules cannot use up the substrate which diffuses in. The substrate then makes use of enzyme molecules immobilized in the inner surface of pores. The latter thus act as a "reserve of fresh activity". The total loss of activity takes more time, giving the impression of enhanced stabilization. "In the absence of diffusional restrictions, activity decays exponentially with time, whereas, when diffusional limitations are present, activity decays linearly with time" [43].

(II) *Reusability*: This is, in fact, a far more important factor both in terms of convenience and cost saving. Tischer and Kasche in fact remark that "… leaving aside the potential advantage of easier removal of the enzyme from the product, immobilized enzymes so far provide no cost benefit. However, cost savings will be achieved by the repeated reuse of the immobilized enzyme" [44]. The two factors are in fact interlinked in as much as no reuse is possible if the immobilized enzyme is not stable under operational conditions!

1.2
Advantages of Soluble Bioconjugates over Insoluble Forms

Enzymes linked to smart polymers combine the benefits of using soluble carriers with the advantage of reusability discussed above. Enzyme immobilization with soluble supports implies homogenous catalysis instead of heterogenous catalysis. This means reduced mass transfer constraints. This factor becomes especially important when one is dealing with insoluble/poorly soluble substrates or macromolecular substrates like proteins, polysaccharides or nucleic acids. Poor solubility of the substrate will also result in fouling of carriers. In the case of packed bed reactors, this will clog the bed and no further catalysis is possible.

Against this background, it is not difficult to understand why soluble supports were tried quite early for enzyme immobilization [45–47]. While such soluble supports quite often enhance stability, these do not constitute reusable biocatalysts. It is in this respect that smart polymers, which are reversibly soluble–insoluble in response to various stimuli, form ideal support material for design of smart biocatalysts. Hard data on the extent to which these solve the problems associated with soluble, but macromolecular, substrates have been recently provided by Arasaratnam et al. [48]. With trypsin immobilized on Eudragit S-100, "for soluble conjugates, the relative amount of catalytically active trypsin inaccessible to soybean trypsin inhibitor was about 25%–30%. When precipitated, more trypsin molecules were shielded by the polymer and became inaccessible to soybean trypsin inhibitor. The higher the trypsin content in the conjugate, the more pronounced the effect: about 51% and 66% of the activities were not inhibited in insoluble conjugates with 5 and 500 mg trypsin/g Eudragit S-100 respectively." Thus, such enzyme conjugates, in reality, do have better accessibility to other macromolecules when they are in soluble form.

Perhaps the first work based on this concept was by Charles et al. [49] who coupled lysozyme to alginic acid. The enzyme assay was with a suspension of *M. lysodectikus* cells. The enzyme lost about 25% activity after seven cycles, losing some activity in each cycle. The latter feature indicates that it is possible that at least some lysozyme (a basic protein) was bound to alginate (a negatively charged polymer) merely by adsorption (see the discussion in Sect. 3).

Two years later, van Leemputten and Horisberger immobilized trypsin on acrolein-acrylate copolymers which precipitate at pH 4–4.5 [50]. The proteolytic activity was quite low and this was attributed to "electrostatic repulsion of the substrate by the negatively charged enzyme complex."

Perhaps, it is because of such mixed results that the approach was not vigorously pursued for a while. Availability of suitable commercially available polymers in recent years has definitely acted as a "catalyst" for this area.

2
Structural and Kinetic Consequences of Immobilization

In order to fully appreciate the advantages of homogenous catalysis (with enzyme bound to soluble matrices) over heterogeneous catalysis (with enzymes bound to insoluble supports), it is necessary to briefly look at a few parameters which become important upon immobilization. For a more comprehensive and rigorous treatment of these parameters, the reader is referred to some earlier reviews [29, 44].

Engasser and Horvath distinguish between intrinsic, inherent and effective rates in case of immobilized enzymes [29].

Intrinsic rate: This can be different from the enzyme in free solution because of conformational changes, matrix interaction and steric effects. The conformational changes in the enzyme molecule occur (to a varying degree) irrespective of the method of immobilization. In the case of covalent coupling, the chemical modification of the essential residues can lead to a drastic reduction in activity. Steric hindrance refers to the blocking of access of the substrate by the matrix. This effect can often be minimized by using a "spacer" between the matrix and the enzyme. The intrinsic rate can only be observed if both macro- and microenvironments of the enzyme are identical vis-à-vis substrate and product concentrations.

Inherent rate: This is the rate which would be observed if there were no diffusional limitations. What distinguishes inherent rate from intrinsic rate is that the former factors in partition effects. These partitioning effects due to electrostatic or other interactions between the matrix and various species render concentrations in microenvironments different from the bulk solution. "Such partitioning effects can cause the pH optimum and/or K_m of the immobilized enzyme to differ from the values obtained for the free enzyme" [29].

Effective rate: This is what one measures in reality. This is different from intrinsic rate and inherent rate as both diffusional limitations as well as partition effects modify them.

It is obvious that, as far as soluble supports are concerned, effective rate may be closer to inherent rate since diffusional limitations are expected to be much lower in the case of a homogenous catalyst. At the same time, there is a price which one may have to pay in terms of enhanced partition effects. Frequently used solid matrices for conventional immobilization are generally hydrophilic supports like agarose. The partition effects tend to be quite low in these cases. Smart polymers used to design smart biocatalysts are reversibly soluble–insoluble. By their very nature, such polymers are designed on the principle of "balance of power" in terms of hydrophilicity vs hydrophobicity. Invariably, these have fairly high nonspecific binding towards various species including proteins. Thus, it is expected that these will show far more partition effects than solid supports. This is one aspect of smart biocatalysts that is yet to be investigated in depth.

In all cases, however, one can define the effectiveness factor:

$$\text{Effectiveness factor, } \eta = \frac{\text{Rate of the reaction catalyzed by the immobilized enzyme}}{\text{Rate of the reaction catalyzed by the same enzyme concentration in free form}}$$

An effectiveness factor of 1.0 means that the activity of all the enzyme molecules has remained unchanged after immobilization, by diffusion limitations and partition effects. It also means that the intrinsic rate, inherent rate and effective rate are all identical for the immobilized preparation!

3
Conjugation Between Enzymes and Smart Polymers

The two main approaches that have been used to create these hybrid biocatalysts are adsorption or covalent coupling of the enzyme to the polymer (Table 1). As in the case of insoluble matrices, adsorption, if it works, is a convenient and economical way of immobilizing enzymes. Another major advantage is that it does not involve harsh and toxic chemicals, which are often required for covalent coupling methods. The trouble is that, even when significant adsorption is possible, adsorbed enzymes often come off the matrix in small amounts. This is illustrated by the experience of Cong et al. [54]. Amylase adsorbed to Eudragit did not leak from the matrix; however, repeated uses led to a continuous loss of enzyme activity. "As amylase is sensitive to acidic pH conditions, the decrease in activity observed was most likely due to the denaturation of the enzyme when low pH was used for precipitation" [54]. Thus, multicovalent linkages between the matrix and the enzyme are necessary to enhance the stability of the enzyme. It is noteworthy that, in the above case, the workers solved the problem by using the covalent coupling method.

As far as the covalent coupling method is concerned, a large number of options for conjugation chemistries are available (Table 2). These have been reviewed extensively in a number of publications [34, 57, 58]. Another publication contains many useful practical hints [59]. It is neither possible nor necessary to cover the voluminous literature on the subject. We will focus on the experience of different workers and their results while conjugating some specific enzymes to some polymers later on in this work. Here, in this section, some general comments are made.

Tyagi et al. investigated the coupling of five different enzymes, viz., wheat germ acid phosphatase, β-glucosidase, β-galactosidase, trypsin and xylanase, to Eudragit S-100 and observed: (a) the activity of the bioconjugate was critically dependent on the physical state of the polymer and the pH at the time of coupling [32]. In some cases, better activity was obtained if carbodiimide coupling was performed with Eudragit in the insoluble form while, in other cases, better results

Table 1 Covalent vs. noncovalent binding

Some of the advantages associated with covalent coupling

- Sometimes it is possible to attach more protein covalently than by adsorption [51]
- Oriented immobilization [52] or site-specific conjugation is possible only with covalent coupling
- As a variety of crosslinkers are available, one can almost always obtain an immobilized product with a reasonable level of activity. Adsorption is less predictable, it is not always possible to find a suitable matrix which adsorbs a specific protein
- One can minimize nonspecific binding and covalently link only the desired protein; this is rather difficult in the case of adsorption [53]

Table 2 Coupling chemistry

Functional group on the matrix	Activating agent	Activation time (h)	Functional group on the ligand to be coupled	Ligand coupling time	Remarks
-OH	Cyanogen bromide	0.2–0.4	-NH$_2$	2–4 h	– The original method has been superceded by a method which allows coupling to occur in alkaline pH and results in low nonspecific interaction [54]; – amine-containing buffers should be avoided during the coupling step; – cyanogen bromide can be replaced by the less toxic 1-cyano-4-dimethylaminopyridinium tetrafluoroborate [55]
	Divinyl sulfone	0.5–2.0	-NH$_2$; -SH; -OH	Fast; slower	Complexes formed are unstable at high pH
	Bisoxiranes	5–18	-OH; -NH$_2$; -SH	15–24 h	– A spacer arm is automatically inserted; – not suitable for base-sensitive ligands
	Tresyl chloride	0.5–0.8	-NH$_2$; -SH	Fast	Reacts faster than tosyl chloride to form stable alkylamine linkage
	Carbonyl-diimidazole	0.2–0.4	-NH$_2$	1–6 d	– The reagent is less toxic than CNBr or bisoxiranes; – coupling forms urethane linkage with no nonspecific interaction
-COOH	Carbo-diimide	0.2–0.4	-NH$_2$	1–3 h	– Avoid acetate and phosphate buffers; – efficiency of carbodiimide coupling increases by adding sulpho-N-hydroxysuccinimide [56]
-NH$_2$	Glutaral-dehyde	1–8	-NH$_2$	6–16 h	– A spacer arm is automatically inserted; – reversible Schiff base formed must be reduced by reductive alkylation for the linkage to be stable

were obtained if coupling was carried out in a homogenous system; (b) with most of the enzymes, a significant amount of enzyme could be bound merely by adsorption. More importantly, after the carbodiimide coupling step, a sufficient fraction of the bound enzyme could be eluted off the matrix, indicating that this was merely adsorbed and not covalently coupled. More recently, Arasaratnam et al. [48] have reported that, in the case of trypsin, the noncovalently bound enzyme could be washed off the matrix Eudragit S-100 by washing with 0.15 M Tris buffer (pH 7.6) containing 3 g/L Triton X-100. One needs to check whether washing with this buffer will help in removing adsorbed proteins in other cases as well.

Fujimura et al. [60] have reported that immobilization with a methacrylate copolymer, MPM-06, in the solution phase with carbodiimide coupling gave much better results (90% yield) as compared to the immobilization at pH 4.5 (only 16% yield) when the polymer was in the insoluble form.

From the viewpoint of conjugation, an early work with copolymers of *N*-isopropylacrylamide with *N*-acryloxysuccinimide or glycidyl methacrylate is worth mentioning [61]. The solubility of the copolymers could be controlled by temperature and salinity of the solution. The pendant groups of the copolymer could easily link with amino group bearing ligands. The copolymer with glycidyl methacrylate was used to immobilize immunoglobulin G (IgG) and alkaline phosphatase by simply mixing the protein and polymer solutions and stirring at 40 °C overnight. The bound enzyme showed up to 87% of activity yield and could be reused 10 times without appreciable loss of activity. The simple and efficient conjugation protocol is an attractive feature of this work and illustrates the potential of designing polymers with suitable inbuilt reactive pendant groups for obtaining smart biocatalysts.

A considerable amount of work in such systems has utilized carbodiimide coupling. In most of the cases, not much effort has been made to optimize conjugation conditions. A few guidelines to enhance the efficiency of carbodiimide coupling need to be kept in mind: The literature suggests that the coupling proceeds quite well at any pH between 4.5 and 7.5 but that buffers that contain free amines, sulfhydryls or carboxyl groups should be avoided [57, 59]. Acetate and phosphate buffers also reduce the reactivity of carbodiimides. Addition of *N*-hydroxysulfosuccinimide is also recommended to enhance coupling efficiency. These aspects have seldom been taken into consideration. For example, phosphate buffer has been employed for carbodiimide coupling with Eudragit [48]. In this respect, it is extremely relevant to refer to another early work by Dominguez et al. [62] wherein it is shown that a considerable amount of optimization is required if one wants to obtain the best possible immobilized biocatalyst. It is obvious that more serious efforts are required while developing strategies for obtaining these smart biocatalysts.

3.1
End-Group Conjugation

It has been found that, even with soluble polymers, steric hindrance by the matrix is not totally abolished. In all such cases, multiple linkages hold the two components together. In order to reduce the steric hindrance from the soluble polymer, an alternative design is to link the polymer via one reactive end group to the enzyme. An example of such bioconjugates are PEG-enzyme conjugates. Gaertner and Puigserver described a PEG-trypsin conjugate that in fact showed a lower dissociation constant with the low molecular weight substrate than native trypsin [63]. Similar behavior was exhibited by PEG-RNase A conjugate [64]. Moreover, in the case of PEG-chymotrypsin, the conjugate did not hydrolyze BSA presumably since the "PEG phase" excluded BSA [65]. Also, in PEGylation of proteins, multiple chains of the polymer become attached to various amino groups present on the protein surface. This still prevents the free accessibility of a macromolecule like an antibody (against the enzyme) binding to the PEG-enzyme conjugate [64].

In recent years, bioconjugates with better performance have been designed by using oligomers of N-isopropylacrylamide [66–68]. For example, Ding et al. [66] described the synthesis and characterization of oligo(N-isopropylacrylamide)-trypsin conjugates. In these conjugates, oligomeric chains of N-isopropylacrylamide are linked to amino groups of the enzyme via dicyclohexyl carbodiimide (DCC) coupling in the presence of N-hydroxysuccinimide. The oligomer chain is attached only by its one reactive end group and the ratio of the oligomer to the enzyme can be varied in a controlled and measurable fashion. These conjugates showed some unusual properties [67]. The esterase activities of the conjugates were greater than those of native trypsin and increased with an increase in the extent of conjugation. Similar results were observed in the case of amidase activity (towards N-α-benzoyl-DL-arginine p-nitroanilide, DL-BAPNA) as well. It was found that K_m values of the enzyme towards DL-BAPNA decreased with increase in oligomer-to-enzyme ratio in the conjugates. In fact, both k_{cat} and k_{cat}/K_m showed similar trends. The exact mechanism by which this increase in activity occurs was not elucidated. It is possible to evaluate the "accessibility" of the active site of trypsin in such conjugates by titrating a polymer-trypsin conjugate with soybean trypsin inhibitor (STI). This inhibitor forms an inactive complex with trypsin by combining in a 1:1 ratio. Various conjugates of trypsin (E) with oligo (isopropylacrylamide) (O) made with O/E ratios of up to 12 showed inhibition comparable to native trypsin when titrated with STI. There was a marginal increase in residual esterase activity with increasing O/E of conjugate. Thus, such single attachment polymer-enzyme conjugates probably constitute the best design for minimum steric hindrance from the matrix. While the trypsin conjugate mentioned above had better thermal stability at 60 °C and retained total activity when stored at 4 °C for up to 25 d (thus being better in both respects when compared to native trypsin), it did worse than native trypsin when subjected to thermal cycling through the lower critical solution temperature (LCST). Thus, the polymer going through temperature-based changes in its configuration resulted in microenvironmental stress for the associated enzyme molecule. At present, there does not seem to be any solution to this "design problem". Earlier, Matsukata et al. [68] compared the performance of PIPAAm in trypsin conjugates when these had been prepared by either single end or multipoint chemistry. It was found that the conjugate prepared by single end chemistry had better stability under repeated temperature recycling.

3.2
Site-Specific Conjugation

Yet another way of designing a smart polymer–protein conjugate is to conjugate the polymer at a specific site on the protein in a predictable way. This approach has been described by Hoffman's group in a series of papers [69–72]. In one case, a single thiol group was introduced by protein engineering in an electron transport protein cyt b_5 [70]. This group was purposely positioned away from the electron-transfer site of the protein. A poly(N-isopropylacrylamide) chain with a maleimide terminal group was conjugated to the thiol group of the protein. The conjugate was formed via the reaction between maleimide and thiol groups and had a ratio of one polymer chain per protein molecule.

A similar approach was used by this group to design a polymer-streptavidin conjugate [71]. Streptavidin is a protein that binds to biotin. The interaction of biotin towards streptavidin is the basis of a host of techniques in the areas of immobilization and analysis [73]. The conjugate in this case showed temperature-dependent binding behavior towards biotin [71]. Site-directed mutagenesis was used to substitute cysteine in place of asparagine at the 49th position. The latter amino acid residue is known to be important in binding of biotin to this protein. Poly(N-isopropylacrylamide) [poly(NIPAAm)] with a terminal vinyl sulfone group was used to conjugate the polymer specifically at the protein-engineered thiol group. At 4 °C (well below the phase transition temperature of 32 °C), the polymer had a hydrated coil conformation and the polymer-streptavidin conjugate bound to biotin. At 37 °C, the polymer assumed a molten globule conformation that made the polymer-streptavidin come off the biotin molecule. It has been found that "biotin binding switching activity" is strongly dependent on the conjugation position on streptavidin [72]. A smart polymer, N,N-dimethyl-acrylamide-co-4-phenylazophenylacrylate, was conjugated via a vinyl sulfone end group to two mutants of streptavidin wherein cysteine was placed at the 116 position (close to the biotin binding site) and the 139 position (which is away from the biotin binding site). Only the former conjugate showed significant thermal response vis-à-vis biotin binding capability. An interesting insight obtained was that "off-rate of biotin was unperturbed and that the thermally triggered release of biotin with the E116C conjugate was due to the blocking the re-association of biotin" [72].

In yet another design, complementarity of two oligo-nucleotide chains was exploited to noncovalently conjugate poly(NIPAAm) with streptavidin [74]. One oligo-nucleotide was linked to protein-engineered streptavidin at a site near its binding site for biotin. The complementary oligo-nucleotide chain was conjugated to poly(NIPAAm). Hybridization of the modified polymer and modified protein led to formation of the bioconjugate. A novel design element in this site-specific conjugation was that, by varying the length of the oligo-nucleotide, the distance of the polymer from the binding site of biotin on the streptavidin could be controlled.

Site-specific conjugation is not limited to thiol groups. Almost any coupling chemistry can be used by a suitable choice of site-directed mutagenesis of a protein. For example, poly(N,N-diethylacrylamide) was conjugated to a closed double lysine mutant of streptavidin [75]. Computer modeling suggested that these protein-engineered lysine groups were more solvent-accessible than other lysine groups inherent to the native molecule. It is expected that the polymer (via its terminal succinimidyl group) is conjugated to either of the two lysines in the resultant conjugate population. An interesting observation was that, while a large biotinylated protein (IgG, molecular weight 150 kDa) did not bind to the conjugate at either below the LCST or above it, a small biotinylated protein (Protein G, molecular weight 6.2 kDa) bound at both below and above the LCST. Furthermore, biotinylated bovine serum albumin with an intermediate molecular weight of 67 kDa "exhibited increased binding as the temperature was raised through the LCST" [75].

Site-specific conjugation has also been carried out to obtain conjugates which show combined temperature and pH sensitivity. A poly(NIPAAm) acrylic acid copolymer (with as little as 5.5 mol% of acrylic acid) was completely soluble at 37 °C and pH 7.4 but insoluble at 37 °C and pH 4.0. A conjugate of this polymer with the E116C mutant of streptavidin showed significantly reduced binding of biotin at pH 4.0, the pH at which the polymer component is expected to have a more compact coil structure [75].

The conjugates of streptavidin discussed above do not strictly constitute examples of smart biocatalysts. However, these are still relevant inasmuch as streptavidin/biotin or enzyme/substrate interactions both essentially operate via similar molecular recognition paradigms. Thus, it should be possible to design such "extra smart" conjugates with enzymes as well wherein the stimuli-sensitive nature of the polymer is not merely for controlling solubility but for building in regulatory features as well.

3.3
Conjugation via Protein Chain Activation

In most cases, conjugation of the enzyme/protein is carried out by activation of the polymer and the free amino groups on the protein surface are utilized to link up with the activated polymer. A recent report uses a reverse strategy with chitosan [76]. Chitosan, being partially deacetylated chitin, has some free amino groups, which are glucosamine residues. In fact, one can control deacylation of chitin and hence produce chitosan with less or more free amino groups [77]. In this work, free carboxyl groups on the enzyme laccase (from *Coriolopsis gallica*) were activated with carbodiimide and the enzyme was conjugated with chitosan [76]. The conjugate had pH-dependent solubility like chitosan, i.e., soluble at acidic pH and insoluble near neutral pH. The conjugate, moreover, was more stable than free laccase at both pH 1 and pH 13. Laccase is an enzyme that oxidizes dihydroxyphenol and thus its stability at extreme pH may not be relevant. However, these results serve as a good illustration of use of chitosan as a smart polymer.

4
Design of Smart Polymers for Creating Hybrid Biocatalysts

The change in solubility of water-soluble polymers with a variable like temperature, pH or concentration of another species is different from small molecules. While the solubility changes monotonously in the case of the latter, it is an 'all or none' phenomenon with the polymers. Thus, a polymer is soluble up to a certain point (in practice, it is a narrow range of values) of temperature, concentration of cosolute, etc., and is insoluble beyond that point. With solutions of amorphous polymers, insolubility implies a liquid/liquid phase separation where the precipitate is actually a highly concentrated polymer solution. If the polymer concentration is low, the precipitated polymer forms a colloidal suspension. Hence, the point at which precipitation starts is called the "cloud point" of the polymer. The solubility of a polymer depends on three interactions: (a) inter- and intrachain interactions among the polymer's molecules, (b) interactions between solvent and other species (if any) present in solution, and (c) interaction between polymer and solvent molecules (and other species). All these interactions are noncovalent in nature: electrostatic, van der Waals forces, H-bonds and hydrophobic effects. The precipitation is, therefore, essentially a result of (a)+(b) dominating over (c)-type interactions. Hence, the design of a smart polymer involves defining medium conditions in a predictable way by which this can be made to take place in a reversible way.

The origin of 'smartness' in the case of reversibly soluble–insoluble polymers (like all other smart materials) lies, of course, in their structures. This is perhaps best understood by discussing a few specific polymers.

Eudragit, a polymer of methacrylic acid and methylmethacrylate, can be dissolved in aqueous solutions when the pH is such that enough free carboxyl groups on the polymer are ionized (pH 5.5 or higher) and hence confer negative charge on the polymer [7–10]. Apart from interaction with water, these charges cause repulsion and the polymer is in a stretched form. At pH below ~4.7, the carboxyl groups become protonated, the H-bonding between polymer and water is no longer possible; the repulsion between intramolecular charges is considerably reduced and the polymer precipitates due to hydrophobic aggregation (Fig. 1). Most of the polymers responding to pH as stimulus follow similar molecular mechanisms.

It is also interesting to note that the precipitation of Eudragit is also promoted by combination of Ca^{2+} and temperature [78]. Ca^{2+} ions crosslink the polymer; higher temperature simultaneously increases hydrophobic forces resulting in polymer precipitation. This precipitation is also reversible; the lowering of temperature and addition of EDTA dissolve the polymer precipitate. Eudragit precipitation with Ca^{2+} can also be promoted by adding water-miscible organic solvents [78]. The latter mode presumably increases the degree of interaction between the metal ions and carboxylate groups of the polymer (the dielectric constant of water is around 80 and that of organic solvents is much lower). This increased interaction leads to a greater crosslinked network of polymer molecules. The result is a rather compact precipitate.

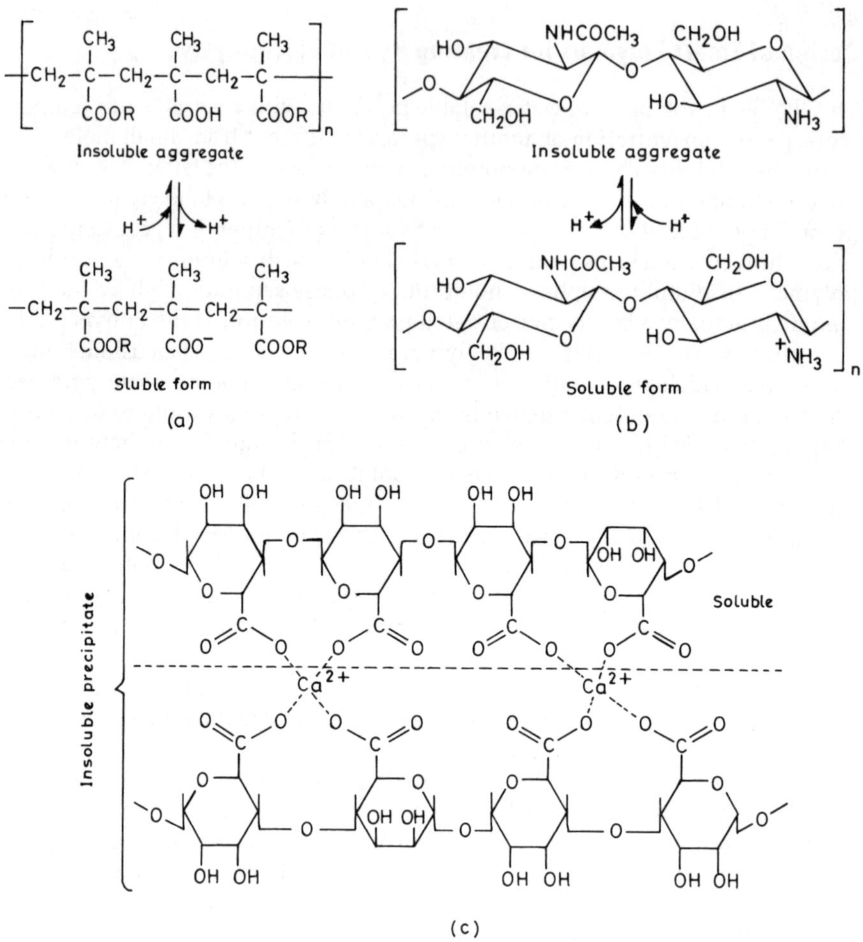

Insoluble aggregate

Insoluble aggregate

Sluble form

Soluble form

(a)

(b)

Insoluble precipitate

Soluble

(c)

Fig. 1a–c Mechanism of reversible solubility of smart polymers. The illustrative examples shown here are a Eudragit (R=H or CH_3). In the case of Eudragit L-100, the ester-to-acid ratio is 1:1 while, in the case of Eudragit S-100, this ratio is 2:1; **b** chitosan; and **c** alginate. The addition of Ca^{2+} leads to complexation of the free carboxyl groups, leading to the precipitation of alginate. The mechanism of solubility of these polymers is described in the text

Eudragit is an enteric polymer. It was designed to respond to pH changes so that, once the coating dissolves, the drug becomes available to the body. Its applications in bioseparation and biocatalysis have been a case of adapting an existing design since it suited the intended applications.

Similarly, naturally occurring alginate, an anionic polysaccharide, has been used as calcium alginate beads for entrapping cells and enzymes. The rapid mixing of Ca^{2+} with alginate gives a precipitate (Fig. 1) that dissolves in the presence of EDTA. Hence the mechanism of reversible solubility is similar to that discussed above for Eudragit. In fact, even in this case, protonation of carboxyl groups below pH 2.0 also leads to precipitation of alginate.

Chitosan, on the other hand, is a cationic polysaccharide [9]. It is obtained by partial deacetylation of chitin, which is poly(N-acetylglucosamine). Chitin constitutes cell walls of fungi and moulds and crab shells. Deprotonation around pH 6.5 results in the removal of the positive charge on the free amino groups of chitosan and the polymer is insoluble in aqueous media at and above this pH (Fig. 1).

It is possible to build in similar features for designing reversibly soluble–insoluble polymers with more or less predictable properties. Synthetic routes are available mostly for temperature-sensitive polymers. Illustrative examples for these polymers are outlined in Scheme 1. The polymer poly(NIPAAm) is an important illustration of this [8–10]. These temperature-responsive polymers have a cloud point above which an aqueous polymer solution separates into a polymer phase and an aqueous phase. The phase separation is believed to be due to the association of the polymer molecules by hydrophobic interaction. It is possible to design NIPAAm (and other temperature-responsive polymers) with lower or higher cloud points [79].

The design principles involved in the case of photoresponsive polymers are given in Sect. 4.1.

While smart polymers have been designed which respond to a variety of stimuli [9, 10] (Table 3): pH, temperature, addition of Ca^{2+}, addition of tetraborate, addition of a lectin, addition of an organic solvent along with Ca^{2+}, their use as a component of biocatalyst conjugates imposes certain constraints on the choice:

- The pH of transition between soluble and insoluble forms should not be an extreme one. Many enzymes denature below pH 4.0 and above 8.0. However, it may be added that, at least in one case, conjugation to the polymer has enhanced the stability of the enzyme towards repeated exposure to cycles of pH changes [13].
- Similar remarks may be made regarding temperature-sensitive polymers. Most of the enzymes denature beyond 40–45 °C; exceptions are enzymes from thermophiles and enzymes stabilized through chemical modification, protein engineering and directed evolution [80, 81]. A worthwhile objective will be to design smart polymers that, upon conjugation with enzymes, could enhance the thermostability of the latter. The limited data available on the consequences of conjugation of enzymes with smart polymers regarding their stabilization will be discussed later in this review while describing the work with specific substrates.

In general, the co-monomer composition will dictate the response of the polymer towards the stimuli. There is a lot more data on this in the case of responsive gels [4]. This experience is beginning to be extrapolated to the design of reversibly soluble–insoluble polymers.

4.1
Photoresponsive Polymers and Their Hybrid Bioconjugates with Proteins/Enzymes

Apart from other stimuli, irradiation in the UV/visible range can also be used to alter the properties of polymers [82]. Properties like volume changes, wettabil-

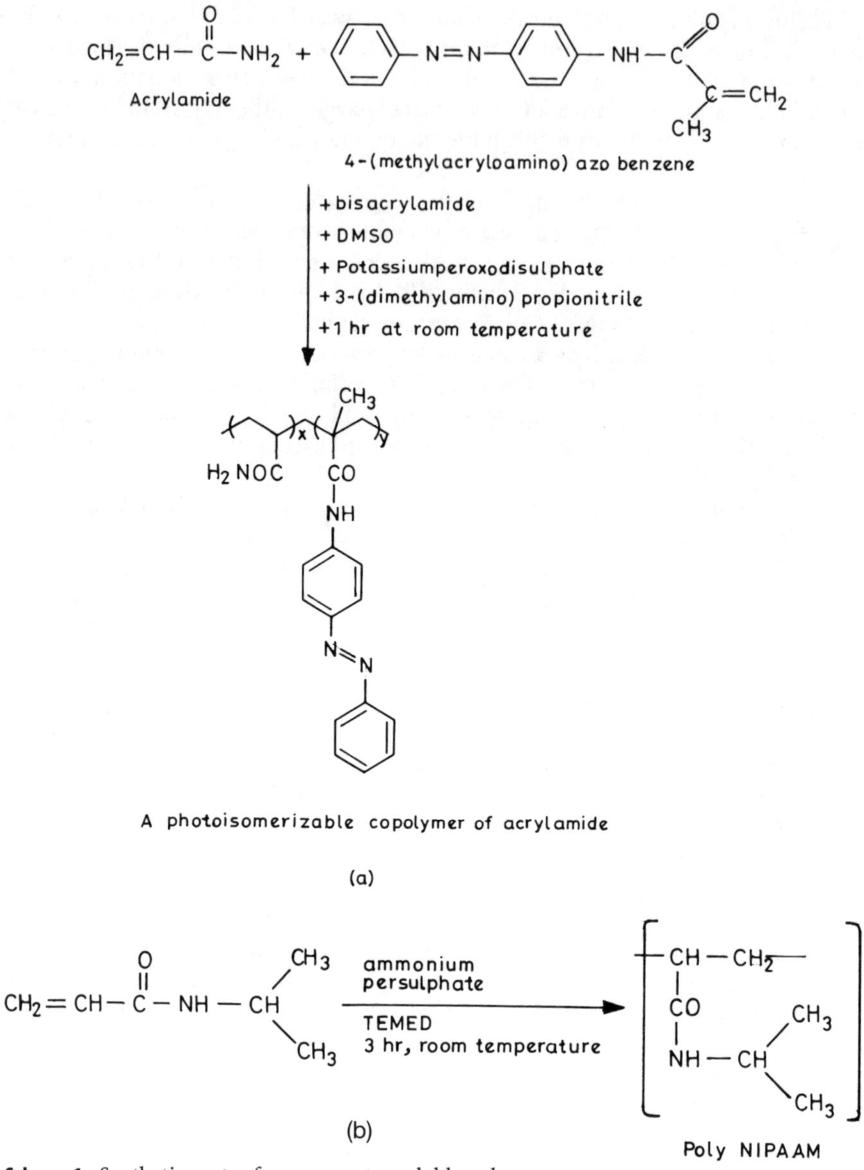

CH2=CH—C—NH2 +
Acrylamide
N=N
NH—C
C=CH2
CH3
4-(methylacryloamino) azo benzene
+bisacrylamide
+DMSO
+Potassiumperoxodisulphate
+3-(dimethylamino) propionitrile
+1 hr at room temperature
CH3
H2NOC
CO
NH
N=N
A photoisomerizable copolymer of acrylamide
(a)
CH2=CH—C—NH—CH
CH3
ammonium persulphate
TEMED
3 hr, room temperature
CH—CH2
CO
NH—CH
Poly NIPAAM
(b)

Scheme 1 Synthetic routes for some water soluble polymers

ity and viscosity can be controlled in such cases [83]. In fact, such photoregulation of biomaterials is of considerable interest in the context of development of bioelectronic devices. As far as proteins/enzymes are concerned, three approaches have been chiefly used to make them photoresponsive [84]. The first approach is the modification of active sites by photoisomerizable units or use of photoisomerizable inhibitors. The second approach is "their modification by

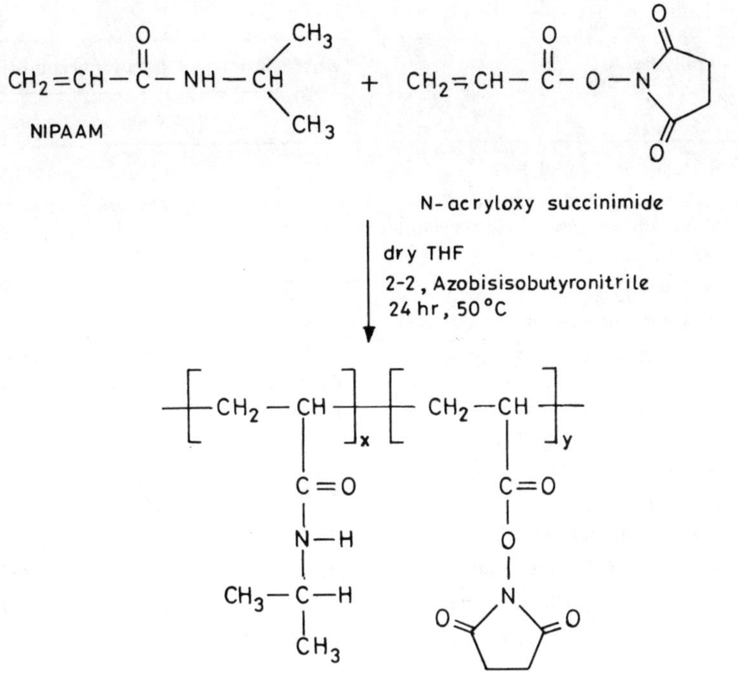

NIPAAM

N-acryloxy succinimide

dry THF
2-2, Azobisisobutyronitrile
24 hr, 50 °C

Copolymer of NIPAAM and NASI

(c)

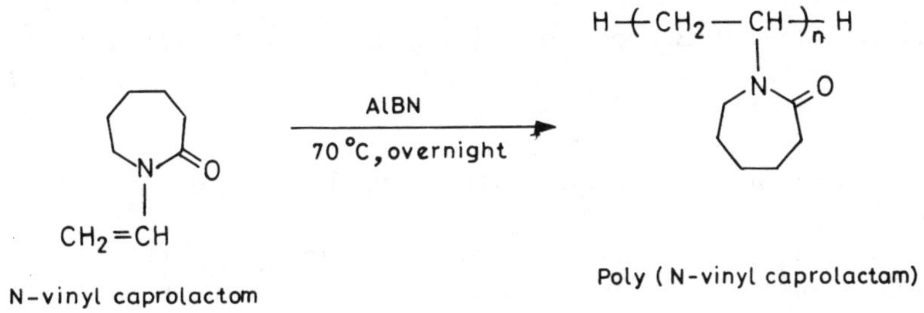

AIBN
70 °C, overnight

N-vinyl caprolactom

Poly (N-vinyl caprolactam)

Synthesis of Poly (N-vinyl caprolactam)

(d)

Scheme 1 (continued)

photoremovable components. Nevertheless, this approach leads to single-cycle photoactive biomaterials, that lack reversible activities." The third strategy is to form a hybrid of the protein with a photoresponsive polymer. One illustration of this approach is the work on encapsulation of the hydrolytic enzyme α-chymotrypsin in the crosslinked copolymer of acrylamide 1-(β-methacryloxy)-ethyl-3,3'-dimethyl-6-nitrospiro (indoline-2,2'[2H-1]benzopyran) with N,N'-

Table 3 Some commercially available smart polymers

Name of polymer	Abbreviation	Operational conditions	
		Soluble	Insoluble
Acrolein-acrylic acid copolymer		>pH 7.0	<pH 4.0
N-Acryloyl-*m*-aminobenzamidine-*N*-Acryloyl-*p*-aminobenzoic acid acrylamide copolymer		>pH 8.0	<pH 4.0
NIPAAm-*N*-acryloxysuccinate copolymer	poly S		>42 °C
NIPAAm-glycidyl methacrylate copolymer	poly G		>37 °C
Methacrylic acid-methacrylate copolymer	MPM-06	>pH 6.0	<pH 4.5
	Eudragit S	>pH 5.5	<pH 4.8
	Eudragit L	>pH 4.3	<pH 4.0
Methacrylic acid-methylacrylate methyl-methacrylate copolymer	MPM-05	>pH 6.0	<pH 4.5
Carboxymethyl ethylcellulose	CMEC	>pH 5.5	<pH 4.5
Cellulose acetate phthalate	CAP	>pH 5.0	<pH 4.0
Hydroxypropyl methylcellulose phthalate	HP-50	>pH 4.5	<pH 3.0
Hydroxypropyl methylcellulose acetate succinate	HP-55	>pH 4.7	<pH 3.8
Alginic acid		>pH 2.0	Ca^{2+}

methylene bisacrylamide as the crosslinking reagent [84]. Upon illumination of the encapsulated enzyme with 400 nm$>\lambda>$300 nm, the copolymer becomes permeable to *N*-(3-carboxypropionyl)-L-phenylalanine-*p*-nitroanilide and this amide is hydrolysed. Illumination with $\lambda>$475 nm turns the polymer impermeable to the substrate. This "on" and "off" cycle could be carried out repeatedly in a reversible fashion. Structurally, it was the transition between spiropyran (off) and merocyanine (on) states of the acrylamide-modifying ligand that made this possible.

More recently, the same photochemistry has been exploited for designing reversibly soluble–insoluble polymers [85, 86]. Polymerization of methacrylate, methacrylic acid and spiropyran-carrying methacrylate gave a smart polymer that precipitated from solution upon irradiation with UV radiation [85]. The precipitate could be resolubilized upon irradiation with visible light. It was found that the photoresponsive nature of the polymer did not change upon its conjugation to the protease subtilisin. The conjugation involved reaction between free amino groups of the protease with carboxyl groups on the polymer by the carbodiimide coupling method. In fact, the smart hybrid biocatalyst exhibited this photoresponse in nonaqueous media (toluene) and could carry out transesterification reactions (see Sect. 5 for further details).

5
Applications

The following sections detail the applications and results reported with smart polymer-enzyme bioconjugates in the case of some biochemically and industrially useful bioconversions. Table 4 gives an illustrative list of the use of some hydrolytic enzymes in various industrially important bioprocesses.

5.1
Starch Hydrolysis

According to a 1998 survey, starch-processing enzymes constitute a global market of US\$ 500 million [87]. This excludes amylases sold for detergents and the textile industries. The major market for starch hydrolysates is in the food industry where they are used as sweeteners and syrups. They are also fermented further to obtain ethanol, acetone, butanol and lactic acid [88]. Enzymatic hydrolysis of starch to glucose requires α-amylase and glucoamylase, whereas production of industrial sweeteners, high fructose corn syrup, requires glucose isomerase as an additional enzyme (Fig. 2). Hoshino et al. [21] immobilized a commercial preparation of amylase, Diabase K-27, to tine types of enteric coating polymers. These water-soluble polymers were: cellulose acetate phthalate, two kinds of hydroxypropyl methyl cellulose phthalate, three kinds of hydroxypropyl methyl cellulose acetate succinate and three kinds of methacrylate copolymers (Eudragit L-100, Eudragit S-100 and MPM06). Best results were obtained with Eudragit L-100. The bioconjugate was completely soluble above pH 5.0. At pH 3.5, the bioconjugate precipitated completely. The specific activity of the immobilized enzyme was 85% of that of the native enzyme. Three approaches were used for starch hydrolysis with this biocatalyst. As these represent generic situations in such systems, it is worthwhile describing them here.

Table 4　Some important macromolecular substrates and their corresponding hydrolases

Enzyme(s)	Substrate(s)	Importance
α-Amylase β-Amylase Glucoamylase Pullulanase	Starch	Food industry, detergents, textile industry
Cellulase, Hemicellulase	Cellulose and lignocellulose	Biomass utilization for production of alcohol
Xylanases	Xylan	Paper pulp industry
Pectinases	Pectin	Increasing yield of fruit juice; prevention of haze formation in wines and fruit juices; extraction of citrus oils and pigments
Proteases	Protein and proteinaceous material	Cheese making; detergents; pharmaceuticals; leather industry

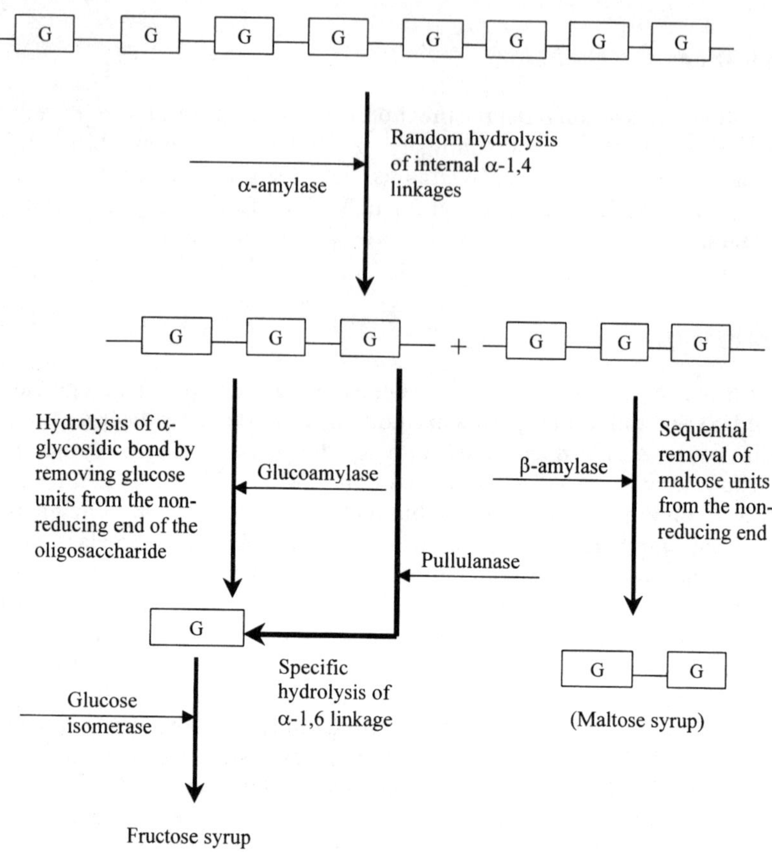

Fig. 2 Mechanism of starch hydrolysis using different enzymes. Depending on the mixture of enzymes used, different products of industrial importance may be obtained

(A) Hydrolysis with removal of residual substrate: In this approach, the smart biocatalyst was separated from the hydrolysis reaction mixture after 24 h and fresh substrate suspension was added to the redissolved biocatalyst. The conversion rate was 100% up to second use and dropped to 55% during the fifth use of the biocatalyst. A major factor was adsorption of the biocatalyst on the unhydrolyzed substrate although some inactivation during recycling was also reported.

(B) Hydrolysis reaction without removal of residual substrate: In this case, the next cycle was started by mixing the solubilized enzyme and the residual substrate with a fresh substrate suspension. Less loss of biocatalyst activity was observed in this case.

(C) Hydrolysis reaction with addition of buffer solution: This was the same as (B) except, instead of fresh substrate, only fresh buffer was added after removal of the hydrolysate by centrifugation. As product inhibition is circumvented, higher conversion rates are obtained. It is reported that the time necessary for obtaining 300 kg of glucose was 120 h for the free enzyme and about 72 h for the smart biocatalyst with approach (C).

Hoshino et al. [89] have also used amylase immobilized to the smart polymer along with immobilized cells of *Lactobacillus casei* for continuous lactic acid production from raw starch. The same amylase was linked to hydroxypropyl methylcellulose acetate succinate by the carbodiimide coupling method and *Lactobacillus casei* was entrapped in kappa-carregeenan. This smart polymer dissolves at pH 5.5 and precipitates at pH 4.0. By optimizing the immobilization conditions, the specific activity of the enzyme was found to be 1.4 times the "best preparation" (obtained with Eudragit earlier).

Cong et al. [54] have used Eudragit L-100 and polyethyleneimine for immobilizing Termamyl (a commercial preparation of alpha amylase) by carbodiimide coupling. Polyethyleneimine can be precipitated as a viscous zone reversibly by the addition of 120 mM phosphate buffer, pH 6.5. About 96% of the enzyme activity was retained in the case of Eudragit L-100, while the binding to polythyleneimine resulted in activation of the enzyme (about 30%), with improvements in both K_m and V_{max} contributing to this. As in other cases [90, 91], the transition pH of the solubility–insolubility of Eudragit L-100 shifted to higher pH upon conjugation with the enzyme. This was attributed to reduction in the number of free carboxyl groups. The enzymes immobilized on Eudragit and polyethyleneimine had pH optima of 5.8 and 6.5 as compared to that of the native enzyme, which was 6.0. Immobilization on Eudragit lowered the temperature optima to 80 °C (from 90 °C in the case of the free enzyme) whereas it increased to 100 °C in the case of polyethyleneimine. Coupling to the polymers resulted in relative stabilization, although the relative stability in the absence of Ca^{2+} was the same for free as well as immobilized enzyme. Approach (B) (outlined in [21]) was used for starch hydrolysis. "Breakdown of starch occurred initially at a very high rate, the major amount of reducing sugar being formed in the first half hour ..." (67% conversion). About 80% conversion was obtained in 2 h. This was similar to that obtained with the free enzyme and the polyethyleneimine conjugate. The enzyme activity remained unchanged after five repeat runs in the case of both Eudragit and polyethyleneimine conjugates [54].

5.2
Hydrolysis of Cellulose

Cellulose, the most abundant organic compound on earth, constitutes a large potential source of feed stock for the fermentative production of fuel alcohol. The intrachain hydrogen bonds result in long and large (250 Å wide) aggregates called microfibrils. These microfibrils enmesh with lignin and hemicellulose to constitute lignocellulosic material which constitutes over 90% of the dry weight of a plant cell [92]. Enzymatically, cellulosic hydrolysis is carried out by three enzymes (Fig. 3). Endoglucanase hydrolyzes some internal bonds in more accessible regions of the microfibrils. Next, cellobiohydrolases attack to generate nonreducing ends to release cellobiose. The disaccharide is hydrolyzed by β-glucosidase to yield glucose.

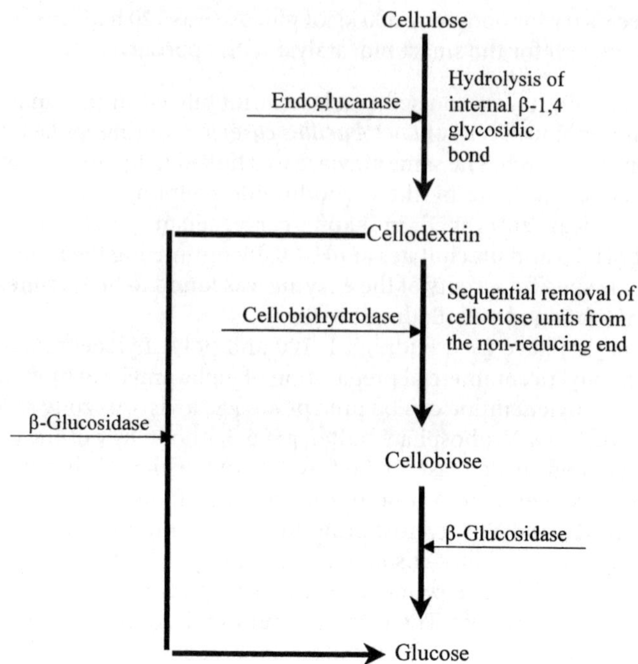

Fig. 3 Mechanism of cellulose hydrolysis using different enzymes. Cellulase is a multienzyme system composed of several enzymes with numerous isoenzymes, which act in synergy. Endoglucanase (EC 3.2.1.4 or beta-1,4-glucan-4-glucanhydrolase) hydrolyzes the beta-1,4-glucan chain of cellulose at random. Cellobiohydrolase hydrolyzes the bonds at the non-reducing end of the crystalline cellulosic chain, producing cellobiose. Beta-glucosidase splits cellobiose into glucose

Taniguchi et al. [90] immobilized a commercial preparation of cellulase from *Trichoderma viride* on the methacrylate polymer, Eudragit L. The bioconjugate was completely soluble above pH 5 and insoluble at pH 4. The specific activities of the immobilized enzyme towards microcrystalline cellulose, carboxymethyl cellulose and cellobiose were 63, 53 and 63% of the free enzyme, respectively. It is believed that as hydrolysis proceeds, cellulose molecules diffuse into the pores of microcrystalline cellulose and become inactivated. It was found that the conjugate of Eudragit-cellulase adsorbed less on the substrate as compared to the free enzyme. This explained the higher activity of the conjugate in the later stages of the reaction. Thus, the size increase upon conjugation, in this case, fortuitously gave a better catalyst. The recycling of the conjugate biocatalyst was carried out in various ways, as described in the case of starch hydrolysis [21]. The most efficient hydrolysis rate was observed when the biocatalyst and unutilized substrate were separated from the product by precipitation, and hydrolysis was recommenced by redissolving the coprecipitate of the biocatalyst and the substrate by adding fresh buffer.

The next challenge in this area will be to use such smart biocatalysts to generate alcohol from cellulose biomass. Among many others that may crop up, two obvious complications will be the inhibitory concentrations of ethanol and control of pH. The latter, of course, could be taken care of by choosing a smart polymer that responds to a stimulus other than pH.

5.3
Protein Hydrolysis

According to an estimate published in 1994 [93], proteolytic preparations account for two-thirds of world market for bulk enzymes. Among the applications listed are detergent manufacture, brewing, cheese making, pharmaceuticals and leather industries.

Fujimura et al. [60] immobilized papain and chymotrypsin on an enteric coating methacrylate–methylmethacrylate copolymer MPM-06. The immobilized papain showed insoluble form below pH 4.8 and was completely soluble above pH 5.8. Immobilization reduced the specific activity of papain by only 20%. About 70%–80% of the initial activity survived after five cycles of its use in solution as amidase. Only marginal improvement (upon conjugation) in thermal stability was observed when evaluated by heating at 85 °C for 15 min. Unfortunately, no data at lower temperatures was provided. The specific activity of the immobilized papain was 62% (of the free enzyme) with casein as substrate and, curiously enough, only 25% with haemoglobin. Even in the case of low molecular weight amide substrate, while immobilized chymotrypsin and trypsin gave specific activity in the similar range of 70%–80%, the results with thermolysin and carboxypeptidase gave disappointingly low specific activities of 30% and 16%, respectively. The immobilized chymotrypsin was used to synthesize a dipeptide Z-Tyr-Arg (NO_2) in the presence of about 45% (v/v) methanol. A constant yield of 67% was obtained in each of the five cycles and the specific activity of immobilized chymotrypsin after repeated use was still 80% of its initial activity.

5.4
Other Miscellaneous Applications

While hydrolysis of cellulose, starch and protein has been the main focus of efforts in this area, sporadic applications in some other systems have also been reported. Dominguez et al. [62] have coupled β-galactosidase from *A. oryzae* to alginate. Alginate is a naturally occurring copolymer of mannuronic acid and guluronic acid, and is generally obtained from sea kelp. It is nontoxic and is widely used in the food industry. Addition of Ca^{2+} precipitates this water-soluble polymer as calcium alginate; removal of Ca^{2+} converts this polymer back to its soluble form. β-Galactosidase hydrolyzes lactose to glucose and galactose [94]. Many adults cannot digest lactose. Hence, production of "low lactose milk" by using β-galactosidase (or lactase as it is often called) is an important industrial process. The enzyme is also used for hydrolysis of lactose present in whey for obtaining ethanol by fermentation of product sugars. Dominguez et al. [62] have optimized the conditions for coupling of the enzyme to alginate by the carbodiimide coupling procedure. About 55% of the activity added to the carrier appeared as bound, 21% remained unbound, remaining activity constituted operational losses. Unfortunately, the enzyme forms were assayed using a synthetic substrate and no data with lactose as substrate are available.

Another interesting application is the immobilization of a commercial endopectinase, Pectolyase Y23, from *A. japonicus,* on Eudragit L-100 [95]. Pectin-hydrolyzing enzymes are used to maximize juice yields from fruits, maceration of fruits and vegetables for preparing nectars, pulpy drinks and baby food, extraction of citrus oils and pigments and to remove haze formation from fruit juice preparations [93]. Immobilization on Eudragit L-100 was carried out both by adsorption as well as the carbodiimide coupling method [95]. In both cases, about 80% of the loaded activity was observed in the conjugate, which survived even after repeated washing with 0.2 M NaCl. Linkage (eitherway) with Eudragit shifted the pH optimum of the enzyme from pH 5.6 to 7.0. For the adsorbed enzyme, both K_m and V_{max} remained in a similar range; however, V_{max} of the covalently linked enzyme was reduced drastically. Immobilization led to enhancement of thermal stability; the half-life of the free enzyme was 1 week at 25 °C, the corresponding values for the adsorbed and the covalently linked enzyme were 20 d and 1 month, respectively. One crucial point, which is missing in this work, is on the recycling of the biocatalyst. As discussed earlier in the case of amylase [54], it is during reuse/recycle that one often finds the superiority of the covalent coupling method over adsorption.

A slightly different concept has been described by Okumura et al. [96]. The α_{S1}-casein polymerized through the formation of intermolecular ε-(γ-glutamyl) lysine crosslinked by transglutaminase was used as a smart biopolymer which could be reversibly precipitated by Ca^{2+} and solubilized by adding EDTA. Conjugates of phosphoglyceromutase, enolase, peroxidase and pyruvate kinase with polymerized α-casein were prepared using a heterobifunctional reagent *N*-succinimidyl 3-(2-pyridylthio)propionate. No improvement in thermal stability was observed. Kinetic parameters were not reported. An interesting observation was that a mixture of phosphoglyceromutase, enolase and pyruvate kinase conjugated to the casein gave the same rate of conversion for 3-phosphoglycerate to pyruvate. "This result indicates that the enzyme conjugates can act, without any diffusion limitations, as free enzymes..." [96].

Finally, mention should be made of *A. niger* xylanase linked to the methacrylate polymer Eudragit L-100 which was used for hydrolysis of xylan [13]. The immobilization of the xylanase activity by adsorption was simultaneously accompanied by removal of cellulase activity. Thus, the immobilized xylanase, which retained 60% of its activity towards xylan, can be used for treatment of paper pulp bleaching in paper industry. The Eudragit-bound xylanase was soluble under the conditions used for xylan hydrolysis. Oat xylan, however, was used as a suspension and hence this is a case of heterogeneous catalysis. Incubation at 65 °C showed that immobilization resulted in significant stabilization as compared to the free enzyme (Fig. 4). The first-order rate constant for thermoinactivation at 65 °C for the free enzyme was calculated as 0.026 min^{-1}, that gave the value of $t_{1/2}$ as 26.6 min. In the case of immobilized enzyme, inactivation was biphasic in nature. The enzyme lost only 22% activity after 90 min with an inactivation rate constant of 1.2×10^{-3} min^{-1}. However, incubation beyond this resulted

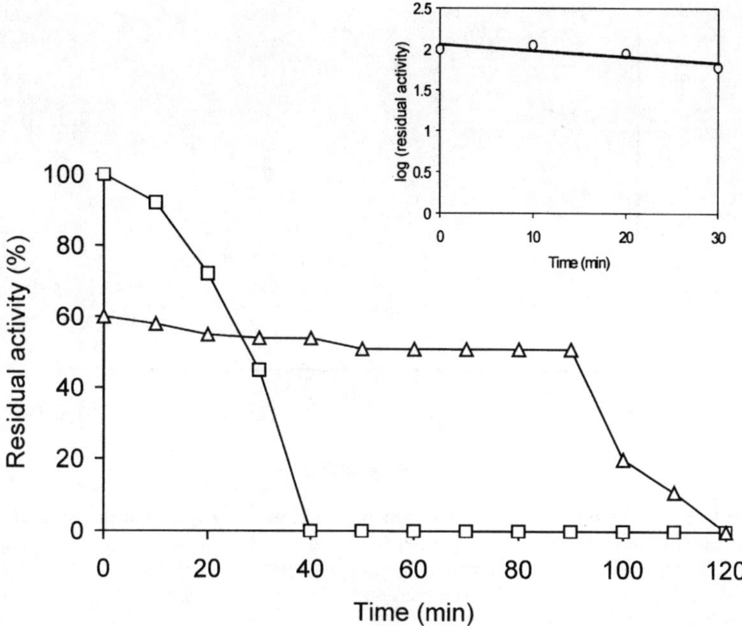

Fig. 4 Thermal stability of free and immobilized xylanase (on Eudragit L-100) at 65 °C. The activity of the free enzyme (□) is taken as 100%. The actual expressed activity of the immobilized enzyme (△), i.e., when the enzyme load, protein-wise, corresponds to a protein equivalent of 100% activity for the free enzyme, is plotted. Inset shows the first-order deactivation kinetics for the native enzyme

in considerable inactivation and the enzyme preparation was completely inactivated after 120 min. No significant changes in pH optimum and K_m were observed upon adsorption to the polymer. The latter indicated that the enzyme-substrate binding continued to be efficient in spite of the macromolecular nature of the substrate. Fluorescence spectroscopy and UV difference spectroscopy were used to probe the changes in the enzyme structure upon immobilization. The spectroscopic data with both techniques reflected significant structural changes. Figure 5 shows the changes in absorbance (ΔA_{272}) as a result of immobilization at different "protein load" upon the polymer. The effectiveness factor η (given in brackets along with each value of A_{272}) has been defined earlier. In order to fully appreciate the data, it is necessary to analyze how the effectiveness factor is expected to vary with different enzyme loads. At low enzyme loading, the effectiveness factor increases with increasing load. While this is a frequently observed phenomenon, it is seldom discussed in the literature. Bosley and Peilow offer an insight into this [97]. According to them, at low loading of enzyme, there is a large excess of surface (of the matrix) available to it. The enzyme molecule attempts to maximize the contact with the surface and undergoes considerable conformational distortion during this process. These authors also suggest that initially the enzyme molecules occupy areas of high affinity hot spots on the matrix surface and this relatively more vigorous interaction leads to greater loss of activ-

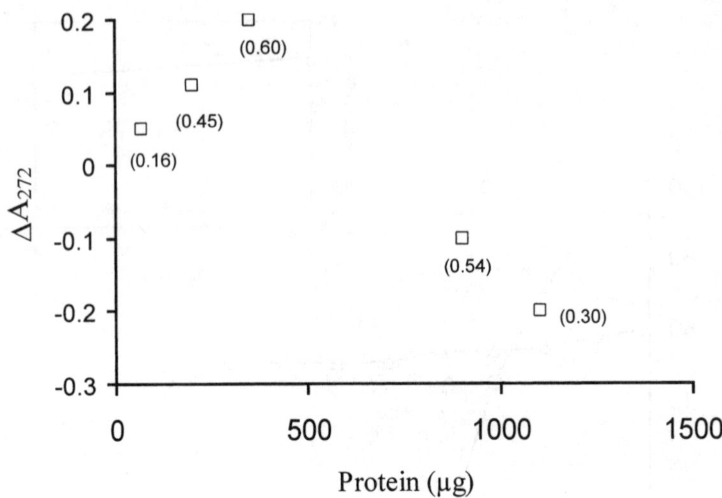

Fig. 5 Dependence of the change in the structure of the immobilized enzyme on increasing the protein load. The effectiveness factor (η) (given in brackets) in each case has been determined as described in the text. ΔA_{272} is calculated by subtracting the absorbance of the free enzyme from the absorbance of the immobilized enzyme at 272 nm. The amounts of protein in both cases have been kept the same in each case

ity. Both factors become less damaging as the enzyme concentration increases. While the discussion by Bosley and Peilow is in the context of immobilization by adsorption, it may be equally valid for covalently coupled enzyme preparations. At some point, the effectiveness factor reaches its maximum value and, if the enzyme loading increases beyond this optimum value, the effectiveness factor decreases. This decrease in effectiveness factor is due to steric hindrance and mass transfer constraints (see discussion in Sect. 1.1).

The main virtue of the data in Fig. 5 is that it is able to correlate a structural factor (as reflected by ΔA_{272}) with the effectiveness factor. Absorbance around 272 nm reflects the microenvironment around aromatic amino acid residues in the protein. A change in absorbance around this region reflects conformational changes. The reversal in the sign of ΔA_{272} correlates well with the reversal in the trend shown by the effectiveness factor with the increasing enzyme load. One needs to establish whether this kind of correlation is a general phenomenon. If so, UV spectroscopy can be a useful tool while designing immobilization protocols for water-soluble polymers. Circular dichroism (CD) spectroscopy revealed that the enzyme undergoes drastic changes in the conformation upon adsorption but retains significant level of activity in spite of that.

Three comments regarding these data may be useful to other workers: (a) the adsorption of the enzyme to the polymer was quite firm since no enzyme activity could be eluted with 1 M NaCl, 50% ethylene glycol or 0.2% Triton X-100. The immobilized enzyme could be reused four times for hydrolysis of xylan without any significant loss of enzyme activity. The strategy used was based on an earlier observation that, in the case of Eudragit polymers, use of higher concentration of polymer resulted in stronger binding of the enzymes [98]; (b) immobilization on water-soluble polymers makes it possible to probe structural changes which occur in an enzyme in a more convenient as well as extensive fashion. We need more data on this as this may bring more rationality to the design of immobilized biocatalysts; and (c) simultaneous purification, especially removal of an 'undesirable' contaminant activity (for a particular application), is a welcome feature of this approach.

Galaev and Mattiasson quote some other polymer-enzyme conjugates which have been described in conferences/patent literature [7]. Coupling of trypsin and cyclodextrin glycosyltransferase to copolymers of *N*-isopropylamide and glycidyl methacrylate constitute examples of thermoprecipitating enzyme derivatives. Bioconjugates of *E. coli* penicillin amidase derivatives and chymotrypsin to poly(*N*-vinyl caprolactam) are two other similar temperature-dependent systems which have been mentioned.

5.5
Nonaqueous Enzymology with Smart Biocatalysts

Enzymes have also been usefully employed in nonaqueous environments for a variety of applications [99–102]. While enzymes have been used in various forms in such media, it is becoming increasingly clear that soluble forms show higher activity than insoluble suspensions [103]. Hence, ideal biocatalyst design for use in organic media again is to link enzymes to smart polymers. This way, the biocatalyst will carry out catalysis in an efficient way and can be recovered for reuse by precipitation in response to the appropriate stimulus.

The earliest reported approach for solubilizing enzymes in dry organic solvents is that of linking enzymes to polyethylene glycol (PEG) [104, 105]. While PEG is not a smart polymer, the PEG-linked enzymes can be recovered by precipitation with a nonpolar organic solvent or used in a membrane reactor [103]. Apart from this historical reason, it is relevant to refer to some data with PEG-linked enzymes since enzymes linked to smart polymers are likely to exhibit similar behavior. Recently, Secundo et al. [106] have shown that PEG-linked lipase displays much higher activity in organic solvents as compared to crude lipase suspension or even crosslinked crystals of the enzyme. Ljunger et al. [107] have found that, while a greater extent of modification facilitates solubility in organic solvents, the enzyme gradually loses activity. Thus, it is necessary to optimize the extent of modification to suit the individual system in terms of specific enzymes and organic solvents likely to be used. Another interesting feature revealed by dynamic light scattering is that PEG-subtilisin in organic solvents is in fact a dispersion of microparticles of at least 300 nm size [108]. It is not unlikely that the various soluble enzyme forms described here may also be of similar kinds. It will be undoubtedly interesting to probe this aspect in all such cases.

The first report of the use of an enzyme linked to a smart polymer in nonaqueous media is that of subtilisin for carrying out transesterification of *N*-acetyl-L-phenylalanine with n-propanol in toluene [85]. The smart polymer used is a photoresponsive copolymer of methacrylate and spiropyran groups. The polymer, soluble in toluene, becomes insoluble when photoirradiated with ultraviolet light since nonpolar spiropyran groups are reversibly converted to ionic mesocyanine form. The bioconjugate had about three to four copolymer chains per subtilisin molecule. The authors also compared the initial reaction rate of the conjugate with PEG-subtilisin and subtilisin powders. The smart conjugate gave a transesterification rate of 1.3×10^2 nmol (mg E)$^{-1}$ min^{-1} whereas PEG-subtilisin and subtilisin powder gave 2.1×10^2 and 1.2 as the corresponding numbers. The authors add that the maximum activity of subtilisin powders did not exceed that of the soluble forms even after the former had been subjected to optimum pH or solutions containing substrate analogs. Such treatments are reported to dramatically enhance enzyme activities in organic solvents [103]. Unfortunately, comparative activity values for these insoluble forms after these improvements are not given. The smart bioconjugate could be reused four times without any impairment in the catalytic efficiency. This recycling was possible since the precipitated bioconjugate could be dissolved again upon exposure to visible light in about 30 min.

Similar "smart performance was also obtained in chloroform although the catalytic activity was slightly lower than that in toluene (data not shown)" [85]. Ito, while reviewing this work, observes that obtaining even better catalytic rates should be possible by trying modification with other polymers [86]. As "proof of the concept" is available, we are likely to see more work on the design of such smart biocatalysts for use in nonaqueous media.

6
Conclusions and Future Prospects

Reusability and generally enhanced stability are the twin assets of immobilized enzymes. Hence, their widespread applications in both synthesis and analysis have occurred despite these being heterogeneous catalysts. The smart biocatalysts reviewed here, to a large extent, combine the virtues of immobilized enzymes with the advantages of being able to use them as homogeneous catalysts. More experience will undoubtedly create better designs in crafting such biocatalysts. It is not unlikely that these smart biocatalysts will replace the conventional form of enzymes in many applications in the coming years.

Before we conclude, it may be good to briefly talk about the possible developments in the near future. There are many ground-breaking developments taking place in the area of smart polymers/hydrogels. Some have already been discussed. One which is more "off the beaten track" as compared to others is the work of Petka et al. [109] which outlines design of reversible hydrogels from proteins produced by recombinant technology. The authors argue that gel formation requires both strong interaction as well as solvent retention function in a polymer. The recombinant proteins were designed as multidomain assemblies in which the two functions were performed by terminal leucine zipper domains and a control flexible water-soluble polyelectrolyte segment. Within a certain pH range, such proteins formed thermoreversible gels. It should, in principle, be possible to extrapolate this idea to a single seamless "conjugate" in which the functions of a smart polymer as well as the enzyme would be in-built in the structure.

7
References

1. Cao W, Cudney HH, Waser R (1999) Proc Natl Acad Sci 96:8330
2. Hoffman AS (1992) Artificial Organs 16:43
3. Galaev IY, Mattiasson B (1993) Enzyme Microb Technol 15:354
4. Kokufuta E (1993) In: Dušek K (ed) Responsive gels: volume transitions II. Springer, Berlin Heidelberg New York, p 158
5. Galaev IY, Warrol C, Mattiasson B (1994) J Chromatogr A 684:37
6. Cooper SL, Bamford CH, Tsuruta T (1995) (eds) Polymer biomaterials: in solution, as interfaces and as solids. VSP, Zeist, The Netherlands
7. Galaev IY, Mattiasson B (1999) Trends Biotechnol 17:335
8. Gupta MN, Mattiasson B (1994) Chem Ind 17:673
9. Gupta MN, Mattiasson B (1994) In: Street G (ed) Highly selective separations in biotechnology. Blackie Academic and Professional, Glasgow, p 11
10. Galaev IY, Gupta MN, Mattiasson B (1996) Chemtech 26:19
11. Agarwal R, Gupta MN (1996) Protein Expr Purif 7:294
12. Roy I, Gupta MN (2000) Curr Sci 78:587
13. Sardar M, Roy I, Gupta MN (2000) Enzyme Microb Technol 27:672
14. Sharma S, Sharma A, Gupta MN (2000) Bioseparation 9:93
15. Teotia S, Gupta MN (2001) J Chromatogr A 923:275
16. Kondo A, Kaneko T, Higashitani K (1994) Biotechnol Bioeng 44:1
17. Kumar A, Kamihara M, Mattiasson B (2002) In: Gupta MN (ed) Methods for affinity-based separation of enzymes and proteins. Birkhäuser Verlag, Basel, p 163
18. Mashelkar RA (1993) J Indian Inst Sci 73:193
19. Park TG, Hoffman AS (1988) Appl Biochem Biotechnol 19:1
20. Park TG, Hoffman AS (1990) Biotechnol Bioeng 35:152
21. Hoshino K, Taniguchi K, Netsu Y, Fujii M (1989) J Chem Eng 22:54
22. Kokufuta E, Aman Y (1997) Polym Gels Networks 5:439
23. Imanashi Y, Ito Y (1995) Pure Appl Chem 67:2015
24. Kost J, Langer R (1999) Trends Biotechnol 10:127
25. Monji N, Hoffman AS (1987) Appl Biochem Biotechnol 14:107
26. Stile RA, Healy KE (2001) Biomacromolecules 2:185
27. Gray HN, Bergbreiter DE (1997) Environ Health Prospect 105:55
28. Lin S-C, Lin K-L, Chiu H-C, Lin S (2000) Biotechnol Bioeng 67:505
29. Engasser JM, Horvath C (1976) In: Wingard LB Jr, Katchalski-Katzir E, Goldstein L (eds) Applied biochemistry and bioengineering (immobilized enzyme principles). Academic Press, New York, p 127
30. Tyagi R, Gupta MN (1994) Process Biochem 29:443
31. Batra R, Gupta MN (1994) Biotechnol Appl Biochem 19:209
32. Tyagi R, Roy I, Agarwal R, Gupta MN (1998) Biotechnol Appl Biochem 28:201
33. Khare SK, Vaidya S, Gupta MN (1991) Appl Biochem Biotechnol 27:205
34. Gupta MN, Mattiasson B (1992) In: Suelter CH, Kricka L (eds) Bioanalytical applications of enzymes, John Wiley, New York, p 1
35. Tyagi R, Gupta MN (1998) Biochemistry (Moscow) 63:334
36. Tyagi R, Batra R, Gupta MN (1999) Enzyme Microb Technol 24:348
37. Margolin AL (1996) Trends Biotechnol 14:223
38. Greenwood JM, Gilkes NR, Miller RC Jr, Kilburn DG, Warren RAJ (1994) Biotechnol Bioeng 44:1295
39. Ong E, Gilkes NR, Warren RAJ, Miller RC Jr, Kilburn DG (1989) Bio/Technology 7:604
40. Sadana A (1991) Biocatalysis: fundamentals of enzyme deactivation kinetics. Prentice Hall, New Jersey
41. Gupta MN (1991) Biotechnol Appl Biochem 14:1
42. Sardar M, Agarwal R, Kumar A, Gupta MN (1997) Enzyme Microb Technol 20:361

43. Cheetham PSJ (1995) In: Wiseman A (ed) Handbook of enzyme biotechnology. Ellis Horwood, London, p 83
44. Tischer W, Kasche V (1999) Trends Biotechnol 17:326
45. Abuchowski A, Davis FF (1981) In: Holcenberg JS, Roberts J (eds) Enzymes as drugs. John Wiley, Chichester, p 367
46. Wongkhalaung C, Kashiwagi Y, Magae Y, Ohta T, Sasaki T (1985) Appl Microb Biotechnol 21:37
47. Lenders JP, Germain P, Crickton RR (1986) Biotechnol Bioeng 27:572
48. Arasaratnam V, Galaev IY, Mattiasson B (2000) Enzyme Microb Technol 27:254
49. Charles M, Coughlin RW, Hasselberger FX (1974) Biotechnol Bioeng 16:1553
50. van Leemputten E, Horisberger M (1976) Biotechnol Bioeng 28:587
51. Douglas AS, Monteith CA (1994) Clin Chem 40:1838
52. Turkova J (1999) J Chromatogr B 722:11–31
53. Shu H-C, Guoqiang D, Kaul R, Mattiasson B (1994) J Biotechnol 34:1
54. Cong L, Kaul R, Dissing U, Mattiasson B (1995) J Biotechnol 42:75
55. March SC, Parikh I, Cuatrecasas P (1974) Anal Biochem 60:149
56. Carpenter A, Purdy WC (1992) J Chromatogr 573:172
57. Hermanson GT, Mallia AK, Smith PK (1992) Immobilized affinity ligand techniques. Academic Press, San Diego
58. Carlsson J, Janson J-C (1989) In: Sparrman M, Janson JC, Ryden L (eds) Protein purification: [rinciples, high resolution methods and applications. VCH, New York, p 275
59. Technical Note # 205, Bangs laboratories Inc. Can be downloaded from labs.com
60. Fujimura M, Mori T, Tosa T (1987) Biotechnol Bioeng 29:747
61. Nguyen AL, Luong JHT (1989) Biotechnol Bioeng 34:1186
62. Dominguez E, Nilsson M, Hahn-Hagerdal B (1988) Enzyme Microb Technol 10:606
63. Gaertner HF, Puigserver AJ (1992) Enzyme Microb Technol 14:150
64. Caliceti P, Schiavon O, Veronese FM, Chaiken IM (1990) J Mol Recognit 3:89
65. Chiu HC, Zalipsky S, Kopeckova P, Kopecek J (1993) Bioconjug Chem 4:290
66. Ding ZL, Chen GH, Hoffman AS (1996) Bioconjug Chem 7:121
67. Ding ZL, Chen GH, Hoffman AS (1998) J Biomed Mater Res 39:498
68. Matsukata M, Aoki T, Sanui K, Ogata N, Kikuch A, Sakurai Y, Okano T (1996) Bioconjug Chem 7:96
69. Morris JE, Hoffman AS, Fisher RR (1993) Biotechnol Bioeng 41:991
70. Chilkoti A, Chen G, Stayton PS, Hoffman AS (1994) Bioconjug Chem 5:504
71. Stayton PS, Shimoboji T, Long C, Chilkoti A, Chen G, Harris JM, Hoffman AS (1995) Nature 378:472
72. Shimoboji T, Ding Z, Stayton PS, Hoffman AS (2001) Bioconjug Chem 12:314
73. Wilchek M, Bayer EA (1990) (eds) Methods in enzymology, vol 184. Academic Press, New York
74. Fong RD, Ding Z, Long CJ, Hoffman AS, Stayton PS (1999) Bioconjug Chem 10:720
75. Hoffman AS, Stayton PS, Bulmus V, Chen G, Chen J, Cheung C, Chilkoti A, Ding Z, Dong L, Fong R, Lackey CA, Long CJ, Miuta M, Morris JE, Murthy N, Nabeshima Y, Park TG, Press OW, Shimoboji T, Shoemaker S, Yang HJ, Monji N, Nowiski RC, Cole CA, Priest JH, Harris JM, Nakamae K, Nishino T, Miyata T (2000) J Biomed Mater Res 52: 577
76. Vazquez-Duhalt R, Tinoco R, D'Antonio P, Topoleski LDT, Payne GF (2001) Bioconjug Chem 12:301
77. Cho YW, Jang J, Park CR, Ko S-W (2000) Biomacromolecules 1:609
78. Guoqiang D, Batra R, Kaul R, Gupta MN, Mattiasson B (1995) Bioseparation 5:339
79. Kuckling D, Adler H-JP, Arndt K-F, Ling L, Habicher WD (2000) Macromol Chem Phys 201:273
80. Gupta MN (1993) (ed) Thermostability of enzymes. Springer, Berlin Heidelberg New York
81. Kuchner O, Arnold FH (1997) Trends Biotechnol 15:523
82. McCurdle CB (1991) (ed) Applied photochromic polymer systems. Blackie Academic and Professional, Glasgow
83. Willner I, Rubin S, Shetzmuller R, Zor T (1998) J Am Chem Soc 115:8690

84. Willner I, Rubin S (1993) Reactive Polym 21:177
85. Ito Y, Sugimura N, Kown OH, Imanishi Y (1999) Nat Biotechnol 31:2606
86. Ito Y (1999) J Polym Sci Polym Edn 10:1237
87. Shanley A (1998) Chem Eng 63
88. Crabb WD, Mitchinson C (1997) Trends Biotechnol 15:349
89. Hoshino K, Taniguchi M, Marumoto H, Shimizu K, Fujii M (1991) Agric Biol Chem 55:479
90. Taniguchi M, Kobayashi M, Fujii M (1989) Biotechnol Bioeng 34:1092
91. Kamihara M, Kaul R, Mattiasson B (1992) Biotechnol Bioeng 40:1381
92. Glazer AN, Nikaido H (1995) Microbial biotechnology. WH Freeman, New York, p 327
93. Walsh G, Headon DR (1994) (eds) Protein biotechnology, John Wiley, Chichester, p 30
94. Huber RE, Khare SK, Gupta MN (1994) Int J Biochem 26:309
95. Dinnella C, Lanzarini G, Ercolessi P (1995) Process Biochem 30:151
96. Okumura K, Ikura K, Yoshikawa M, Sasaki R, Chiba H (1984) Agric Biol Chem 48:2435
97. Bosley JA, Peilow AD (2000) In: Gupta MN (ed) Methods in non-aqueous enzymology. Birkhäuser Verlag, Basel, p 52
98. Kumar A, Agarwal R, Batra R, Gupta MN (1994) Biotechnol Tech 8:651
99. Gomez-Puyou MT, Gomez-Puyou A (1998) Crit Rev Biochem Mol Biol 33:53
100. Roberts SM (1999) (ed) Biocatalysts for fine chemicals synthesis. John Wiley, Chichester
101. Gupta MN (2000) In: Gupta MN (ed) Methods in non-aqueous enzymology. Birkhäuser Verlag, Basel, p 1
102. Vulfson EN, Halling PJ, Holland HL (2001) (eds) Enzymes in nonaqueous solvents: methods and protocols. Humana Press, New Jersey
103. Adlercreutz P (1996) In: Koskinen AMP, Klibanov AM (eds) Enzymatic reactions in organic media, Blackie Academic and Professional, Glasgow, p 9
104. Inada Y, Takahashi K, Yoshimoto T, Kodeva Y, Matsushima A, Saito Y (1988) Trends Biotechnol 6:131
105. Inada Y, Furukawa M, Sasaki H, Kodeva Y, Hiroto M, Nishimura H, Matsushima A (1995) Trends Biotechnol 13:86
106. Secundo F, Spadaro S, Carria G, Overbecke PLA (1999) Biotechnol Bioeng 62:554
107. Ljunger G, Adlercreutz P, Mattiasson B (1993) Biocatalysis 7:279
108. Khan SA, Halling PJ, Bosley JA, Clark AH, Peilow AD, Pelan EG, Rowlands DW (1992) Enzyme Microb Technol 14:96
109. Petka WA, Harden JL, McGrath KP, Wirtz D, Tirrell DA (1998) Science 281:389

Received: September 2001

Adv Biochem Engin/Biotechnol (2004) 86: 191–213
DOI 10.1007/b12443

Disease Profiling Arrays: Reverse Format cDNA Arrays Complimentary to Microarrays

B. Zhumabayeva · A. Chenchik · P. D. Siebert · M. Herrler

BD Biosciences – Clontech, 1020 East Meadow Circle, Palo Alto, CA 94303, USA
E-mail: bdzhumabayeva@clontech.com

Abstract The identification of differentially expressed genes enables us to understand the molecular mechanisms associated with disease, conditions of stress, drug treatments and developmental processes. Microarrays provide a powerful tool for studying these complex phenomena. Verification of differentially expressed genes and correlation with biological function, which is usually done by northern blot analysis, RNase protection assay or RT-PCR, is the bottleneck in all these protocols.

We developed a new type of cDNA array for high-throughput expression profiling of multiple tissues and blood samples (i) for confirmation analysis of statistically significant number of clinical samples (ii) from limited amount of starting material, and (iii) with detailed clinical data from each individual. In contrast to traditional cDNA arrays, these arrays are spotted with a complex cDNA representing the entire mRNA message expressed in a given tissue or blood sample. cDNAs for these arrays were generated using SMART technology and accurately represent the original mRNA population, producing specific and quantitative signals during hybridization.

cDNAs on Disease Profiling Arrays were derived from total RNAs of diseased and normal tissues or different blood fractions of patients. These cDNAs were spotted onto nylon membranes along with positive and negative controls. The arrays were then hybridized with gene-specific probes. Hybridization results revealed disease-related as well as patient-specific gene expression patterns between different disease types for these genes.

These studies demonstrate that Disease Profiling Arrays are suitable for high-throughput studies comparing the relative abundance of a target gene, for simultaneously detecting differentially expressed genes in a wide variety of tissues and blood samples, and can be used down-stream from cDNA microarrays for confirmation analysis.

Abbreviations

RT-PCR	Reverse transcription polymerase chain reaction
PCR	Polymerase chain reaction
SMART	Switching mechanism at the 5′end of mRNA template
RNA	Ribonucleic acid
mRNA	Messenger ribonucleic acid
G3PDH	Glyceraldehyde-3-phosphate dehydrogenase
VEGF	Vascular endothelial growth factor
TAT	Tyrosine amino transferase
TE	Tris-EDTA
dNTP	Desoxyribonucleotide triphosphates
RNase	Ribonuclease
SSC	Sodium chloride/sodium citrate buffer
SDS	Sodium dodecyl sulfate
C_0t-1	cDNA enriched for repetitive DNA sequences
CD14	CD14 positive cells
CD19	CD19 positive cells
CD3	CD3 positive cells
MN	Mononuclear cells
PMN	Polymorphonuclear cells
TL	Total leukocytes
LN	Lymph node
ND	Normal donor
AML	Acute Myelogenous Leukemia
CML	Chronic Myelogenous Leukemia
HD	Hodgkin's Disease
NHL	Non-Hodgkin's Lymphoma
SLE	Systemic Lupus Erythematosus
LA	Lupus Anticoagulans
RA	Rheumatoid Arthritis

TA	Takayasu Arteritis
MS	Multiple Sclerosis
ITP	Idiopathic Thrombocytopenic Purpura
VWD	Von Willebrand Disease
cDNA	Complementary desoxyribonucleic acid
SAGE	Serial analysis of gene expression
CPA	Cancer Profiling Array
DPA	Disease Profiling Array
UBI	Ubiquitin
23kD	23-kD highly basic protein
GSN	Gelsolin
GP	Glutathione peroxidase

1
Introduction

The development and progression of cancer and other life-threatening diseases are accompanied by complex changes in patterns of gene expression. Identifying differences that exist at the level of gene expression in response to various stimuli, treatments, or conditions is a common starting point in understanding the molecular mechanisms underlying biological processes of diseases. Techniques currently used to identify differentially expressed genes include (i) subtractive hybridization [1–3], (ii) differential display [4, 5], (iii) serial analysis of gene expression [6] and (iiii) cDNA microarrays [7, 8]. Among these techniques, microarrays have been well received and widely used in recent years by researchers, because they allow high-throughput screening of thousands of samples in a relatively short time. These techniques usually require verification steps to prove differential expression of identified genes. Approaches such as RT-PCR or real time RT-PCR are superior sensitive techniques and commonly used for detecting differentially expressed genes. However, it can be laborious and time consuming when many genes and samples are analyzed simultaneously. Acquiring sufficient RNA when working with clinical samples such as biopsies, microdissected tumors, and laser-captured cells is very difficult. The use of such tissues is becoming more common in clinical research, and methods for amplifying small amounts of RNA while maintaining the original mRNA representation are extremely valuable.

We have developed a new type of array, a reverse format to cDNA microarrays, as a complementary tool down-stream from microarrays for screening of hundreds of clinical samples in a high-throughput manner.

These arrays are generated using the same technology which has been used for membrane-based RNA dot blots, cDNA microarrays, and cDNA library screening. cDNA samples for Disease Profiling Arrays are spotted onto nylon membranes. Then, membranes are treated accordingly for further use in hybridization experiments. cDNA samples for Disease Profiling Arrays are generated from limited amounts of hard-to-obtain clinical samples by highly-efficient PCR-based SMART technology.

This amplification method is used to obtain sufficient cDNA material for high-throughput gene – disease association studies.

In contrast to cDNA microarrays, cDNAs on Disease Profiling Arrays represent the entire mRNA message expressed in a given tissue or blood sample, but not an individual gene.

Generation and subsequent SMART amplification of cDNA is a highly practical method for overcoming the obstacle of isolating and using limited RNA material from biological samples. However, amplification of cDNA requires the presence of primer binding sites at both cDNA ends. These primer binding sites can be attached by (i) homopolymer tailing on the 3′ end of the first-strand cDNA [9–11], (ii) single-stranded anchor ligation to the 5′ end of mRNA [12, 13] or to the 3′ end of the first-strand cDNA [14], and (iii) double-stranded adaptor ligation to the 5′ end of double-stranded cDNA [15]. Each of these approaches requires several inefficient and complex manipulations.

SMART PCR cDNA synthesis provides an elegant method for amplifying cDNA samples. SMART cDNA technology combines, in one step, first-strand cDNA synthesis and attachment of anchor sequences at the 5′ and 3′ ends of single stranded cDNA.

We report that SMART PCR-generated cDNA from normal and diseased tissues (i) retain the original mRNA transcript profile, and, therefore, (ii) is a suitable tool for high-throughput tumor and disease expression profiling.

2
Applications of Disease Profiling Arrays

Disease Profiling Arrays provide data on the differential expression of a specific gene in multiple patient samples in a high-throughput format. Therefore, DPAs allow comprehensive studies in combination with other techniques such as subtraction, differential display, SAGE, or microarrays, which allow the identification of differentially expressed genes between two different mRNA populations.

DPAs allow analysis of an individual gene in hundreds of clinical samples representing patients with different disease types, different disease stages in a single experiment. This is in contrast to cDNA microarrays where one can use only one or (at most) two samples to analyze thousands of individual genes. Detailed clinical data, including patient age and gender, tumor size, extent of local invasion and sites of metastasis, can be corroborated with the change in gene expression patterns. DPAs are useful to validate microarray data due to the significant number of samples on the array. Therefore, the combination of microarrays and DPAs provides a valuable tool for comprehensive gene-expression studies.

In the case of CPA, multiple tumor types originating in breast, uterus, colon, stomach, ovary, lung, kidney, cervix, rectum, thyroid gland, prostate, pancreas, and small intestine can be screened simultaneously. Thus, a single hybridization experiment with an individual gene-specific probe allows a comparison of hybridization patterns of 13 different tumor types. In addition to comparing multiple tumor types, the Cancer Profiling Array also allows comparison of hybridization patterns between normal and tumor samples of the same individual patient.

The Blood Cancer Disease Profiling Array includes cDNAs derived from four different blood cancers: Acute Myelogenous Leukemia, Chronic Myelogenous Leukemia, Hodgkin's Disease, and Non-Hodgkin's Lymphoma.

The Autoimmune Disease Profiling Array consists of six different disease types: Systemic Lupus Erythematosus, Lupus Anticoagulans, Rheumatoid Arthritis, Takayasu Arteritis, Multiple Sclerosis, and Idiopathic Thrombocytopenic Purpura.

The Blood Cancer and Autoimmune Disease Profiling Arrays allow a survey of gene-expression patterns of individual cDNAs in several types of blood fractions such as CD14, CD19, CD3, mononuclear cells, polymorphonuclear cells, total leukocytes and in the case of HD and NHL, in additional lymph node samples. Both arrays include cDNAs from healthy donors and from patients with Von Willebrand Disease as controls.

Disease Profiling Arrays are less sensitive than real time RT-PCR. Despite their reduced sensitivity, DPAs provide the benefits of a high-throughput format for gene expression profiling, making this technology superior to RT-PCR. The use of SMART cDNA amplification technology opens a unique opportunity for generation of hundreds of micrograms of cDNA from nanograms of total RNA samples (laser-captured cells) which can be used for production scale preparation of Disease Profiling Arrays.

Disease Profiling Arrays provide quick access to multiple patient samples of normal and diseased tissues or blood fractions while also providing valuable information on the temporal expression of a distinct gene within an individual. Though there are other more sensitive methods available, i.e. real-time PCR, DPAs offer a high-throughput patient profiling format where hundreds of patient samples may be simultaneously screened.

3
The Principle of SMART cDNA Amplification Technology

We describe here the main principle of SMART technology and its applications. SMART is a core technology for generating Disease Profiling Arrays.

Many gene cloning and analysis methods, such as cDNA library construction and cDNA subtraction, require large quantities of mRNA. However, there are many cases where mRNA samples are limited. Examples include biopsy samples, unstable cell lines, or tissues at different developmental stages. In such cases, amplification of cDNA in a sequence-independent manner is needed. Generation of cDNA with a high representation of mRNA sequences is critical for further molecular biology applications.

We have developed an efficient cDNA amplification technique that is based on the switching mechanism at the 5′ end of RNA templates (Fig. 1) [16]. This technique combines, in one step, first-strand cDNA synthesis and attachment of anchor sequences at the 5′ and 3′ ends of single-strand cDNA. SMART is achieved by utilizing two intrinsic properties of Moloney Murine Leukemia Virus reverse transcriptase: (i) the ability to add non-templated nucleotides (predominantly

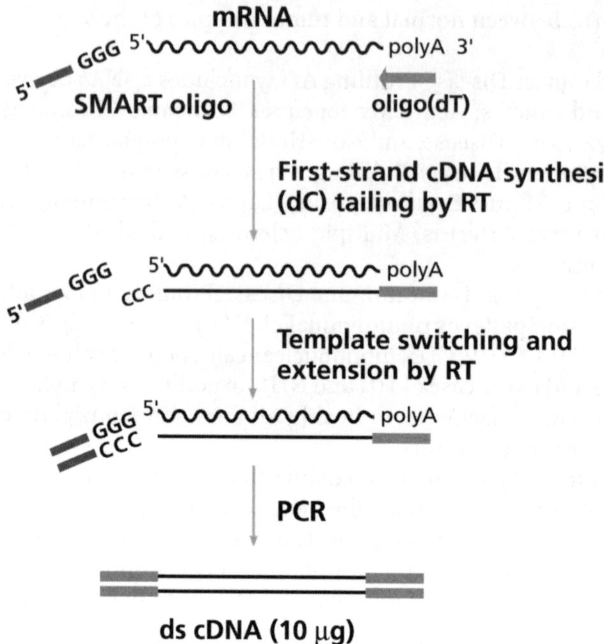

Fig. 1 Mechanism of SMART cDNA synthesis

dC) to the 3′ end of the first-strand cDNA and (ii) the ability to switch templates [16–18]. In SMART cDNA synthesis, the SMART oligonucleotide, which has an oligo (G) sequence at its 3′ end, anneals to the dC-rich cDNA tail and serves as an extended template for RT. When RT reaches the 5′ end of the mRNA, the enzyme switches templates and continues cDNA synthesis until the end of the SMART oligonucleotide. The resulting first-strand cDNA contains the complete 5′ end of the mRNA as well as sequences that are complementary to the SMART oligonucleotide. The SMART anchor sequence then serves as a universal PCR primer binding site for amplifying the entire cDNA population. SMART PCR generates a high yield of full-length double-stranded cDNA products with a high representation of the starting mRNA population.

3.1
Advantages of SMART cDNA Amplification

The SMART PCR method has several advantages over other PCR-based methods. SMART-amplified cDNAs are beneficial to investigators, who have a limited access to various biological samples, especially, clinical samples such as biopsies, laser-captured cells, or microdissected tumors.

– A major advantage of SMART technology is the simplicity of this technique. It eliminates enzymatic steps such as ligation, or homopolymer tailing, or any intervening purification steps required for the subsequent reactions to come.

Reverse transcription and template switching are catalyzed by a single enzyme, reverse transcriptase. The coupled cDNA synthesis and template switching step takes only one hour and is completed in a single tube.
- The combination of SMART technology with long distance PCR generates cDNA of up to 6 kb and with approximately 90–95% gene representation, yielding a pool of cDNA that reflects the sample's original complexity.

SMART cDNA amplification enriches for the full-length cDNA due to its template switching mechanism.

- The simplicity of this technology allows generation of large amounts of high-quality full-length enriched cDNA for a variety of molecular biology applications from as little as 50 ng of total RNA starting material. SMART technology has also been optimized to generate cDNA from as little as 100 cell equivalents or ~2 ng of total RNA. This protocol extends the possible applications of SMART by allowing samples, such as laser-captured microdissected samples, with extremely small amounts of RNA, to be used for gene expression profiling.

SMART cDNA can be used for:

1. cDNA library constructions [16, 19];
2. cDNA cloning [20];
3. as probes for microarrays [21–23];
4. "virtual" Northern [16, 24, 25];
5. high-throughput tissue, disease, and gene expression profiling [26, 27].

3.2
Representation of the Original mRNA Complexity

Generation of PCR-amplified cDNAs with a high representation of mRNA transcripts is critical for gene expression profiling as well as for other molecular biology applications. In order to avoid biased amplification of cDNAs, we restricted PCR amplification to the linear phase, which is necessary to generate highly representative cDNA [28]. In our experience, the optimal number of PCR cycles tends to vary with different templates, amounts of starting material, and type of thermal cycler. Each cDNA synthesis reaction should be monitored on an agarose gel to avoid overcycling of PCR products. In our experiments, the optimal number of cycles is typically one cycle less than is needed to reach the exponential plateau.

To determine whether SMART PCR-generated cDNA accurately represents the original mRNA population, we first amplified cDNAs using SMART technology from five human total RNA samples purified from liver, placenta, brain, kidney, and HeLa cells. Fifty nanograms of purified cDNA were spotted on nylon membranes along with 1 μg of each corresponding total RNA. Arrayed membranes were hybridized with labeled cDNA probes for glyceraldehyde-3-phosphate dehydrogenase (Fig. 2A), and vascular endothelial growth factor (Fig. 2B). Signal intensities on membranes were then quantified by phosphorimaging. Both experiments reveal a strong correlation between the relative signal intensities of

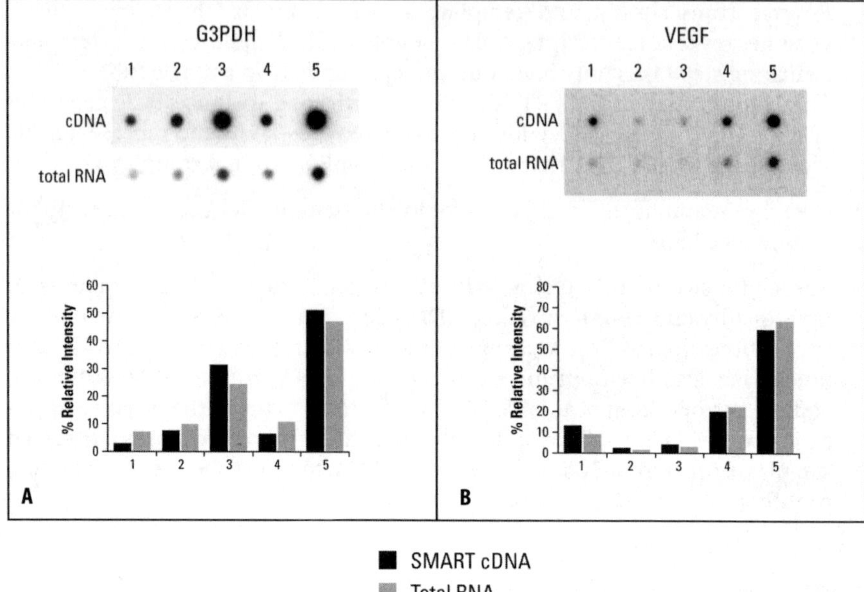

Fig. 2 Array of SMART cDNAs and original total RNA samples probed with G3PDH (*Panel A*) and VEGF (*Panel B*). 50 ng samples of cDNAs and 1 µg of corresponding total RNAs from five different tissues were spotted on nylon membranes. The membranes were hybridized with radiolabeled probes for G3PDH (*Panel A*) and VEGF (*Panel B*). Hybridization signals were detected by phosphorimaging. Due to the different hybridization characteristics of cDNA and RNA, hybridization signal values for both G3PDH and VEGF were quantified and represented as a percentage of the total signal from all five spots for either cDNA or RNA (bottom rows). *1-* liver, *2-* placenta, *3-* brain, *4-* kidney, and *5-*HeLa cells

total RNA and corresponding SMART cDNA in all 5 different tissues (Fig. 2, bottom rows). The results show that 50 ng of SMART cDNA produces the same gene expression profile across the samples as 1 µg of the original total RNA samples, while yielding a higher sensitivity of detection.

3.3
SMART cDNA Produces Tissue-Specific and Quantitative Hybridization Signals

The specificity of SMART PCR amplification for an entire cDNA population is important for generating high-fidelity and representative cDNA molecules. Non-specific amplification can easily be determined when a single gene is amplified, but the issue is more complex for an entire cDNA population.

To evaluate whether SMART cDNA yields signals that are both specific and quantitative, we performed the following experiment. We synthesized SMART cDNA from total RNA derived from the following 12 human tissues: spleen, placenta, kidney, lung, adult liver, testis, whole brain, fetal liver, thymus, pancreas, prostate, and HeLa cells. 50, 30 and 10 ng of each purified cDNA were arrayed on

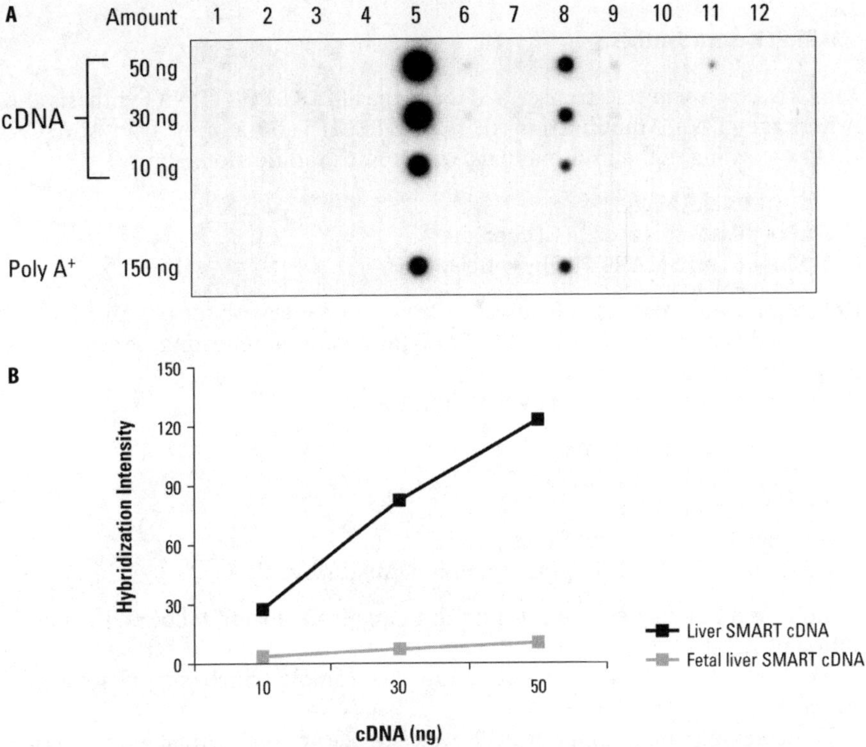

Fig. 3 Array of different amounts of SMART cDNA hybridized with a TAT probe. (**A**) cDNAs were prepared from 12 different human total RNA samples. 10, 30, and 50 ng of each purified cDNA were arrayed on a nylon membrane along with 150 ng of corresponding poly A⁺ RNA. The membrane was hybridized with a radiolabeled probe for TAT and visualized by phosphorimaging. (**B**) Linear correlation between the intensity of each hybridization signal and the amount of spotted cDNAs for adult liver and fetal liver. *1* – spleen, *2* – placenta, *3* – kidney, *4* – lung, *5* – adult liver, *6* – testis, *7* -whole brain, *8* – fetal liver, *9* – HeLa cells, *10* – thymus, *11* – pancreas, and *12* – prostate

a nylon membrane. The membrane was hybridized with a cDNA probe for the liver-specific gene tyrosine aminotransferase [29]. Figure 3 shows that TAT specifically hybridizes only with cDNAs originated from adult and fetal liver (Panel A).

Quantitative analysis of the hybridization signals generated by the various amounts of cDNAs spotted on a membrane revealed a linear correlation between signal intensity and cDNA quantity indicating that the cDNA amounts applied to the membrane did not reach a saturation level (Fig. 3B). This experiment also showed the higher sensitivity of cDNA in comparison with 150 ng/dot of corresponding poly A⁺ RNA.

3.4
SMART PCR cDNA Synthesis

Total RNAs were reverse-transcribed using the SMART PCR cDNA Synthesis and Advantage cDNA Amplification kit (CLONTECH Laboratories, Palo Alto, CA, USA) according to the user manuals, with some modifications:

- 1 µg of total RNA,
- 1 µl of 10 µM oligo(dT) (CDS primer),
- 1 µl of 10 µM SMART II Oligonucleotide

were mixed with deionized RNase-free water for a total volume of 5 µl.

Then, this mixture was heated to 72 °C for 2 min for annealing, cooled on ice for 2 min, and spun briefly.

The reaction was followed by the addition of

- 2 µl of 5X Reaction Buffer,
- 1 µl of 20 mM dithiothreitol,
- 1 µl of 10 mM dNTP,
- 0.5 µl of 100 U/µl of PowerScript RT,
- 0.5 µl of 28 U/µl ANTI-RNase (Ambion, Austin, TX, USA).

Samples were incubated at 42 °C for 1 h in an air incubator, followed by inactivation of RT at 72 °C for 7 min.

Then, 40 µl of TE buffer was added to each sample, which was subsequently stored at –20 °C.

To determine the number of PCR cycles necessary for optimal amplification of cDNA,

- 1 µl from each first-strand cDNA reaction mixture was combined with
- 10 µl of 10X Advantage Polymerase Buffer,
- 1 µl of PCR primer,
- 2 µl of 10 mM dNTP, and
- 1 µl of the Advantage cDNA Polymerase Mix.

Samples were then amplified in an MJ Research Thermal Cycler (MJ Research, Watertown, MA, USA) using the following program:

- 1 cycle at 95 °C for 1 min, then
- 15 cycles at
- 95 °C for 15 s,
- 65 °C for 30 s, and
- 68 °C for 6 min.

After 15 cycles, 15 µl of the reaction mixture was transferred to a fresh 0.5-ml tube and subjected to 3 additional cycles, while the remaining 85 µl of the PCR mixture was kept at 4 °C. After 3 additional cycles, 5 µl of the reaction mixture was aliquoted for gel-analysis while the remaining 10 µl of the reaction mixture was subjected to another 3 cycles for a total of 21 cycles. To determine the optimal number of cycles, 5 µl of each of the reaction mixture (i.e., 15-, 18- and 21-cycle PCR products) was analyzed on a 1.1% agarose gel in Tris-Acetate

EDTA buffer. Based on results of the agarose gel analysis, samples were subjected to additional cycles as necessary.

4
Strategy of Generating Disease Profiling Arrays

The main strategy of generating DPAs using SMART technology is outlined in Figure 4. cDNAs for DPAs were synthesized from total RNAs isolated from diseased and their corresponding normal tissues or different blood fractions. Then, first-strand cDNA is amplified using an enzyme especially formulated for efficient, accurate amplification of cDNA templates by long-distance PCR. For best results, we first optimized the PCR cycling parameters to ensure that the ds DNA remains in the linear phase of amplification. Then, amplified cDNA was purified and the concentration was measured. Initially, we spotted a constant amount (15 ng) of each cDNA on a nylon membrane and hybridized with a set of different housekeeping genes for normalization as described under Sect. 4.2. After normalization, we spotted the normalized amount of cDNA for further analysis.

4.1
Collection of Blood Fractions and Isolation of Total RNA

Blood samples were fractionated in Vacutainer Brand Tubes for preservation of whole blood for total leukocytes and Vacutainer CPT tubes with citrate for all other cell types (BD Biosciences, San Jose, CA, USA). Blood samples were processed within 3 h of collection. The purity of each cell fraction was determined cytologically and immunohistochemically and found to be not less than 96% in all cases.

Total leukocytes were isolated as follows.

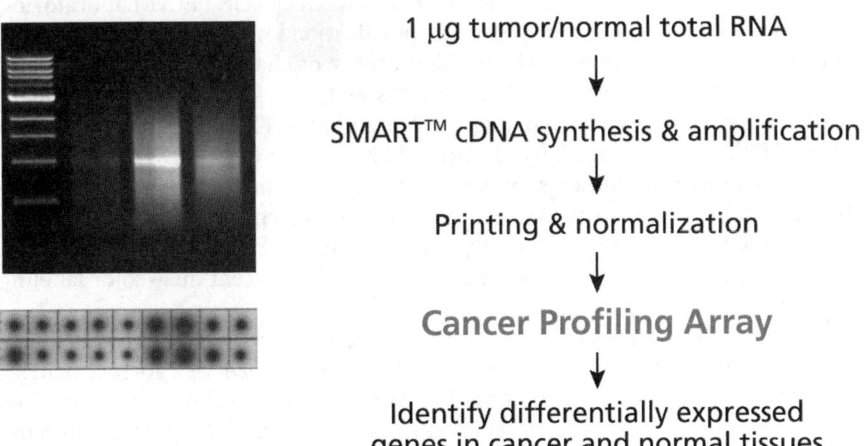

Fig. 4   The Strategy of generating disease profiling arrays

a) The platelets and blood cells were separated from the plasma by centrifugation at 1,100 rpm for 15 min at room temperature.

b) The blood cell suspension was transferred to a clean tube and erythrocytes were lysed at room temperature with erythrocyte lysis buffer for 5 minutes.

c) Total leukocytes were pelleted by centrifugation at 1,400 rpm for 15 min at room temperature.

d) Leukocytes were washed once with 20 ml of 1X PBS and 2 mM EDTA and then counted.

e) Leukocytes were pelleted by centrifugation at 1,400 rpm for 15 min at room temperature.

f) Leukocytes were resuspended in 330–950 µl of RLT Lysis Buffer (QIAGEN, Valencia, CA, USA), exact volume dependant on cell number.

g) Lysed cells were transferred to an eppendorf tube and flash frozen with liquid nitrogen for storage at –70 °C until RNA purification. MN cells were isolated according to the accompanying Vacutainer CPT tube protocol. PMN cells were isolated from the pellets after MN cell isolation. Erythrocytes were lysed with erythrocyte lysis buffer at room temperature. Cells were washed twice with 1X PBS and lysed in RLT Lysis Buffer. CD3, CD14, and CD19 positive cells were isolated using Miltenyl Biotec magnetic microbeads on MS⁺ separation columns according to the manufacturer's protocol (Miltenyl Biotec, Germany). Total RNAs were isolated using an acid guanidium thiocyanate-phenol-chloroform method [30]. RNA samples were treated with DNase I to eliminate genomic DNA contaminations. All total RNA samples were then tested for RNA integrity on a formaldehyde gel, and for genomic impurities by RT-PCR using primers designed for intron sequences.

4.2
Preparation of Membrane-Based cDNA Arrays and Normalization of cDNAs

Amplified cDNAs were purified from unincorporated nucleotides, and primer-dimers using the NucleoTrap PCR Purification Kit (CLONTECH Laboratories, Palo Alto, CA, USA). The principle of this purification kit is based on an especially activated matrix that binds DNA in the presence of chaotropic salts, and allows the purification of large DNA fragments without shearing. A subsequent washing step removes all impurities. DNA is eluted in TE buffer. One microgram of total RNA yields up to 10 µg of purified SMART cDNA.

We used nylon membranes for the printing of amplified cDNA, as nylon membranes were superior to any tested glass surface in terms of sensitivity. Disease Profiling Arrays do not require costly scanners and can be used in any laboratory. Glass-based arrays, on the other hand, allow for fluorescent dual-color labeling and detection.

For normalization of cDNA, we printed 15 ng of cDNA on Nytran Nylon membranes (Schleicher and Schuell, Keen, NH, USA) in a 384-well format using a BIOMEK 2000 Laboratory Automation Workstation (Beckman Instruments, Fullerton, CA, USA). Three arrayed membranes were probed separately with the following housekeeping genes: ubiquitin, β-actin, and 23-kD highly basic protein [31–35]. The hybridization signals were quantified by phosphorimaging.

The term "normalization" refers to a procedure using the average gene expression level of a set of housekeeping genes to standardize the amount of cDNAs spotted onto nylon membranes. Normalization, based on the relative abundance of housekeeping genes, is necessary for comparative studies, but particularly challenging in cancer and other disease tissues, because of significant changes in gene expression in diseased cells [36]. The right selection of housekeeping genes for the normalization procedure was one of the most challenging tasks in developing Disease Profiling Arrays. We tested more than 20 different housekeeping genes for variability in expression. Among those genes we were able to select three genes, which showed 0.5–2 fold variability in expression between tumor and normal samples.

As a first step of normalization, a constant amount (15 ng) of each cDNA is spotted onto 48 membranes.

- Each membrane is then hybridized to one of the housekeeping gene probes.
- The hybridization signal produced by each dot is quantified by phosphorimaging.
- The percentage that each dot contributes to the total signal of all dots on the array is calculated.
- The average percentage for all three housekeeping genes is calculated for each dot.
- A normalization factor is calculated.
- The original cDNA amounts are adjusted to determine the normalized amounts to be spotted onto membranes.

Calculating the value of each dot as a percentage compensates for differences in intensity of the signal for each housekeeping gene.

Normalized amounts of cDNAs were then spotted on nylon membranes along with nine cancer cell line cDNAs (50 ng/spot); negative controls (50 ng/spot) including yeast total RNA (NC-1), yeast tRNA (NC-2), *Escherichia coli* DNA (NC-3), poly r(A) (NC-4), human C_ot-1 DNA (NC-5), human genomic DNA (NC -6), and ubiquitin cDNA (10 ng/spot) as a positive control.

4.3
Hybridization of cDNA Arrays

cDNA probes were labeled using DECAprime II DNA Labeling Kit (Ambion, Austin, TX, USA) and [α-^{32}P] dATP (Amersham Pharmacia Biotech, Piscataway, NJ, USA) according to the user manual. The labeled probes were then purified from unincorporated nucleotides using Chromaspin 100 columns (CLONTECH Laboratories, Palo Alto, CA).

cDNA membranes were prehybridized and then hybridized overnight at 68 °C with radiolabeled cDNA probes in a hybridization bottle containing ExpressHyb Hybridization Solution (CLONTECH Laboratories, Palo Alto, CA, USA), 100 µg/ml sheared salmon sperm DNA, and 1–2×10^6 dpm/ml labeled probe. cDNA membranes were washed three times with 2×SSC, 1% SDS for 30 min, once with 0.2×SSC, 0.5% SDS for 30 min, then rinsed in 2×SSC, and exposed to BIO-MAX MS X-ray film with an intensifying screen (Eastman Kodak Co., Rochester,

NY, USA). Signal intensities were calculated for individual spots using a STORM-860 Phosphorimager (Molecular Dynamics, Sunnyvale, CA, USA).

5
Identification of Differentially Expressed Genes Using Disease Profiling Arrays

5.1
Cancer Profiling Array

The Cancer Profiling Array offers a new approach for profiling gene expression. This array provides quick access to multiple samples of normal and cancerous tissues from individual patients so that confirmatory evidence for the cancer-specific changes in the expression pattern of a particular gene is easily obtained in a single experiment.

The CPA contains pairs of cDNA spots generated from tumor and corresponding normal tissue samples from individual patients spotted side-by-side on a nylon membrane (Fig. 5). The array contains multiple cDNA pairs from 13 different tissues, including breast, uterus, colon, stomach, ovary, lung, kidney, rectum, thyroid gland, prostate, small intestine, pancreas, and cervix. All cDNA samples are normalized to the expression of three housekeeping genes: ubiquitin, beta-actin, and 23-kD highly basic protein. After normalization, the differential expression pattern of a particular gene is easily observed among both normal and tumor samples. The array provides a thorough survey of gene expression among hundreds of cancer patients, and thus, produces statistically relevant data. Detailed clinical data, including patient age and gender, tumor size and stage, extent of local invasion and sites of metastasis is provided for all samples included on the array. In this way, any changes in gene expression can be corroborated with clinical data.

We used the Cancer Profiling Array to screen for the cancer-specific gene gelsolin. Gelsolin is an actin filament regulatory protein that has been shown to be down-regulated in invasive human breast ductal carcinoma [37]. Based on its potential involvement in cancer, we examined gelsolin expression using the CPA to demonstrate its utility in cancer research. Following an overnight hybridization with the radio-labeled probe for gelsolin, we washed and exposed the CPA to an X-ray film (Fig. 6B). Subsequently, we stripped the gelsolin probe off from the membrane and, using the same protocol, hybridized the array again with a ubiquitin probe (Fig. 6A). The hybridization pattern for ubiquitin illustrated that both normal and tumor tissue expression levels are comparable, as expected for this housekeeping gene. In contrast, hybridization with a probe for gelsolin demonstrated that gelsolin was differentially regulated among normal and tumor tissues. Upon closer examination of the data (Fig. 6B), 49 out of 50 paired breast samples, 41 out of 42 paired uterine samples and 10 out of 14 paired ovarian samples showed at least three-fold down regulation of gelsolin expression in tumors when compared to their corresponding normal tissues. In contrast, both tumor and corresponding cDNAs from colon and lung tissues showed only slight down-regulation of gelsolin expression. The data suggest that down-regulation of gel-

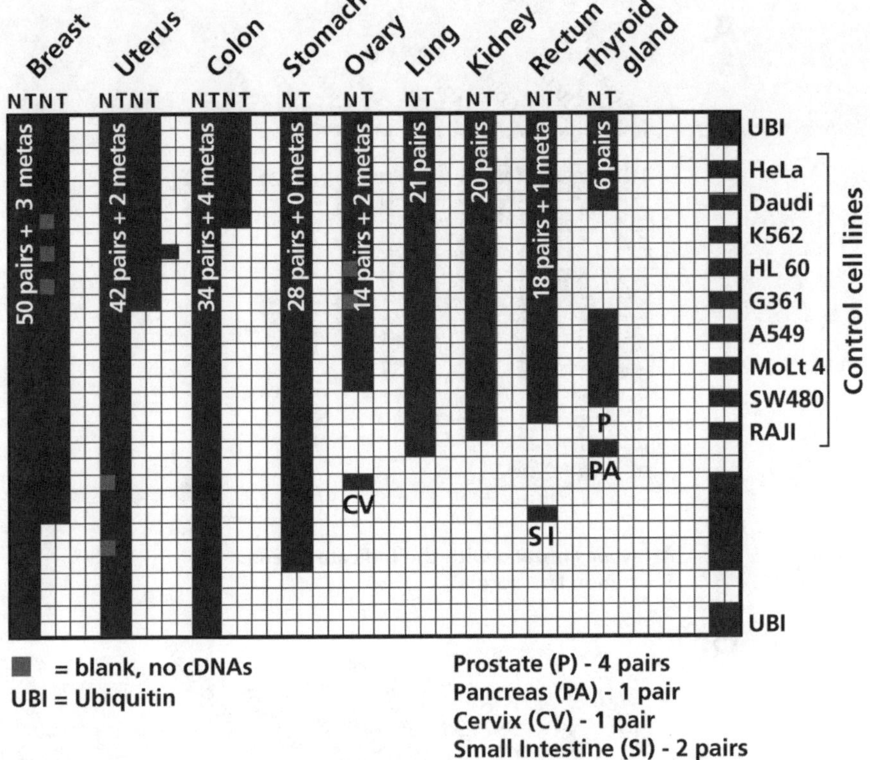

Fig. 5 Schematic overview of the Cancer Profiling Array. The CPA contains 241 cDNA pairs – each representing tumor and normal tissue from an individual patient. N = Normal, T = Tumor, CV = Cervix, PA = Pancreas, P = Prostate, UBI = Ubiquitin, TG = Thyroid Gland

solin expression in these tumors may be correlated to cancer in female reproductive tissues.

We also hybridized the CPA with a probe for Glutathione peroxidase (Fig. 6C). The expression of glutathione peroxidase was predominantly detected in normal kidney samples with down-regulation in kidney tumors (Fig. 6C). The hybridization also indicates down-regulation of GP expression in breast, colon, and lung tumors. No GP expression was detected in cancer cell lines. Down-regulation of GP was previously detected in lung tumors by subtractive hybridization (unpublished data). In fact, our cDNA array reveals the down-regulation of GP in lung tumors of several independent patients. In addition, our hybridization data reveal down-regulation of GP in each of kidney, breast, and colon tumors. RT-PCR results for selected lung and breast cDNA samples confirmed the hybridization results for glutathione peroxidase (Fig. 6D).

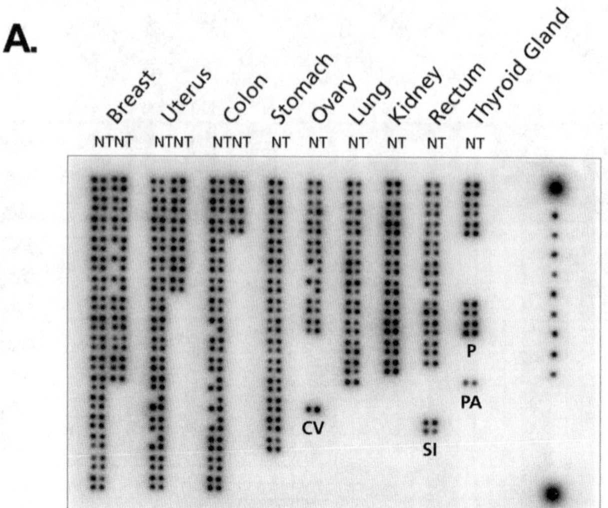

Array was probed with the indicated gene.
T = tumor. N = normal.

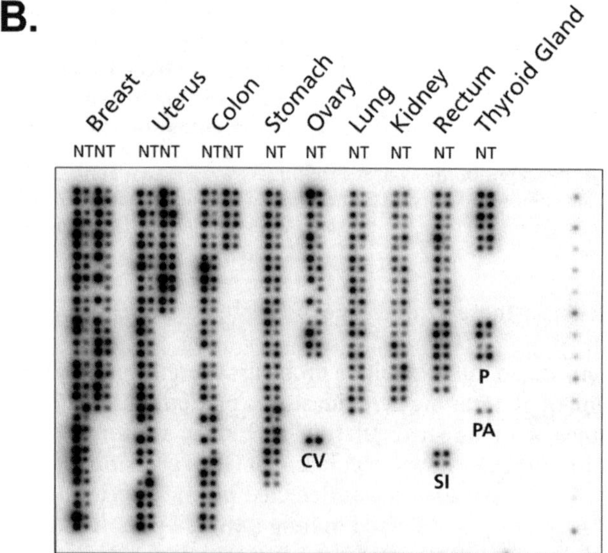

Array was probed with the indicated gene.
T = tumor. N = normal.

Fig. 6 Cancer Profiling Arrays probed with UBI, GSN, and GP. The normalized membranes were hybridized with radiolabeled probes for ubiquitin (**A**), gelsolin (**B**), and glutathione peroxidase (**C**). Four cDNA pairs from lung and two cDNA pairs from breast were used in RT-PCR to confirm the differential expression of GP (**D**). Signals were visualized by phosphorimaging. M-1-kb DNA ladder size markers; N-cDNA from normal tissue; T-cDNA from tumor tissue.

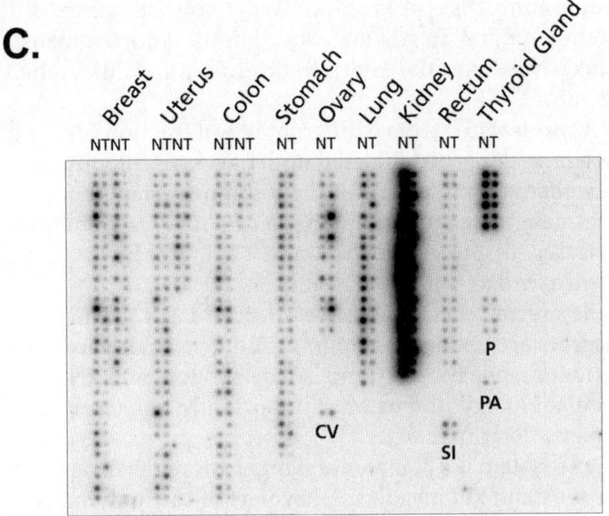

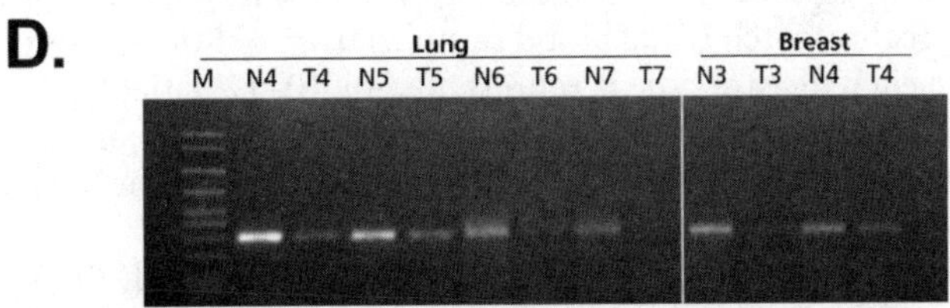

Array was probed with the indicated gene.
T = tumor. N = normal.

Fig. 6 (continued) Cancer cell line cDNAs (from left to right HeLa cells, Daudi, K-562, HL-60, G361, A549, MOLT-4, SW480, and Raji); UBI-positive control cDNA for ubiquitin (10 ng per spot) on the right top and bottom corners of the membrane; The matched tumor and normal cDNAs are derived from the following tissues: breast, uterus, colon, stomach, ovary, cervix, lung, kidney, rectum, small intestine, thyroid gland, prostate, and pancreas

5.2
Blood Cancer and Autoimmune Disease Profiling Arrays

The Blood Cancer Disease Profiling Array includes cDNAs from four different blood cancers: Acute Myelogenous Leukemia, Chronic Myelogenous Leukemia, Hodgkin's Disease, and Non-Hodgkin's Lymphoma.

The Autoimmune Disease Profiling Array consists of six different disease types: Systemic Lupus Erythematosus, Lupus Anticoagulans, Rheumatoid Arthritis, Takayasu Arteritis, Multiple Sclerosis, and Idiopathic Thrombocytopenic Purpura.

Total RNA was isolated from 6 different blood fractions derived from one individual patient as described in detail under Sect. 4.1. Blood cells were fractionated to study changes in the state of the immune system that are considered to take place in malignant cells from leukemias and lymphomas as well as autoimmune diseases. In the case of HD and NHL, total RNA from lymph node biopsies was isolated as well when available.

Blood cells are the most accessible source of RNA when blood and autoimmune diseases are under study. A few milliliters of drawn blood allow one to perform various diagnostic tests, and isolate enough total RNA to be used in different molecular biology studies. Most importantly, blood cells carry out defense functions against foreign invaders.

The immune system is a complex and highly developed system, yet its mission is simple: to seek and kill invaders. A key part of the immune system's role is to differentiate between invaders and the body's own cells – when it fails to make this distinction, a reaction against "self" cells and molecules results in autoimmune diseases. Understanding the mechanism of autoimmune diseases still remains challenging. The Autoimmune Disease Profiling Array can be used as a tool for clinical research using blood samples from 6 different autoimmune diseases where each disease type is represented by 10–12 patients.

The Blood Cancer Disease Profiling Array focuses on the cancer of blood cells.

Cancer of blood cells is characterized by replacement of the bone marrow with malignant, leukemic cells.

We selected 6 different blood fractions to generate both Blood Cancer and Autoimmune Disease Arrays: CD3, CD14, CD19, MN, PMN, and TL.

It is well established that all these cells differentiate from pluripotent stem cells into three lineages: myeloid, erythroid and lymphoid cells.

Myeloid stem cells give rise to MN and PMN, while lymphoid stem cells give rise to T and B cells. The CD3 antigen is expressed by thymocytes, peripheral T cells and natural killer cells. The CD19 antigen is present on marrow pre-B cells and mature B cells. By selecting these two cell types, we have access to study gene expression levels of T- and B-cells.

Finally, CD14 is expressed in all monocytes, representing the myeloid lineage.

Total RNA was prepared from total leukocytes. Overall, six different blood fractions give a broad representation of different blood cells, which could be directly related to blood cancer or autoimmune disorders. The schematic overview of both DPAs is outlined in Figure 7 and 8. All cDNA samples were normalized to the expression of two housekeeping genes: ubiquitin and β-actin. Figure 9 shows comparable expression levels of ubiquitin in both DPAs as expected after normalization. Figure 10 shows the expression profile of 23-kD highly basic protein when hybridized to the Autoimmune Disease Profiling Array. Interestingly, the housekeeping gene 23 kDa basic protein showed comparable expression levels when hybridized with the CPA. An entirely different expression

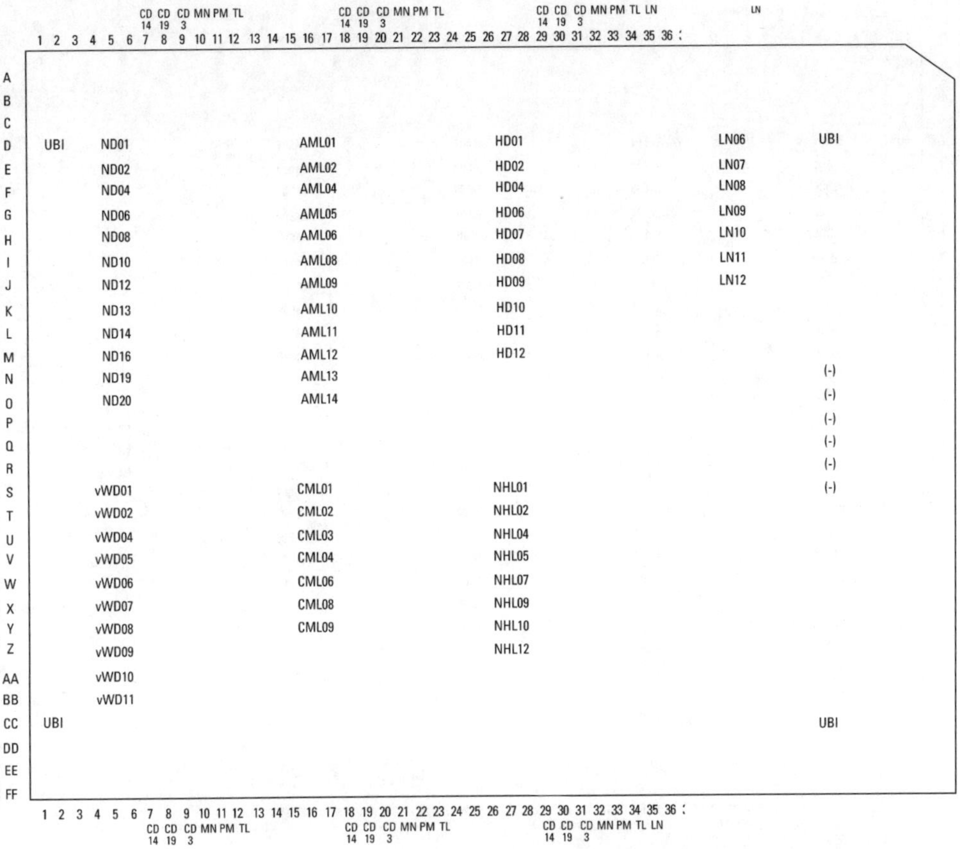

Fig. 7 Schematic overview of the Autoimmune Disease Profiling Array. The Autoimmune Disease Profiling Array contains cDNA samples representing different blood fractions from 6 autoimmune diseases: SLE, LA, RA, TA, MS, and ITP. First quadrant on the top left corner represents cDNA samples from normal donors: *ND*. Each row represents fractionated blood samples of an individual patient. Each column represents different blood fractions: CD14, CD19, CD3, MN, PMN, and TL. UBI = Ubiquitin; (-) = 6 negative hybridization controls

profile was shown for blood fractions for the 23-kDa protein. Upon closer examination, we can see that the 23-kD is highly expressed in CD19, CD3, and MN, but its expression is diminished in PMN cells. In summary, we conclude that Blood Cancer and Autoimmune Disease Profiling Arrays can be used as an innovative tool to study blood cancers and autoimmune disorders by means of gene expression profiling of hundreds of clinical samples.

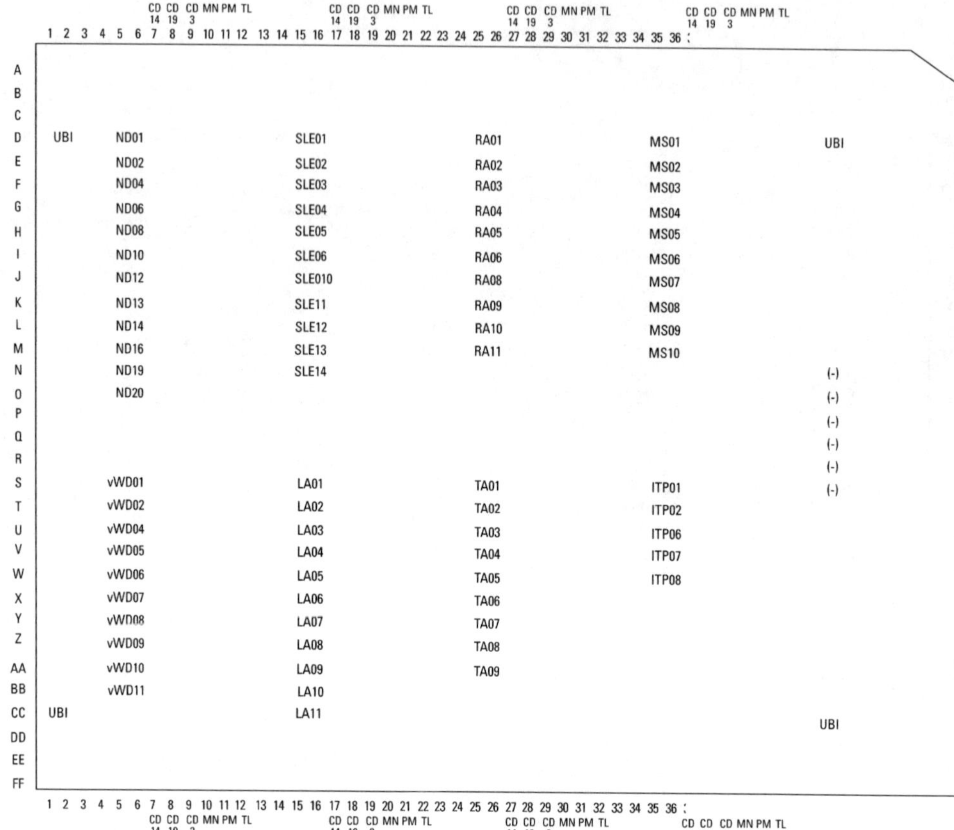

Fig. 8 Schematic overview of the Blood Cancer Disease Profiling Array. The Blood Cancer Profiling Array contains cDNA samples representing different blood fractions from 4 blood cancer diseases: AML, CML, HD, and NHL. First quadrant on the top left corner represents cDNA samples from normal donors: ND. Each row represents a fractionated blood samples of an individual patient. Each column represents different blood fractions: CD14, CD19, CD3, MN, PMN, TL, and LN. UBI = Ubiquitin; (-) = 6 negative hybridization controls

6
Conclusion

We conclude that SMART PCR-generated cDNA could be beneficial to investigators who only have access to small amounts of tissue or blood samples, for (i) quick detection of differentially expressed genes between tumor and normal tissues, (ii) to confirm differentially expressed genes identified by other techniques, i.e., microarrays, and (iii) to screen for differentially expressed genes in a large number of patient samples. These advantages of SMART cDNA open a wide range of possibilities for investigators in high-throughput gene – disease association studies in conjuction with other technologies.

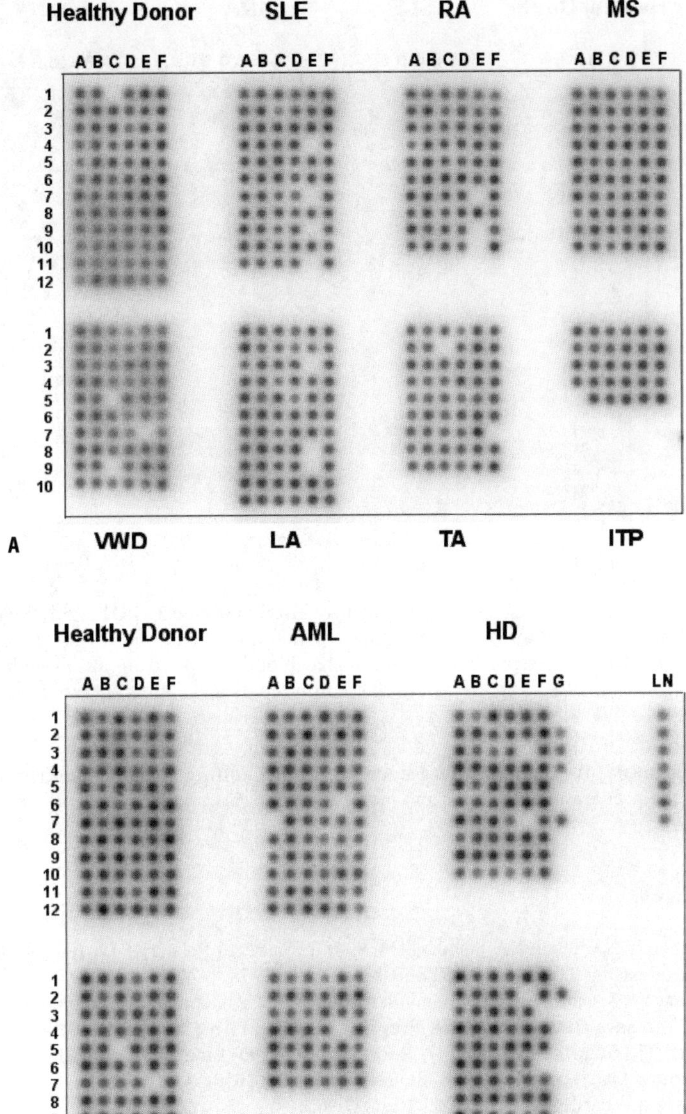

Fig. 9 Blood Cancer and Autoimmune Disease Profiling Arrays hybridized with Ubiquitin. The normalized membranes were hybridized with radiolabeled probes for ubiquitin. **A** – Autoimmune Disease Profiling Array; **B** – Blood Cancer Disease Profiling Array. Each row represents a fractionated blood sample from an individual patient (numbers indicate different patients). Each column represents a different blood fraction

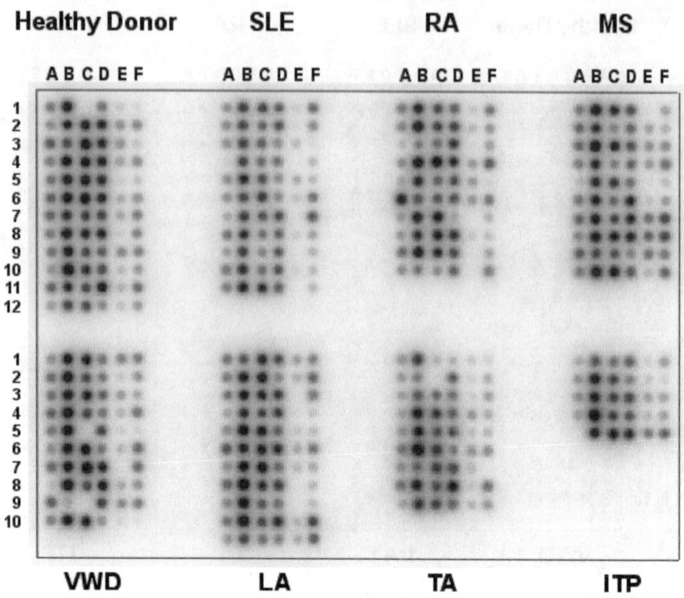

Fig. 10 Autoimmune Disease Profiling Array hybridized with the 23-kDa highly basic protein. The normalized membranes were hybridized with radiolabeled probes for 23-kDa highly basic protein. Each row represents a fractionated blood sample from an individual patient (numbers indicate different patients). Each column represents a different blood fraction

Acknowledgement We thank Myrna Faulds for critical editing of the manuscript and Anna Sayre for the preparation of graphs and pictures.

7
References

1. Diatchenko L, Chenchik A, Siebert PD (1998) In: Siebert PD, Larrick JW (eds), Gene cloning and analysis by RT-PCR. Eaton Publishing, Natick, MA, p 213
2. Gurskaya NG, Diatchenko L, Chenchik A, Siebert PD, Khaspekov GL, Lukyanov KA, Vagner LL, Ermolaeva OD, Lukyanov SA, Sverdlov ED (1996) Anal Biochem 240:90
3. Schoen TJ, Mazuruk K., Chader GJ, Rodrigues IR (1995) Biochem Biophys Res Com 213:181
4. Ikomonov OC, Jacob MN (1996) BioTechniques 20: 1030
5. Liang P and Pardee AB (1992) Science 257:967
6. Velculescu VE, Zhang L, Vogelstein B, Kinzler KW (1995) Science 270:484
7. DeRisi J, Penland L, Brown PO, Bittner ML, Meltzer PS, Ray M, Chen Y, Su YA, Trent JM (1996) Nat Genetics 14:457
8. Lockhart DJ, Dong H, Byrne MC, Follettie MT, Gallo MV, Chee MS, Mittmann M, Wang C, Kobayashi M, Horton H, Brown EL (1996) 199:1675
9. Akowitz A, Manuelidis L (1989) Gene 81:295
10. Belyavsky A, Vinogradova T, Rajewsky K (1989) Nucleic Acids Res 17:2919
11. Domec C, Garbay B, Fournier M, Bonnet J (1990) Anal Biochem 188:422
12. Kato S, Sekine S, Oh SW, Kim NS, Umezawa NA, Yokoyama-Kobayashi M, Aoki T (1994) Gene 150:243
13. Maruyama K, Sugano S (1994) Gene 138:171
14. Apte AN, Siebert PD (1993) BioTechniques 15:890

15. Frohman MA, Dush MK, Martin GR (1988) Proc Natl Acad Sci USA 85:8998
16. Chenchik A, Zhu YY, Diatchenko L, Li R, Hill J, Siebert PD (1998) In: Siebert PD, Larrick JW (eds) Gene cloning and analysis by RT-PCR. Eaton Publishing, Natick, MA, p 305
17. Clark JM (1988) Nucleic Acids Res 16: 9677
18. Hu WS, Temin HM (1990) Science 250:1227
19. Zhu YY, Machleder EM, Chenchik A, Li R, Siebert (2001) BioTechniques 30:892
20. Matz M, Shagin D, Bogdanova E, Britanova O, Lukyanov S, Diatchenko L, Chenchik A (1999) Nucleic Acids Res 15:1558
21. Gonzales P, Zigler JS, Epstein DL, Borras T (1999) BioTechniques 26:884
22. Livesey FJ, Furukawa T, Steffen MA, Church GM, Cerko CL (2000) Current Biology 10:301
23. Wang E, Miller LD, Ohnmacht GA, Liu ET, Marincola FM (2000) Nat Biotechnology 18:457
24. Endege WO, Steinmann KE, Boardman LA, Thibodeau SN, Schlegel R (1999) BioTechniques 26:542
25. Franz O, Bruchhaus I, Roeder T (1999) Nucleic Acids Res 27:e3
26. Zhumabayeva B, Diatchenko L, Chenchick A, Siebert PD (2001) BioTechniques 30:158
27. Wiechen K, Diatchenko L, Agoulnik A, Scharff KM, Schober H, Arlt K, Zhumabayeva B, Siebert PD, Dietel M, Schafer R, Sers C (2001) Am J Pathol 159:1635
28. Lukyanov K, Diatchenko L, Chenchik A, Nanisetti A, Siebert PD, Usman N, Matz M, Lukyanov S (1997) Biochem Biophys Res Com 230:285
29. Schweizer-Groyer G, Groyer A, Cadepond F, Grange T, Baulieu EE, Pictet R (1994) Nucleic Acids Res 22:1583
30. Chomczynski P, Sacchi N (1987) Anal Biochem 162:156
31. Adams MD, Kerlavage AR, Fields C, Venter C (1993) Nat Genetics 4:256
32. CLONTECH Laboratories (1998) CLONTECHniques XIV(1):8
33. CLONTECH Laboratories (1999) CLONTECHniques XIV(1):30
34. Liew CC, Hwang DM, Fung YW, Laurenssen C, Cukerman E, Tsui S, Lee CY (1994) Proc Natl Acad Sci USA 91:10645
35. Spanakis E, Brouty-Boye D (1994) Nucleic Acids Res 22:799
36. Goldsworthy SM, Goldsworthy TL, Sprankle CS, Butterworth BE (1993) Cell Prolif 26:511
37. Asch HL, Winston JS, Edge SB, Stomper PC, Asch BB (1999) Breast Cancer Res. Treat 55:179

Received: March 2002

Adv Biochem Engin/Biotechnol (2004) 86: 215–253
DOI 10.1007/b12444

Hematopoietic Stem Cells: Clinical Requirements and Developments in Ex-Vivo Culture

Asok Mukhopadhyay[1] · T. Madhusudhan · Rajat Kumar[2]

[1] National Institute of Immunology, Aruna Asaf Ali Marg, 110067 New Delhi, India
E-mail: ashok@nii.res.in
[2] Army Hospital Research and Referral, Department of Hematology, 110010 New Delhi, India
E-mail: rajatkr@hotmail.com

Abstract Ex vivo expansion of hematopoietic stem cells is a promising technology for many potential applications from marrow reconstitution to gene therapy. Considerable progress has been made during the past ten years in understanding the biology of hematopoietic stem cells and its ex vivo expansion; despite this, the cultured cell is still between pre-clinical and phase I clinical trials. This review summarizes recent progress in the ex vivo expansion of hematopoietic stem cells and its requirements for clinical applications. The second section covers hematopoiesis and the bone marrow microenvironment. The third and fourth sections deal with therapeutic applications of stem cells and transplantation requirements, respectively. Biological alteration of expanded stem cells, molecular control of hematopoiesis, characterization of cells, and bioreactors for culture of stem cells and its operational parameters are the subjects of the fifth section. The next section covers pre-clinical and clinical studies on expanded stem cells. Ex vivo expansion of stem cells in three-dimensional culture system is the subject matter for the last section.

Keywords Hematopoietic stem cells · Ex vivo expansion · Clinical applications · Engraftment

Abbreviations

BFU-E	Burst forming unit erythrocyte
BFU-Mk	Burst forming unit megakaryocyte
BM	Bone marrow
CAFC	Cobblestone area forming cells
CB	Cord blood
CFC	Colony forming cell
CFU-E	Colony forming unit erythrocyte
CFU-Eo	Colony forming unit eosinophil
CFU-GEMM/	Colony forming unit granulocyte, erythrocyte,
CFU-Mix	monocyte, megakaryocyte
CFU-Mk	Colony forming unit megakaryocyte
CFU-S	Colony forming unit spleen
CLP	Common lymphoid progenitors
CMP	Common myeloid progenitors
CMV	Cytomegalo virus
ECM	Extra cellular matrix
Epo	Erythropoietin
FL	Fetal liver
FltL	Flt-3 ligand
G	Granulocyte
G-CSF	Granulocyte colony stimulating factor
GM-CSF	Granulocyte macrophage colony stimulating factor
GVHD	Graft vs host disease
HLA	Human leukocyte antigen

HPP-CFC	High proliferative potential colony forming cell
HPP-CFU	High proliferative potential colony forming unit
HSC	Hematopoietic stem cell
HUVEC	Human umbilical vein endothelial cell
IGF-1	Insulin like growth factor-I
IL-1, 3, etc.	Interleukin-1, 3, etc.
IMDM	Iscove's modified Dulbecco's medium
INFγ	Interferon γ
LFA-1	Leukocyte functional antigen-1
LIF	Leukocyte inhibitory factor
LTC-IC	Long term culture initiating cell
LTR	Long term repopulating
M	Monocyte
M-CSF	Macrophage colony stimulating factor
MIPα	Macrophage inhibitory protein
Mk	Megakaryocyte
N	Neutrophil
PB	Peripheral blood
PHPC	Primitive hematopoietic progenitor cell
SCF/SF	Stem cell factor/steel factor
SDF-1	Stroma derived factor-1
STR	Short term repopulating
TGFβ	Transforming growth factor β
TNFα	Tumor necrosis factor α
Tpo	Thrombopoietin
VLA-4/5	Very late antigen-4/5

1
Introduction

Bone marrow (BM) is the principal site for blood cell development in humans. The production of mature blood cells is a continual process that is the result of proliferation and differentiation of stem cells, oligopotent progenitor cells and mature cells (Fig. 1). A single stem cell has been proposed to be capable of more than 50 generations (doublings) and has the capacity to generate 10^{15} cells to support up to 60 years of life [1]. The identification, cloning and production of recombinant hematopoietic growth factors/cytokines together with the identification and purification of hematopoietic stem and progenitor cells has enriched our understanding of hematopoiesis. As a result of these fundamental discoveries, many investigators are performing ex vivo manipulation of hematopoietic stem cells (HSC) for potential therapeutic applications, from myeloablative situation to hematological disorders, and from gene therapy to regeneration of specialized tissues.

Dexter et al. [2] described the first in vitro murine BM stem cell culture system. Later, Moore and Sheridan [3] and Gartner and Kaplan [4] adopted the Dexter culture system for human BM culture. During the past two decades several instances of outstanding progress have taken place on ex vivo culture of

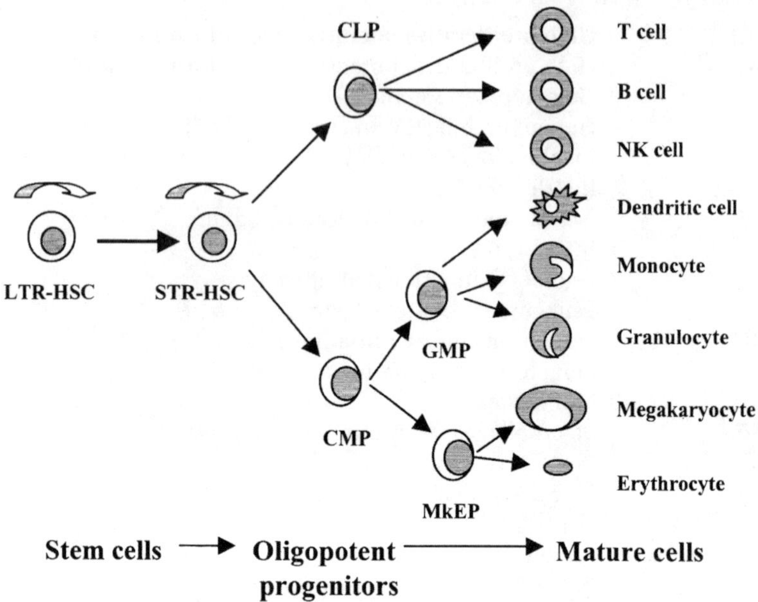

Fig. 1 Hematopoietic system hierarchy. Dividing pluripotent stem cells (LTR and STR) may undergo self-renewal to form daughter cells without loss of potential, and also experienced a concomitant differentiation to form oligopotent daughter cells (CLP, CMP, and CMP lead to GMP and MkEP) with more restricted potential. Continuous proliferation and differentiation results in the production of many mature cells. Macrophages are obtained from monocytes and similarly platelets are derived from megakaryocytes (not shown in the figure) (adopted from [18])

HSC. It seems likely that, after human skin and cartilage, BM will be the next tissue that will be reconstituted to the point of clinical efficacy. Despite astounding advances in culture of HSC and a fair degree of success in clinical studies [5, 6], ex vivo expanded hematopoietic cells still remain as a potential source of stem cells for transplantation. The main reason for this is a curious lack of consensus among the investigators that ex vivo expanded HSC could be used in clinical applications. Few investigating groups believed that HSC expand in vitro at the cost of its engraftment potential, so there is no reason why the expanded cells should be used in clinical applications. However, many investigators, by pre-clinical and clinical studies, demonstrated that ex vivo expanded HSC are well tolerated by the patients; it has marrow repopulating ability and can be differentiated into each type of blood cells.

Several reviews in the recent past have covered various aspects of the ex vivo expansion of HSC [7–12] and its clinical implications [13, 14], but none of these reports addressed the basic question – why ex vivo expanded HSC has not yet reached clinics for regular therapeutic usage. In this review we will discuss the technological breakthrough for culture of HSC, biological alterations, and the clinical applications of the expanded cells.

2
Hematopoiesis

Hematopoiesis, the regulated production of eight different blood cells derived from a common small population of pluripotent stem cells, occurs mainly in the bone marrow of adult mammals. An average human produce nearly 400 billion mature blood cells per day to maintain homeostasis [7]. During embryonic development, the site of hematopoiesis is in the yolk sac, in the second trimester it is shifted to the liver and spleen, and in the third trimester hematopoiesis is abundant in the central and peripheral skeleton [15]. In adulthood the main sites of hematopoiesis are in vertebral bodies, sternum, ribs, and pelvis [15].

2.1
Hematopoietic System: Cells and its Lineages

Studies of Wu et al. [16, 17] lead scientists to believe that during hematopoiesis a small population of primitive, self-renewing cells divide to generate progeny cells, each progressively more restricted in developmental potential, until mature effector cells are formed. On the basis of differentiation status, hematopoiesis is divided into three broad stages [18], as shown in Fig. 1. Accordingly, cells are characterized phenotypically and grouped into three distinct subpopulations, as stem cells, oligopotent progenitor cells, and mature cells. Hematopoietic stem cells (HSC) are divided into two groups on the basis of duration of support for blood cell formation. Long-term repopulating (LTR) cells are pluripotent in nature giving rise to cells of both lymphoid and myeloid lineage throughout the life [19, 20], whereas short-term repopulating (STR) cells provide support for several weeks [21, 22]. Further to the downstream are lineage-restricted oligopotent progenitor cells capable of generating both lymphoid and myeloid progeny. These progenitor cells are therefore capable of undergoing proliferation, differentiation, which develop into only one or the several of the mature cell types. Common lymphoid progenitors (CLP) are the precursors of T, B, and NK cells [23], whereas common myeloid progenitors (CMP) gives rise to granulocyte (G), monocyte (M), erythrocyte (E), and megakaryocyte (Mk) precursors [24].

CMP cells are also designated by the term colony-forming unit (CFU) because of their ability to form colonies of mature cells in semisolid agar or methylcellulose cultures. As these progenitor cells are the precursors of granulocyte, monocyte, erythrocyte, and megakaryocyte cells, they are also termed CFU-GEMM or CFU-Mix. In semisolid medium four distinct colonies of these precursor cells can be easily identified. In adult, these myeloid progenitor cells are located in the bone marrow (BM) with small populations in the spleen and in the circulation; therefore BM is considered the major site of myeloid blood cell production.

Under normal conditions, the vast majorities of stem cells are in the G_0/G_1 state (quiescence), and believed to repair damaged DNA and maintain genetic integrity [25]. To fulfill the continuous demand for mature cells, progenitor cells must in turn be continuously generated from LTR-HSC, which have the capacity to persist throughout the life span of the individual.

2.2
BM Microenvironment

BM is the principal site for hematopoiesis; it is a connective tissue located in the medullary cavities of bones. Functionally, BM plays two major roles, as the hematopoietic tissue and as a major reticuloendothelial organ [26, 27]. In mammals, hematopoiesis occurs in the extravascular spaces between marrow sinuses, in a specialized area, known as niche. The stem cells niche is yet to be described in the ultrastructure level, but it is believed that niches provide microenvironment to meet certain functional requirements of hematopoiesis. The microenvironment is characterized by the local geometry and cellular arrangement, by stroma cells, by the chemical nature of the microenvironment in terms of extracellular matrix (ECM) components and the growth factors (cytokines). The stroma cells include endothelial cell that line the marrow sinuses, adventitial reticular cells, macrophages, and adipocytes. The BM stroma cells synthesize growth and differentiating factors, and ECM components. The stroma cells and the ECMs form three-dimensional scaffolding upon which the HSC lodge. The stroma cells through their intimate physical contacts with the hematopoietic cells, through the ECM components and the growth factors they secrete create an intricate hematopoietic inductive microenvironment, which promotes and regulates hematopoiesis [28]. It is this type of inductive microenvironment is termed hematopoietic niche.

Endothelial cells have been observed to produce constitutively G-CSF and GM-CSF [29]. They are also capable of transporting plasma derived substances through their cytoplasm into the hematopoietic space. Adventitial reticular cells are fibroblastoid cells, that ectopically synthesize a number of growth factors, e.g., G-CSF, M-CSF, GM-CSF, IL-1, IL-6, SCF, FltL, LIF [30, 31, 32, 33]. Adventitial reticular cells are capable of taking up fat in the presence of corticosteroids and converted into adipocyte [34]. In young age marrow is red in color and cellular, with the age hematopoietic activity is reduced and the marrow is composed of large proportion of fat cells (yellow marrow) [35]. Macrophages are critical in controlling erythropoiesis, which proceeds in isolated islets surrounding an associated macrophage nurse cell in the BM [36]. Macrophages are considered to be a major source of cytokines under normal conditions; these are, GM-CSF, IL-1 Epo [37]. They also produce hematopoiesis inhibitory cytokine like, TNFα and IFNγ [36, 37].

In BM, ECM components are mainly produced by endothelial and adventitial reticular cells. The ECM of BM consists of collagen types I, II, IV [38], laminin, fibronectin [39], vitronectin [40], thrombospondin [41], hemonectin [42], and proteoglycans with glucosaminoglycan side chains, such as heparan sulfate and hyaluronic acid [43, 44]. ECMs provide support and cohesiveness for the marrow structure. There is a growing body of evidence indicating that ECM is important for the regulation of hematopoiesis. Glucosaminoglycan side chains of heparan sulfate can sequester and compartmentalize certain cytokines in local areas and present them to HSC [45]. This is the mechanism for presenting high, localized concentrations of growth factors that are protected from proteolysis. Another important function of ECMs is to provide anchorage for immature hematopoietic cells.

3
Clinical Applications of Hematopoietic Stem Cells

Bone Marrow Transplantation (BMT) is a life saving procedure for a number of malignant and non-malignant life threatening diseases [46]. Ever since the first successful BMT in the late 1960s, an increasing number of BMT and related procedures are being performed worldwide. Allogeneic stem cell (unrelated human leucocyte antigen, HLA identical donor) transplantation is now a curative treatment modality for a number of disorders, including malignant diseases, dyserythropoiesis, bone marrow aplasia, immunodeficiency states, and a number of inherited disorders [47].

The indications for hematopoietic stem cell transplantation [47] can be conveniently divided into two groups: (a) Malignant disorders: like leukemias, lymphomas, multiple myeloma and solid tumors like breast cancer, testicular cancer. In all these indications, the cure or palliation is by the high doses of chemotherapy or radiation therapy, while the transplant serves to rescue the patient from the myelotoxic effects of the anti-cancer therapy. In allogeneic type of transplants, there is an additional immunological advantage of graft vs cancer effect, which contributes to the disease relief; (b) Non-malignant diseases: like aplastic anemia, thalassemia, Gaucher's disease, etc. In these conditions the abnormal marrow is deliberately destroyed and replaced by the healthy donor marrow. In this setting autologous (patient's own stem cells) transplantation cannot be effective for obvious reasons.

Bone marrow and stem cell transplantation is curative in many potentially fatal conditions. Nevertheless, for those who survive for two years after an allogeneic transplant, the five-year survival is 89% [48]. Although BMT does not confer a normal life span, the slightly increased risk of death with the passage of years may not be so bad when one considers the alternative [49]. The graft vs tumor effect seen after allogeneic transplants for human malignancies represents the clearest example of the power of the human immune system to eradicate cancer [50, 51]. It is likely that in future the high dose preparative regimens used in allogeneic bone marrow and stem cell transplants will be replaced with less toxic therapy leading to a safer transplant procedure. Autologous transplants are mainly done for lymphomas, solid malignancies like neuroblastoma and germinal cancers, hematological malignancies like multiple myeloma [52], and in patients with acute leukemia who do not have an HLA identical donor. The advantage of autologous transplant over allogeneic transplant is that there is no graft vs host disease (GVHD), and once engraftment occurs then graft rejection is unlikely. Thus there is a significant decrease in the complication rate as compared to the allogenic transplantation; however risk of tumor relapse is higher as compared to the allogeneic transplantation.

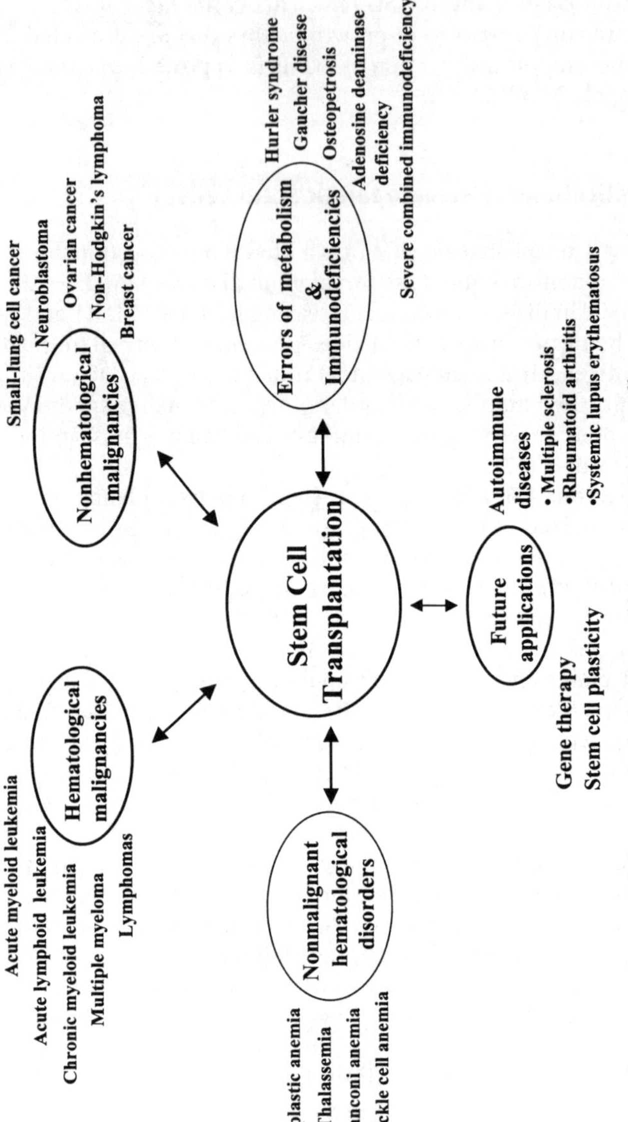

Fig. 2 Current and potential therapeutic applications of hematopoietic stem cells

Various current and future applications of stem cell transplantations are shown in Fig. 2.

4
Sources of Stem Cells and Transplantation Requirements

HSC from four different sources are being used in stem cell transplantation; these are bone marrow (BM), peripheral blood (PB), umbilical cord blood (CB), and fetal liver (FL). The first three sources of stem cells could be from patients (autologous), or from matched donors (allogeneic). Till now, BM and PB stem cells transplantations are most popular. The FL stem cells transplantation has been performed on a limited number of patients with aplastic anemia or acute leukemia, but only transient engraftment has been demonstrated [53]. Various stem cell sources differ in their reconstitutive and immunogenic characteristics, which are based on the proportion of early pluripotent and self-renewing stem cells to lineage-committed late progenitor cells and on the number and characteristics of accompanying "accessory cells" contained in stem cell allografts [54].

BM is collected under anesthesia through repeated puncture as aspiration from the iliac crest (pelvic bone) and some time from the sternum (chest bone). In a typical BM transplant for an adult, 1–2 l of marrow is aspirated for $10–20\times10^9$ mononucleated cells, out of which $CD34^+$ (surface antigen present in the primitive and progenitor stem cells) cell content is about 1% [55]. Due to low mortality rate (<5%), autologous BM transplantation is more popular. BM transplantation from allogeneic HLA-identical requires intensive immunosuppressive therapy to prevent graft rejection. Depletion of T cells from the donor BM significantly reduces the incidence of acute GVHD. Autologous transplantation using purified $CD34^+$ cells from BM in patients with breast cancer or neuroblastoma demonstrated rapid and complete hematopoietic reconstitution without graft failure and with normal immune function [56]. In the case of aplastic anemia, large BM cell doses ($>3\times10^8$ per kg body weight) enhanced engraftment and survival [57]. In autologous bone marrow transplantation, the marrow must be free of disease at the time of harvesting, and it cannot be done for conditions where the marrow is primarily affected like in aplastic anemia or thalassemia.

In 1981, peripheral blood as a source of stem cell for transplantation was introduced by Korbling et al. [58]. Since then many groups performed PB stem cell transplantation, and now it is the preferred source of stem cells. The percentage of $CD34^+$ cells among circulating total nucleated cells at steady state in healthy donors is about 0.06%; it is approximately 1.0% in the BM. With cytokine treatment, there is a temporary shifting of hematopoietic progenitor cells from bone marrow into the circulating blood. When healthy donors are treated with recombinant human granulocyte colony-stimulating factor (G-CSF; 12 µg per kg body weight per day) over 3 days, with another G-CSF dose given on the fourth day before the stem cell apheresis, a process that separates the nucleated cells from red blood cells and platelets. The mean PB $CD34^+$ cells density in the blood increases from 3.8×10^9 cells l^{-1} to 61.9×10^9 cells l^{-1}, a 16.3-fold increase over the baseline. The increase in early $CD34^+CD38^-$ progenitor cells is, at 23.2-fold, even greater [59]. Patients with solid tumors, high dose chemotherapy with cyclophosphamide and GM-CSF or G-CSF are used to mobilize progenitor stem cells ($CD34^+$) in the PB [60]. Good mobilization of PB stem cells means more than 2% $CD34^+$ expressing cell or 50 CFU-GM per10^5 cells that is associated with rapid hematopoietic recovery of neutrophils (>500 per µl) and platelets (>50,000 per µl) within a median of 15 days in myeloablative patients [61]. For consistent engraftment, the minimum numbers of nucleated cells are found to be between 3 and 6×10^8 cells per kg body weight [62]. It has been believed that a minimum of 2×10^6 CD34$^+$ cells per kg are necessary to achieve rapid, consistence, and sustained engraftment [62].

BM and PB allografts either are transfused fresh into the recipient or are cryopreserved in liquid nitrogen or electric deep freezers before transfusion. The latter approach has the advantage of being performed independently of the transplantation procedure, though cryopreservation and thawing can reduce the CD34$^+$ cell number in the transfusate by as much as 10–20%, mainly because of cell trapping during the thawing process [59].

Umbilical CB stem cells are an exciting alternative to BM and PB stem cells for transplantation. This is due to abundantly available and risk free collection, low chances of contamination with CMV and Epstein-Barr virus, transplantation associated with low incidence of acute and chronic graft vs host disease compared to adult bone marrow allografts; cord blood banks can provide HLA matched stem cells on demand. Cord blood transplants were initially done utilizing placental blood collected from a sibling, though delays in myeloid engraftment were noted, the probability of event-free survival was about 70%. The cord blood is enriched in stem and progenitor cells (CD34$^+$ cells ~0.8%); however a small number of cells collected from single cord limits its current use [63]. The total number of CD34$^+$ cells obtained in an optimally collected cord blood donation is about 10×10^6 [64]. Thus the dose of nucleated cell in CB transplantation is limited, and the number of cells infused in recipients is about 1 log less than in BM and PB transplantations. There is concern about engraftment on account of this lower dose. The critical factors for engraftment are the number of cells infused and HLA compatibility; for HLA identical transplants lower number of cells is sufficient for engraftment, which is reversed in case of HLA mismatches [65]. In general, the time taken for engraftment is much longer in CB transplantation when compared with BM or PB transplantation. The median time taken to reach an absolute neutrophil count of 500 per µl has been 30 days and 56 days for platelet count to reach to 20,000 per µl [66]. The major factor in engraftment is considered to be the number of nucleated cord blood cells infused per kilogram of the recipient's body weight. An analysis of 133 cord blood transplants listed in the Eurocord Registry also revealed myeloid engraftment failures occurring in 47% of patients over 15 years of age who had received less than 3.7×10^7 nucleated cells per kg body weight [67]. There are different recommendations regarding the minimum nucleated cell dose for transplant, which ranging from 1 to 4×10^7 per kg body weight [65, 68]. The dose of cord blood to provide durable engraftment in an adult has not been firmly established; however the acceptable minimum nucleated cord blood cell dose is 1×10^7 per kg body weight [69].

The unique properties of fetal cells, including their considerable capacities of proliferation and differentiation as well as their ability to become tolerant to host antigens [70], and to develop normally in a foreign host, have prompted investigations for using FL stem cells for transplantation. In the case of patients with inborn errors of metabolism, a beneficial effect has been noted in FL transplantation, as both stem cells and prehepatocytes got into the host [71]. The ages of fetal donors are ranged from 8 to 22 weeks postfertilization in transplantation to patients with inborn errors of metabolism. The FL is gently disrupted, and single cell suspension is prepared in RPMI-1640 medium [72]. The average number of viable nucleated cells recovered from a single fetal liver is 8×10^8, but it varies with the age of the fetus [72].

5
Ex Vivo Expansion of HSC

5.1
Why Ex-Vivo Expansion is Needed

The reconstitution of a functional hematopoiesis has been long desired as HSC has several advantages and hence need to be cultivated ex vivo. A functional ex vivo human hematopoietic system could be served as an analytical model to study basic biology of hematopoiesis and to examine causes of many hematological disorders. Furthermore, such a test system would provide alternative to animal studies, with the added benefit of generating data from human cells that may be more relevant. Above all, a large numbers of transplantable cells could be cultivated ex vivo, and offers tremendous advantage over the traditional sources of stem cells as mentioned below:

1. Currently, BM harvest is a lengthy inpatient procedure, requiring upto 16 h to collect 1000 ml aspirant using multiple needle sticks. PB collection is an out patient procedure requires several visits with a total duration of 30–40 h and requires several needle sticks. A limited trials consisting of 19-patient using ex vivo-expanded BM cells derived from small aspirant of approximately 40 ml of marrow showed significant engraftment [73]. Thus, ex vivo culture procedure reduces the need for large-scale harvesting of marrow or multiple leukaphereses, and offers a less expensive alternate.
2. For optimum engraftment, current transplantation protocol requires at least 2.5×10^6 CD34$^+$ cells per kg of patient body weight [74]. CD34$^+$ cells alone are not adequate for long-term grafting; the transplant should contain a sufficient number of long term culture initiating cell (LTC-IC). In ex vivo culture, LTC-IC can be expanded several folds [75, 76] to ensure hematopoiesis to support multiple cycles of high-dose chemotherapy.
3. Ex vivo expansion leads to supplementing stem cell graft with more mature precursors to shorten or potentially prevent chemotherapy-induced pancytopenia.
4. Studies have shown that ex vivo culture conditions favor the maintenance of normal cells over leukemic cells. Thus when using patient-derived hematopoietic tissue, ex vivo expansion offers a number of opportunities to reduce the tumor burden of the transplant. Moreover, the expansion process leads to passive purging of some tumors from normal stem cells [77].
5. For adoptive immunotherapy against specific tumor, both antigen-specific cytotoxic T lymphocyte (CTL) and helper T-cells can be expanded in vitro or in vivo with the help of antigen-primed dendritic cells. It has become possible to generate adequate numbers of such dendritic cells by ex vivo expansion of CD34$^+$ cells from as little as 1 ml of BM [78].

6. An optimally collected umbilical cord blood donation generates approximately 10×10^6 CD34$^+$ cells, which is just 10–15% of the optimal dose for an adult. Again, the hematopoietic recovery using cord blood stem cells is approximately 25 days for neutrophils and 43 days for platelets, which are significantly slower than that obtained in BM and PB transplantation [79]. In ex vivo expansion, it is possible to generate sufficient cells of desired combination from a single cord blood to reconstitute an adult following high-dose chemotherapy.

7. The demand of pathogen-free mature blood cells, such as, red blood cells, platelets, and granulocytes are quite large all over the world. Ex vivo manipulation of HSC for lineage specific growth may lead to the revolution in routine clinical use of blood cells.

8. Ex vivo expanded HSC could be used as a delivery vehicle of various genes against potential target diseases.

5.2
Molecular Control of Hematopoiesis

It is generally believed that stroma cells produce many growth factors/cytokines for the proliferation of hematopoietic stem cells in vivo [26] and in vitro [80–83]. In fact, many cytokines and/or their transcripts have been detected in primary stroma cells and in stroma cell lines [84–87]. Growth factors that act on hematopoietic stem cells depend on the hierarchy of the cells in the differentiation tree. In vitro studies have show that more than a dozen of cytokines involved in the process of hematopoiesis, some are primarily responsible for self-renewal of the progenitor cells, some function as differentiating factors (Table 1), and others act as inhibitors. On the basis of the first two biological functions (i.e., self-renewal and differentiation), growth factors are classified into three main groups:

1. Potentiating group: few growth factors/cytokines potentiate activity of other growth factors, such as, SCF (mast-cell growth factor, steel factor, c-kit ligand), FltL, LIF, IL-6, and IL-11. SCF and FltL showed marked activity on hematopoietic stem cells, and are considered as stem cell growth factors. SCF and FltL prevent apoptosis and are primarily involve in replication of primitive progenitor cells [102, 103]. The potentiating effects attributed to FltL include (a) expansion of uncommitted progenitor cells (CD34$^+$CD38$^-$) in presence of SCF and IL-3, (b) an increase of HPP-CFU and CFU-GM number when incubated with IL-3, IL-6 and Epo [104], and (c) formation of CFU-GM from CD34$^-$Lin$^-$ cells in presence of GM-CSF [105]. FltL knock-out studies have shown that both pro-B and pre-B cell numbers drastically reduced in the bone marrow as compared with the control mice [106]. FltL has been shown to synergize with IL-7 in inducing proliferation of B cells [107]. SCF acts synergistically with G-CSF, GM-CSF, IL-3, and Epo to stimulate the growth of human progenitor cells [108].

2. Multilineage group: growth factors essentially act on CFU-GEMM, CFU-GM, BFU-Mk, and BFU-E, such as, IL-3 and GM-CSF. IL-3 also prevents apoptosis of the committed progenitor cells [109].

Table 1 Cytokines used in ex vivo culture of HSC

Source of cells	Cytokines	Major effect	Reference
A. Human			
1. Bone marrow CD34+HLA-DR−c-kit+ cells	CL	Sustain more pre-HPC	[88]
	IL-3	Promotes HPC survival	
	CL+IL-3	More HPC and pre-HPC due to suppression in apoptosis	
2. Bone marrow CD34+CD33− cells	MIPα+IL-3	3.5–7-fold increase in LTC-IC	[89]
3. Bone marrow CD34+ cells	SCF+IL-9	9-fold expansion of CD34+CD33−DR− cells	[90]
4. Bone marrow mononuclear cells	SCF+IL-3+GM-CSF	7.5-fold expansion of LTC-IC	[74]
5. Bone marrow CD34+ cells	SF+G-CSF	Little effect on CFU-GEMM	[91]
	SF+IL-3	Expand CFU-GEMM	
6. Bone marrow CD34+CD38− cells	SCF+IL-3+FltL	45-fold expansion of LTC-IC	[75]
7. Peripheral blood	SCF+IL-3+IL-6+IL-1β +EPO	Preserve LTC-IC and expand progenitor cells	[92]
8. Peripheral blood	SCF+G-CSF+IL-3+FltL	2–5-fold expansion of LTC-IC	[93]
9. Cord blood CD34+ Rh-123^{lo}	SCF+EPO	94-fold expansion of CFC	[94]
10. Cord blood CD34+ cells	MGF+IL-3 +IL-6+EPO	55-fold expansion of BFU-E	
	MGF +IL-6 +FP+G-CSF +M-CSF	70-fold expansion of CFU-C	[95]
11. Cord blood	SLF+PIXY321	10- and 90-fold expansion of CFU-GEMM and CFU-GM respectively	[96]
12. Cord blood CD34+CD38−	SCF+FltL+IL-3+IL-6+GM-CSF	160-fold expansion of LTC-IC	[76]
13. Cord blood CD34+ cells	FltL+TPO	2–5-fold expansion of LTC-IC	[97]
14. Cord blood CD34+ cells	SCF+FltL+TPO		[98]
B. Murine			
15. Bone marrow cells	SF+IL-12	Support development of committed myeloid and multipotent lymphohematopoietic progenitors	[99]
16. 5-FU treated Bone marrow cells	SCF+IL-3+IL-1	Expand HPP-CFC	[100]
17. Bone marrow CD34+ cells	SCF	Stem cells proliferate without differentiation	[101]
	SCF+IL-3	Induce differentiation of stem cells	
	SCF+IL-3 +GM-CSF	Further differentiate and reduces CFU-S	

3. Unilineage group: growth factors act on later hematopoietic progenitors (CFU-E, CFU-Mk, and CFU-Eo), such as Epo, G-CSF, IL-5, M-CSF, and Tpo.

From time to time studies revealed that many hematopoietic growth factors act synergistically to expand LTC-IC and also number of CFC [74–76, 88–101]. It is note-worthy to mention that the function of these growth factors are not restricted to any particular source of hematopoietic stem cells (Table 1).

Amongst the growth factors belong to the unilineage group, Tpo is most important. Tpo, the physiological regulator for platelet production, stimulates megakaryocytopoiesis [110]. Maximum formation of megakaryocyte-committed colonies has been observed in presence of Tpo plus either SCF or IL-3 [111]. Most interestingly, Mpl (Tpo receptor) is expressed predominantly in stem cells (CD34$^+$ cells), other than in megakaryocyte [112]. This finding, together with the functions of SCF and FltL, prompted a more aggressive exploration of the capacity of these three cytokines (together) to sustain and augment the number of primitive progenitor stem cells in ex vivo culture [98, 113, 114].

Many investigators have used IL-1α in stroma-supported culture of hematopoietic stem cells. Whether it has any direct role in hematopoiesis is not clear, but it is revealed that IL-1α stimulates expression of various growth factors, as mentioned above, in HUVEC and in bone marrow-derived mesenchymal stroma cells [115, 116]. LIF is another multi-potent cytokine, constitutively expressed in bone marrow stroma cells to support growth of long-term repopulating stem cells. LIF works indirectly by promoting cytokine expression by the bone marrow stroma cells [117]. The expression of LIF in stroma cells is found to be upregulated by various hematopoiesis modulating cytokines, such as, IL-1α, IL-1β, TGFβ, and TNFα [118]. LIF is believed to have some role in preventing differentiation of hematopoietic stem cells, as it suppresses differentiation of embryonic stem cells [119].

Similarly, there are quite a few cytokines produced by the cells of the immune system that either inhibit proliferation of primitive stem cells, or suppress CFC of the myeloid and erythroid lineages, or induce apoptosis, or restrain cells from differentiation in vitro (Table 2). Jacobsen et al. [127] described TGFβ and TNFα as inhibitors of FltL-induced proliferation of stem cells. TGFβ completely inhibits FltL-dependent colony formation (half maximal inhibition at 0.2 ng ml^{-1}), and TNFα (half maximal inhibition at 2 ng ml^{-1}) inhibits 60–95% [127]. These effects were observed in combinations of FltL and either SCF, IL-3, IL-6, IL-11, IL-12, or G-CSF. In the case of FltL and SCF together, TNFα inhibits colony formation by 80% and shifted the colonies predominantly from granulocytes to macrophages [127]. A member of the chemokine family, MIPα was also found to inhibit differentiation of the most primitive stem cells, at the same time stimulate proliferation of more mature cells [128].

Table 2 Cytokines inhibit hematopoiesis

Source of cells	Cytokines	Major effect	Reference
Human			
1. Bone marrow CD34$^+$ cells	H-ferritin, MIP-1α MIP-2β, PF4, IL-8, TGFβ	Suppressive effects on stem/progenitor cells	[120]
2. Bone marrow CD34$^+$CD38$^-$ cells	IFNγ	Inhibits proliferation, but preserves viability of cells, secondary CFC inhibited	[121]
3. Bone marrow CD34$^+$ cells	IFNγ, TNFα	Inhibit colony formation in methylcellulose culture	[122]
4. Cord blood CD34$^+$ cells	MIP-1α	Inhibits expansion of primitive CD34$^+$, but not in mature myeloid and erythroid progenitors	[123]
	TGFβ, TNFα	Inhibit primitive as well as CD34$^+$ myeloid and erythroid progenitors	
Fetal liver CD34$^+$ cells	MIP1-α, TGFβ, TNFα	Matured erythroid cells are mostly inhibited	[124]
5. Cord blood CD34$^+$ cells	IL-13	Promotes MK colony formation, but inhibits GM and erythroid progenitor cells	
6. Bone marrow CD34$^+$CD33$^+$ cells	IL-13	Macrophage colony formation is significantly suppressed	[125]
Murine			
7. Bone marrow	TNFα	Increases primitive progenitors (CFU-S) cells, but eliminates CFU-C and significantly lower HPP-CFC	[126]

5.3
Ex Vivo Expansion Procedures

5.3.1
Bioreactor/Culture Systems

In 1973, Dexter et al. [2] were first to reconstitute in vivo hematopoietic microenvironment in vitro to culture progenitor stem cells. After one-and-a-half decades, Iscove et al. [129] described a deliberate attempt for 8–12-fold expansion of murine pluripotent stem cells in vitro during four days culture. In two decades of research, it is now evident that expansion of clonogenic progenitor cells is quite feasible in almost any type of ex vivo culture system [88–101, 130–143]. These studies demonstrated that, over a reasonable time period, the number of progenitor cells expand several folds in culture. In some cases, progenitor cell numbers decline after attaining maximum density; whether such decline results from adverse effects of in vitro expansion or is due to natural senescence of more mature progenitor cells is difficult to understand. Different bioreactor configurations that have been used for expansion of human stem cells are shown in Table 3. In a recent review, Noll et al. [144] described biochemical engineering aspects for cultivation of hematopoietic stem cells. How stem cell cultivation concept should be developed matching with the application, and various culture parameters that influence different cell types, were described in this review [144]. Here we have mainly discussed culture of stem cells in presence or in absence of stroma cells, in stagnant or stirred medium, in serum-replete or serum-free medium, and in presence of different combinations of hematopoietic growth factors. The ex vivo cultures were initiated either with purified stem cells (CD34$^+$), or mononuclear cells obtained from bone marrow, peripheral blood, and cord blood. It may be noted that for HSC expansion, flat-bed grooved [135, 136] and Aastrom cell production system [71, 140, 141] were developed, on the basis of the same principle. These bioreactors have few limitations: (a) difficulty in obtaining representative samples from the system, (b) supporting relatively low cell densities as compare to the clinical requirements.

5.3.2
Expansion of HSC in Stroma- and Nonstroma-Based Culture

Primarily ex vivo expansion processes can be divided into two groups, stroma-dependent and stroma-independent. In the traditional cultivation process, culture flasks in direct contact with stroma cells, as described by Dexter et al. [2], were used as stroma cells and were known to produce ECM components and secrete growth factors to support primitive hematopoietic progenitor cell (PHPC). To understand whether stroma cells are directly involved in hematopoiesis in vitro, Verfaillie [145] physically isolated PHPC from the stroma layer by a 0.45-μm microporous membrane and cultured them for eight weeks. By this study, it was demonstrated that direct contact between bone marrow storma and stem cells was not essential for either differentiation or conser-

Table 3 Ex vivo culture system

Nature of cells	Culture system	Results	In vivo studies
Cord blood cells	T-75 culture flask, batch mode of operation [134]	25-fold expansion of promyelocytes 3-fold more progenitor cells than stroma free culture	
Cord blood cells	Flat-bed grooved multi-pass perfusion bioreactor [135]	18- and 5.3-fold expansion of CFU-GM and CFU-GEMM, respectively. LTC by 3-fold	
Bone marrow cells	Flat-bed grooved single-pass perfusion bioreactor [136]	20 to 25-fold expansion of total mononuclear cells 30-fold expansion of CFU-GM	
Bone marrow cells	Microcarrier suspension culture[137]	9-fold expansion of total mononuclear cells 1.8-fold expansion of CFU-GM	
Peripheral and cord blood cells	Stirred tank bioreactor, batch perfusion, cell dilution feeding [138]	200 to 400-fold total cell expansion, 13 to 33-fold expansion of CFU-GM depending on source of cells	
Bone marrow cells	Air-lift bioreactor [137]	12-fold expansion of total mononuclear cells no expansion of CFU-GM	
Peripheral and cord blood CD34+ cells	Spinner flask with batch perfusion [139]	3.4 to 17-fold and 8 to 33-fold expansion of CFU-GM and BFU-E, respectively depending on source of cells	
Cord blood CD34+CD38- cells	Serum free suspension culture [76]	160-fold expansion of LTC-IC	NOD/SCID CRU expansion 2-fold
Cord blood/Bone marrow/ Peripheral cells	Aastrom cell production system-disposable reactor cassettes, fully automated, closed system single pass perfusion [72, 140, 141]	10-fold CD34+ and 100-fold CFU-GM expansion in CB cells	Phase I clinical trials
Peripheral blood/ CD34+ peripheral blood cells	Gas permeable culture bag [142, 143]	66- and 1324-fold expansion of CFU-GM and nucleated cells	Clinical study Conducted

vation of PHPC but was essential for the regulated production of mature blood cells. In other studies Burroughs et al. [146] and Verfaillie et al. [147] demonstrated that various marrow stroma diffusible factors are responsible for maintaining LTC-IC. However, Breems et al. [148] showed that graft quality of mobilized peripheral blood CD34$^+$ cells was preserved when cultured in direct contact with stroma cells. From time to time, many investigators advocated in favor of stroma cells as feeder layer for maintenance of hematopoietic stem cells. Recently, Brandt et al. [149] demonstrated the maintenance of CD34$^+$CD38$^-$ phenotype among proliferating human HPC in three weeks porcine endothelial cell line-supported ex vivo culture, that were multilineage engraftable in human fetal bones implanted in SCID mice (SCID-Hu). Despite these positive effects of stroma cells, it has been shown by various researchers that the function of stroma cells can be partly substituted by either stroma conditioned medium or recombinant growth factors. Using cord blood stem cells (CD34$^+$CD38$^-$), it has been shown that LTC-IC can be expanded to 60-fold in suspension serum-free culture using FltL, SCF, IL-3, IL-6, and G-CSF [76]. Though use of storma cells in ex vivo culture is advantageous for maturation and attainment of functional properties in progenitor cells, stroma may pose several disadvantages in clinical applications: (1) the use of allogeneic source of stroma cells could be potentially harmful due to additional loading of miss-matched tissue, i.e., graft vs host disease (GVHD) and donor-mediated infection, (2) ex vivo expansion of PB/CB stem cells for autologous transplantation requires an additional invasive harvest of patient bone marrow to produce a preformed stroma layer, and (3) due to the adherent nature of progenitor stem cells harvesting of cells may require harsh treatment that could be detrimental to the desired cells. Clinical problem in using stroma-based culture could be solved to a large extent by using endogenous stroma from unprocessed BM (for autologous transplantation). The Aastrom Biosciences bioreactor system currently under evaluation in phase-1 trials is based on endogenous stroma [72]. For cell sources without endogenous stroma, e.g., PB, CB, or CD34$^+$ selected cells, an alternative is to use irradiated stroma cell lines. In the case of CB stem cells, cord vein endothelial cells could be a potential source for stroma.

5.3.3
Operational Parameters in Ex Vivo Expansion

The success of the animal cell culture process primarily depends on the design of the bioreactor and its operation strategy. Various cell lines/strains currently used in traditional animal cell culture processes follow simple growth kinetics and are adopted in vitro. On the other hand, hematopoiesis is a complex process and it is kinetically regulated by numerous biological factors that reside in the BM microenvironment. Being a primary tissue, isolated hematopoietic stem cells gradually lose phenotypes in vitro, if suitable conditions do not prevail during culture. The list of variables that characterize the in vivo state is extensive, and many of these remain unknown. A few variables that are found important in ex vivo culture of hematopoietic stem cells are discussed below.

5.3.3.1
Inoculum Density and its Composition

The most important parameter in stem cell culture is the composition of cells. The purpose of ex vivo expansion is to produce more cells that are engraftable. The outcome of long-term culture depends upon the self-renewal potential of primitive subsets within the cell population. If a starting inoculum is composed of few such primitive progenitors and more committed progenitor cells, the outcome of the culture will be unsatisfactory. The composition of inoculum sometimes controls the type of media to be used. Both PB and CB mononuclear cells differ significantly from BM mononuclear cells in that only the later contains stroma cells. As such, effects of medium conditioning will be quite different in stroma-free cultures. Similarly, CD34$^+$ cell cultures will be different from mononuclear cells as the former is much homogeneous than the later. When using high seeding density or whole cells (bone marrow type), it is important to meet the additional metabolic demands through feeding of medium [150]. The seeding density should be commensurate with surface area of the bioreactor where they are cultured, as the number of cells produced in unit area is fixed. It has been shown that the expansion of total cell number is inversely proportional to the seeding density [136]. A few studies have indicated optimal expansion using low number of enriched population (CD34$^+$) of stem cells [94, 95, 97, 151]. Comparative expansion of CD34$^+$ cells from purified BM, CB, and PB in liquid culture in the presence of cytokines is shown in Fig. 3. The numbers of cells in these cultures increased from 10- to 75-fold and 150- to 250-fold after 10 days and 21 days culture, respectively, depending on the source of CD34$^+$ cells [152]. This study indicated that by 2 weeks of culture, about 100-fold expansion of cells is readily achievable in either of this cell type.

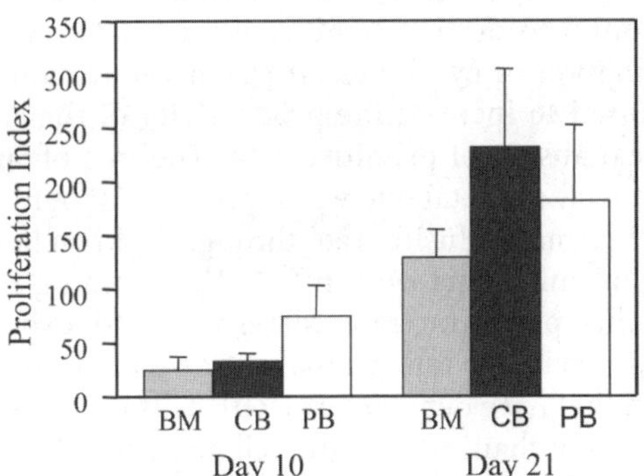

Fig. 3 Proliferation of CD34$^+$ cells from different sources. Purified CD34$^+$ cells derived from BM, CB, and G-CSF mobilized PB were cultured in suspension in presence of SCF, IL-3, GM-CSF, and G-CSF. Samples were drawn at 10 and 21 days of culture, and fold increase in cell numbers were determined by Coulter counter (taken from [152])

5.3.3.2
Media Composition and Perfusion Rate

Standard practice for culture of hematopoietic stem cells is using Iscoves modified Dulbeccos medium (IMDM), or McCoys 5A medium supplemented with 20–30% serum (1:1 mixture of fetal calf and horse) [135, 137, 138]. It is unnecessary to mention that a combination of early acting and lineage specific cytokines were used in the culture medium. The right combination of cytokines and their adequate amounts are essential for a successful ex vivo culture (see Sect. 5.2). Most cytokines are endogenously produced by the stroma cells. If stem cells are cocultured with stroma cells, it is not necessary to add large amounts of exogenous cytokines; instead culture conditions can be manipulated to increase secretion of cytokines by stroma cells [153, 154]. The presence of serum in the culture medium may cause concern for the clinical use of expanded progenitor stem cells. Thus, to facilitate human applications CB and BM stem cells were expanded in absence of serum, but in presence of exogenous cytokines. The expanded cells were successfully engrafted in xenograft models [155, 156]. Serum-free media promote greater expansion of the erythroid and megakaryocytic lineages, as serum contains TGFβ, a potent inhibitor of these cells [157]. In contrast, serum-containing media are superior for expansion of the granulocyte and monocyte lineages [158, 159]. Autologous serum, however, can partly alleviate this shortcoming of serum-free media [158], and further modifications to serum-free medium eventually will solve many problems pertaining to serum [160].

Frequent replacement of medium has been found helpful for self-renewal of hematopoietic stem cells. In static culture, cell concentration is limited due to the build-up of toxic metabolites. By continuous perfusion it is possible to eliminate such toxic substances and to provide essential growth factors to the cells; thus cells can multiply for a longer duration. In fact, primitive LTC-IC (LTR cells) were found to decline significantly in static cultures, even in the presence of combination of cytokines [161]. Therefore, continuous perfusion appears to be required to increase the productivity of the culture in terms of maintenance and expansion of primitive cells. The rate of supply of nutrients and the rate of removal of metabolic waste products should be similar to that occurring in vivo. Plasma perfusion rate through BM has been reported to be approximately 0.1 ml ml^{-1} marrow min^{-1} [162]. Assuming BM cell density of 5×10^8 cells ml^{-1}, this perfusion rate corresponds to using 0.29 ml serum 10^{-6} cells day^{-1}. This exchange rate compares to a daily feeding of 20% serum-containing medium that have densities of 1×10^6 cells ml^{-1}. The media perfusion rate will proportionately changed with the cell densities. The effect of perfusion rates in long-term BM culture has been extensively studies by Palsson and his group [74, 163]. It was shown that a continuously perfused (22.5 ml day^{-1}) 15-ml bioreactor can produce 1.14×10^6 CFU-GM from 30×10^6 BM mononuclear cells in 14 days culture [74]. More importantly, LTC-ICs were expanded to 7.5-fold [74].

5.3.3.3
Dissolved Oxygen Tension

It has been shown earlier that the oxygen tension in BM is between 10 and 50 mmHg [164, 165], which means 5% oxygen saturation in culture is equivalent to oxygen tension in the BM. Several studies conducted over the past 20 years showed that lower oxygen tension significantly enhances both number and size of the progenitor cells [134, 166–168]. Initially the beneficiary effect of lower oxygen tension was thought to be due to lower production of oxygen radicals; further studies however revealed a more complex scenario [169]. Low oxygen tension could stimulate complex intracellular interactions leading to the enhancement of the expression of cytokines by stroma cells. The stimulation of erythropoiesis under low oxygen tension was due to enhanced secretion of Epo by macrophages [170].

5.3.4
How Large Will Be the Clinical-Scale Bioreactor

The size of a clinical-scale bioreactor is decided by the therapeutic dose, and the total requirement for single transfusion. Most conservatively, suppose for a transplantation CB nucleated cells dose of 3×10^7 cells per kg body weight is required, which should be composed of 100 CFU-GM per 10^5 cells and 1×10^6 CD34$^+$ cells per 3×10^7 nucleated cells [63, 67, 68]. Therefore, for a patient of 70 kg body weight, total numbers of cells required will be 2.1×10^9, 2×10^6, and 70×10^6 for nucleated, CFU-GM, and CD34$^+$ cells, respectively. If a nucleated cell density of 3×10^6 cells ml^{-1} is obtained in a suspension culture, the culture volume will be 700 ml, which means total volume of the bioreactor is 1000 ml (assuming 70% working volume). This bench-top bioreactor could be easily obtained as per FDA specifications. The above calculation is based on the assumptions that in cord blood, under most optimum conditions, nucleated cells expands 20- to 30-fold, CD34$^+$ cells 7- to 10-fold, and CFU-GM cells 30- to 50-fold [94–96, 135].

For adhered BM cell culture, 5×10^5 cells could be recovered per square centimeter [7]. With a typical nucleated cell dose of 3×10^8 cells per kg body weight [57], for a 70-kg patient the total culture area of the bioreactor will be 42,000 square centimeters for 210×10^8 nucleated cells. For this number of cells, the number of T-150 flasks needed is 280, and the number of 300 square centimeters gas-permeable culture bags needed is 140. This large surface area is not difficult to accommodate in special bioreactor systems with high specific surface area. It has been shown earlier that a 15-ml continuous perfusion bioreactor produces 1.14×10^6 CFU-GM from 3×10^7 nucleated BM cells in 14 days culture [74]. The total cell numbers and CFU-GM were increased 10-fold and 21-fold, respectively [74]. Assuming a linear scale-up ratio of 1:70, a 1050-ml bioreactor should be sufficient to generate about 210×10^8 nucleated cells having 7.5×10^7 CFU-GM.

5.4
Biological Alteration of Ex Vivo Expanded HSC

There is ample evidence to believe that ex vivo expansion causes substantial changes in the biology of stem cells, which may lead to the changes in BM repopulating ability and long-term hematopoiesis in host [171]. Two major relevant issues to be discussed in this section are potential for self-renewal of stem cells and changes in the expression of adhesion molecules.

5.4.1
Self-Renewal vs Differentiation of Stem Cells

There are three modes of cell division (Fig. 4A) by which an animal cell may give rise to its clone of descendants: (a) *proliferation mode* – under which a cell of type X divides symmetrically to produce two equal daughter cells of type X, both of which also divide symmetrically; (b) *stem cell mode* – cell of type X divides asymmetrically to give rise to two unequal daughter cells, of which one is the mother type X and the other is of another type Y (differentiated); the daughter cell X further divides asymmetrically; (c) *diversifying mode* – where a cell of type X divides into two unequal daughters of type Y and Z; both are differentiated and neither ever gives rise to maternal cells of type X. In general, more than one of these modes of cell division usually occurred in any lineage of differentiated cells [172]. In hematopoietic system, cells are divided following the last two modes of division; thus progenitor stem cells never run-off; at the same time terminally differentiated cells are produced. There is some concern in ex vivo hematopoietic stem cell expansion systems that the source of progenitor stem cells is exhausted in long-term culture. If true self-renewal of stem cells is achievable, it would be possible to generate increased numbers of stem cells that themselves could be further renewed following different modes of cell expansion, as shown in Fig. 4A. On the other hand, it has been proposed that within the stem cell population the proliferation potential of stem cell is inversely related to its generation-age, while its cycling rate is directly related to its generation-age [173–175]. In other words, with increasing generation number (generation-age) the population doubling time of stem cells is reduced at the cost of its proliferation potential. In addition to the generation-age hypothesis, in vivo studies also showed that the length of the cell cycle of murine HSC is about 40 days. With the same cells or comparable human cells when cultured in vitro, the length of cell cycle is surprisingly reduced [176, 177]. It is perhaps this sudden acceleration in cell cycling rate that presumably modulates the outcome of HSC division in culture. In vivo self-renewal and differentiation of stem cells are tightly regulated by multiple feedback and reciprocal intracellular interactions that facilitate in response to variable physiological needs, as for example more production of blood cells after infection or injury. Such a unique complex process as hematopoiesis is governed by various intrinsic regulators include the growth factors responsible for setting up asymmetric cell divisions, nuclear factors controlling gene expression and chromosomal modifications in stem and nonstem daughters, and clocks that may set the number of

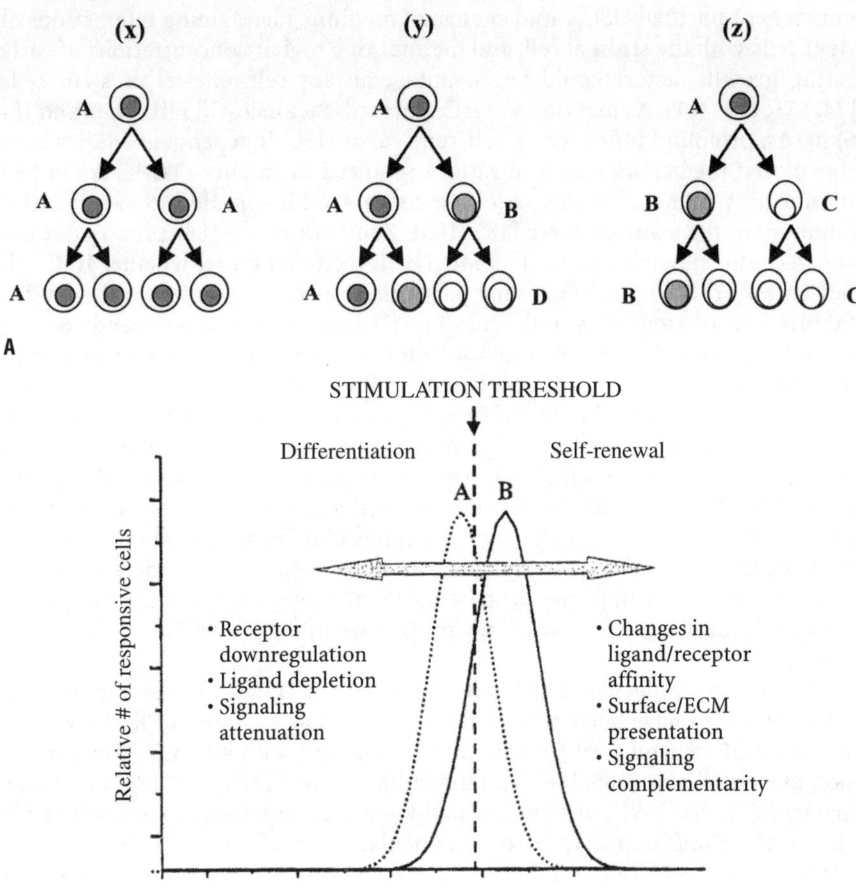

Fig. 4 **A** Three principal modes of cell division. (x) proliferation mode-cell divides in symmetry, i.e., mother cells (A) give rise to same daughter cells (A) ; (y) stem cell mode-cell divides in asymmetry, i.e., mother cells (A) give rise to one daughter (A) and another daughter (B) differing from parental cells; (z) diversifying mode-mother cells (A) never produces daughter cells of same type. B, C, and D are differentiated cells. **B** A ligand-receptor signaling threshold model for control of stem cell differentiation. Examples of different mechanisms by which proliferating cells can move from a net loss in the number of undifferentiated cells (scenario A), to a net gain of undifferentiated cells (scenario B), as well as the reverse of this process (taken from [184])

rounds of division within the transit amplifying population [178]. On the other hand, ex vivo hematopoiesis is a unidirectional process, the kinetics of which is regulated by merely a set of growth factors.

Identification of ex vivo conditions that will support a similar degree of HSC self-renewal activity as compared to in vivo has proven to be a major challenge in large-scale culture of HSC for transplantation. Some degree of success has already been made by many investigators, who have shown that culture with low numbers of purified HSC, rapid change of medium, maintaining interactions of stem cells with the stromal cell, and maintaining higher concentrations of early acting growth factors could be advantageous for self-renewal of stem cells [74, 163, 179–181]. As mentioned earlier, growth factors (SCF, FltL, IL-3, and IL-6) have a profound influence on self-renewal of HSC. Independent studies have shown that the cytokine concentrations required to amplify LTC-IC are in fact significantly higher than the cytokine concentrations needed to expand CFC numbers in the same cultures [182, 183]. Zandstra et al. [184] proposed a ligand-receptor signaling threshold model believed to modulate whether HSC will self-renew or differentiate into any particular lineage. According to the model, the fraction of stem cells undergoing a self-renewal division depends on the numbers of signaling ligand-receptor complexes as compared to a given threshold level. If a cell is expressing sufficiently high numbers of the relevant receptors and is also exposed to sufficiently high concentrations of cognate ligands, the numbers of ligand-receptor signaling complexes are expected to be above a certain threshold levels which improve the probability of self-renewal of progenitor stem cells (Fig. 4B). On the other hand, if a cell expresses few receptors, or its receptors are downregulated by some molecular changes inside the cell, or if the concentrations of cognate ligands are lower, so that the numbers of ligand-receptor signaling complexes drop below the threshold levels, the consequence are a lowering HSC self-renewal and an increase in the probability of differentiation.

Few specific molecules that influence the fate of HSC (either self-renewal or differentiation) have been recently identified. It has been found that the decision for HSC self-renewal appears to be associated with intrinsic over expression of HoxB4 gene [185, 186], and the induction of Notch [187, 188] and Sonic hedgehog (Shh) [189] cell signaling pathways. It is imperative to investigate the effects of various extrinsic factors, particularly cytokines on the induction of above signaling pathways and gene expression in ex vivo culture of progenitor stem cells.

5.4.2
Changes in the Expression of Adhesion Molecules

The ability of HSC to reconstitute normal BM following transplantation into surrogate recipients relies on the potential of these cells to home and to anchor within the BM microenvironment. Homing is an intricate process, through which HSC migrate to the BM microenvironment and presumably anchor to the stroma cells and to the ECM components through the interactions of adhesion molecules expressed on the HSC and the cognate ligands on the BM microenvironment. Several insights regarding the molecular pathways for homing processes have been revealed in the literature. Using blocking antibodies to specific adhesion molecules, or its deficient animals (knock-out) it has been shown that not a single combination of adhesion molecule and its cognate ligand is exclusive for reconstitution of bone marrow with donor cells. Table 4 shows various complementary molecules for the homing processes of the progenitor stem cells. In the recent past, Whetton and Graham [201] reviewed potential

Table 4 Various adhesion molecules expressed in the hematopoietic stem and progenitor cells, and their cognate ligands expressed/produced by the stroma cells

Adhesion molecules expressed on HSC		Phenotypes of stem cells	Cognate ligands
Name	Association		
CD11a (LFA-1)	CD18	Human CD34$^+$CD38$^-$, Murine BM [190]	CD54 (ICAM-1), CD102 (ICAM-2)
CD18 (integrin-β2)	CD11a,b,c	Human CD34$^+$CD38$^-$	
CD29 (integrin-β1)	CD49d,e,f	Murine Sca-1$^+$c-kit$^+$Lin$^-$	Fibronectin [191]
CD43 (leukosialin)		Human CD34$^+$ Murine Sca-1$^+$Lin$^-$ [193]	CD54 (ICAM-1) [192]
CD44 (Pgp-1)		Human CD34$^+$	CD62E (E-selectin) [194], Glycosaminoglycan hyaluronate [195], Collagen
CD49d (VLA-4 α)	CD29	Human CD34$^+$, Murine HSC	CD106 (VCAM-1) [192] CD106 (VCAM-1) [196, 197], CD62P/CD62E (P/E-selectin) [196]
CD49d (VLA-4 α)	CD29	Murine HSC	Fibronectin [198]
CD49e (VLA-5 α)	CD29	Human CD34$^+$, Murine Sca-1$^+$Lin$^-$, Murine BM [190]	Fibronectin [199], Laminin
CD49f (VLA-6 α)	CD29		Laminin
CD62L (LAM-1)		Human CD34$^+$CD38$^-$	TSP
CXCR-4		Murine HSC	SDF-1 [200]
CD117 (c-kit)		Murine HSC	SCF

adhesion molecules on HSC that are responsible for homing and mobilization of donor cells in the stem cell niches.

It is crucial that some of these adhesion molecules should be adequately expressed on the donor cells for proper grafting to the recipient BM. However, it is becoming clear from the ex vivo expansion of CD34$^+$ cells using cytokines that the expression of some of these adhesion molecules are extensively changed [202, 203]. Such changes were not restricted to the expression of adhesion molecules only, but their functions were also modified, as for example the affinities of CD49d (VLA-4) and CD49e (VLA-5) to their cognate ligands were reported to be upregulated by treatment with cytokines [204]. However, study indicated a lower engraftment potential of ex vivo expanded murine HSC in irradiated host [205, 206]. Some investigators correlated the expression of adhesion molecules on human CD34$^+$ and on murine progenitor cells with the status of the cell cycle [207–209]. In vivo engraftable stem cells derived from fresh bone marrow are mitotically quiescent [210]. Upon ex vivo culture with cytokines, a part of these cells exit from G$_0$ phase and loses BM repopulating potential [210]. On the basis of the above results and those obtained by Becker et al. [211], these investigators concluded that a casual relationship exists between loss of engraftment potential and alterations in the expression and function of adhesion molecules on cultured progenitor stem cells. Whether ex vivo expansion of progenitor stem cells leads to the downregulation of adhesion molecules that associates with impaired engraftment has been examined recently. Two independent studies have shown that ex vivo expansion of murine bone marrow stem cells resulted decline in the expression of CD29 (integrin-β1) [191, 212]. Incidentally, CD29 is necessary for homing mediated by $\alpha 4\beta 1$ (VLA-4) and several other integrins [201, 213, 214]. Thus, it is evident from the foregoing studies that the expression profiles of various adhesion molecules in the ex vivo expanded cells should be available to predict the potential for grafting.

Contrary to the above results, a decade ago it was demonstrated in ex vivo culture that IL-3 upregulates CD11a (LFA-1) expression on CD34[+] cord blood stem cells [215]. A decade later, Dravid and Rao [98] have shown that SCF, FltL, and Tpo-induced expansion of CD34[+] cord blood stem cells is not experienced with the downregulation of VLA-4 and LFA-1. The most important finding in recent years was SCF and IL-6 mediated transplantation of human stem cells in xenograft model. These cytokines were found to increase surface expression of CXCR-4 on primitive human stem cells (CD34[+]CD38[-]), thereby facilitating its migration toward SDF-1 for in vivo homing and repopulation [216]. Another interesting observation that appeared a few years back was that human CD34[+]Thy-1[+] cells expanded 150-fold in stroma-dependent culture in the presence of LIF, SCF, IL-3, IL-6, and GM-CSF without losing its phenotype. These ex vivo expanded CD34[+]Thy-1[+] cells gave rise to multilineage differentiation, including myeloid, T, and B cells, when transplanted into SCID-hu mice [217]. These conflicting results on ex vivo expanded HSC vis-à-vis expression of adhesion molecules and in vivo grafting of expanded cells thus warrant further investigation on the effect of cellular proliferation on grafting of progenitor stem cells in vivo.

5.5
Characterization of Engraftable Stem Cells

Characterization refers to the assessment of hematopoietic stem cells, whether it is directly derived from the donors or expanded ex vivo, that are graftable to the hosts (patients) to support short- and long-term requirements. Grafting primarily depends on the immunological competence (HLA-matching) between donor and the host tissues. Since this is beyond the scope of the review it will not be discussed; however, other different parameters that govern a successful grafting will be mentioned briefly. Many in vitro assays that are used to characterize stem cells to be grafted and produce different blood cells are shown in Fig. 5. All these tools, as shown in Fig. 5 may not be used in clinical practice, but are important for research and developmental purpose.

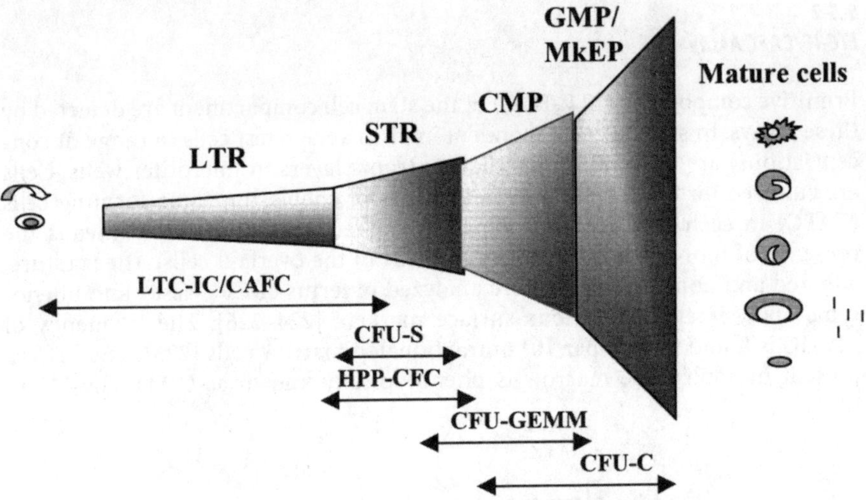

Fig. 5 Functional assays for hematopoietic cells. Schematic representation of the different in vitro assays employed in various stages of hematopoietic cells development (stem cells to mature cells)

5.5.1
Phenotypic Analysis of HSC

The most popular assay for HSC is its phenotypic analysis. Using flow cytometry analysis it has been revealed that most active murine and human LTR cells are CD34[-] [218, 219]. However, for all practical reasons CD34[+] is considered the most popular surface marker of hematopoietic stem cells. Different characteristics of surface markers on stem cells that are engraftable for long-term are shown in Table 5. It is important to be noted that stem cells population is not described based on one particular surface marker. As for example, in case of murine bone marrow stem cells, the marker Sca-1 is not unique, as it is also expressed by some activated immune cells. At the same time, c-kit is a transmembrane tyrosine kinase receptor for SCF/SF and is considered to play an important role in the early stage of hematopoiesis [223]. Therefore, murine stem cells are confidently characterized as Sca-1[+]c-kit[+] cells.

Table 5 Phenotypic characteristic of HSC and their marrow repopulating ability

Source	Repopulating ability	Surface marker
Murine BM	LTR	Sca-1[+]c-kit[+]CD38[+]CD34[-]Lin[-] [220]
Human fetal BM	LTR	CD34[+]Thy-1[+]Lin[-] [221]
Human CB	LTR	CD34[+]CD38[-] [222]

5.5.2
LTC-IC/CAFC Assay

Primitive components (LTR-HSC) of the stem cell compartment are detected by these assays. In stroma cell dependent in vitro assays, test cells (a range of concentrations) are overlaid on irradiated stroma layers in microtiter wells. Cells are cultured for 6–8 weeks, and frequencies of cobblestone area-forming cells (CAFC) in each well are determined. A higher number of CAFC means the presence of more primitive progenitor cells in the overlaid cells. After culture, adhered and nonadhered cells are analyzed in terms of CFU assay and phenotypic analysis against various surface markers [224–226]. The frequency of LTC-IC is found to be 2 per 10^4 unfractionated marrow cells [225]. The LTC-IC present in adult bone marrow is phenotypically known as $CD34^+Thy\text{-}1^+Lin^-$ [221].

5.5.3
Colony Forming Unit Spleen (CFU-S) Assay

Primitive components (STR-HSC) of the stem cell compartment are detected by these assays. A fixed number of test cells are injected through the tail vein of lethally irradiated mice. Mice are sacrificed at days 8 and 12 ($CFU\text{-}S_8/CFU\text{-}S_{12}$), spleens are isolated and numbers of macroscopic colonies in each spleen are determined [227 and reference cited therein]. Cells taking 12 days to generate a spleen colony are designated as more primitive than those taking only 8 days, as cells forming $CFU\text{-}S_{12}$ are capable of self-renewal because they formed CFU-S [228]. In adult mice CFU-S can be detected in BM, spleen and circulation at a proportion of 0.05%, 0.002%, and 0.0001%, respectively [229, 230].

5.5.4
In Vitro Colony Assay (CFU-C)

Hematopoietic progenitor cells are indirectly studied by examining the ability of cells to form discrete colonies from individual cells. Colony forming cells are committed cells providing immediate protection of hosts from the damage caused by radiation or chemotherapy. A fixed number of cells are plated with a specialized semi solid medium (methycellulose/soft agar) containing various growth factors. After incubation of cells for 10–12 days, specific progenies are identified in the form of colonies [231 and reference cited therein]. The CFU-GEMM assay detects committed progenitor cells that produce single colonies containing granulocytic, erythroid, monocytic, and megakaryocytic lineages in the presence of cytokines/growth factors. Other forms of colonies normally detected in this assay are CFU-GM, CFU-G, BFU-E, CFU-E, and CFU-Meg. The CFU-GEMM assay measures cells with limited replating efficiency. In contrast, high proliferative potential CFC (HPP-CFC) assay detects a cell capable of generating large macrophage colonies and progenitors of granulocyte, erythrocytes, and megakaryocytes that can repopulate mice in vivo [232, 233]. HPP-CFC are replatable, indicating a degree of self-renewal. In case of human BM,

HPP-CFC (CD34$^+$HLA-DR$^-$Rh^{dull}) can be maintained and expanded to a limited extent in presence of SF and PIXY (IL-3/GM-CFS fusion protein) [234].

5.5.5
Marrow Reconstitution Assay

The ultimate test of stem cells is its engraftable potential in bone marrow. Hematopoietic stem cells can be analyzed in vitro and in vivo to judge the engraftable potential. In vitro assays include phenotypic analysis of surface adhesion molecules/receptors, such as, CD44, VLA-4, VLA-5, CXCR-4, etc., can be considered as a first step to judge the engraftable potential of stem cells [195, 200, 214, 216]. In vitro adhesion assay using preformed stroma cells reasonably predicts grafting and marrow reconstitution of donor's cells. It has been shown that stroma adhered cells are reconstituted in a syngenic mice [235]. Later, we have found that in vitro adhered cells on preformed stroma layer are comparable with in vivo grafted cells on bone marrow, in terms of LTC-IC (unpublished data).

Establishment of in vivo assay for HSCs is a more difficult task to accomplish. However, recent advances in the transplantation of primitive human hematopoietic populations into surrogate hosts has provided an opportunity to study long-term repopulating ability of these cells in vivo. The ultimate assay would be engraftment of long-term self-renewing multilineage cells which are genetically marked either through DNA manipulation, or detected allogenic differences with the help of antibodies. In vivo models, i.e., in utero transplantation to sheep [236], and mouse models using severe combined immunodeficient-human (SCID-hu) [217, 237], SCID-NOD (nonobese diabetic) [238] are popular.

6
Pre-Clinical and Clinical Studies on Ex Vivo Expanded Stem Cells

Ex vivo expansion of human hematopoietic stem cells has been practiced for more than two decades. During the past few years, considerable progress has been made on expansion of these cells; however, the benefits of research are yet to reach the bed-site. The major concern for using ex vivo expanded stem cells for clinical transplantation has been the doubt that ex vivo expanded progenitor cells may result in loss of capacity for early engraftment. It has been shown that grafting of expanded CD34$^+$ cord blood cells was delayed in NOD-SCID mice. It was presumed that more mature cells were produced at the expense of primitive cells. This was attributed to reduced capacity for post-transplant homing and/or proliferative growth in the host [239]. In another study, five patients received both cultured cord blood cells as well as haploidentical sibling progenitor cells, and the transplant outcome showed no demonstrable engraftment of the cultured cord blood cells [240]. This report, however, does not suggest any difference of kinetics of grafting between the cultured and the fresh cells as no genetic marker was used to distinguish between these two cell types. Therefore, in such a situation one should be careful to interpret the results. The results of inconclusive clinical studies and the ethical dilemma of using ex vivo expanded stem cells prevent clinicians from large scale trials of transplantation using ex vivo expanded stem cells.

The most significant result of transplantation of ex vivo expanded human cells was obtained with the xenograft model. It has been reported that stroma and cytokine supported culture of CD34$^+$Thy-1$^+$ cells for five weeks could be engrafted in SCID-Hu mice that leads to the generation of myeloid, B, and T-cells [217]. A similar result was obtained by Goltry et al. [241] who concluded that in ex vivo culture, bone marrow stroma cells increase expansion potential of cord blood stem cells which may ultimately accelerate hematopoietic reconstitution. In other preclinical studies faster recovery of neutrophils was observed in irradiated baboons receiving ex vivo expanded peripheral blood CD34$^+$ syngenic cells and growth factors, as compared to the control animals receiving normal cells [242].

During the past several years, few clinical studies have been pursued using ex vivo expanded hematopoietic stem cells; none of these reports showed any adverse effects of these expanded cells. Schpall et al. [243] conducted clinical trials on ex vivo expanded cord blood stem cells among various cancer patients receiving high dosages of chemotherapy. A part of cord blood (60%) was infused on day 0 without manipulation; the remaining 40% (only CD34$^+$ fraction) was expanded for ten days in the presence of cytokines and reinfused. All patients receiving expanded cord blood stem cells showed neutrophil and platelet engraftments at median times of 25 and 58 days, respectively. These data suggest two potential benefits of expanded cord blood cells, first lack of graft failure and second faster neutrophil engraftment as compare to the previous study with normal cord blood cells [244]. It is necessary to mention at this point that a few more groups also evaluated transplantation using previously expanded cord blood cells to increase the number of hematopoietic progenitor cells and thus to enhance engraftment. However, the engraftment times were reported to be similar to those with unexpanded cells [245–247].

One of the first clinical trials was performed with allogeneic bone marrow transplant for malignant hematological diseases. Patients who received both fresh and expanded bone marrow cells demonstrated faster recovery (median time 17 days) of platelets as compared to the control group (median time 23 days) [248]. Ex vivo expansion of hematopoietic stem cells for transplantation was not limited to the bone marrow, or to the cord blood. Three independent studies showed that ex vivo cultured mononuclear, or CD34$^+$ fraction peripheral blood progenitor cells were well tolerated by the patients with decreasing time to neutrophil recovery as compared to the control patients [249–251]. Thus, it can be concluded that the diverse results of transplantation of ex vivo expanded stem cells are attributed to (a) limited study and (b) different protocols used for the expansion of cells.

7

Ex Vivo Expansion of HSC on Three-Dimensional Supports

Ex vivo expansion of hematopoietic stem cells are primarily carried out in Dexter's type culture (two-dimensional) with support of stroma cells, or in liquid suspension culture without stroma cells. Neither process supported sustained self-renewal of human pluripotent stem cells [252]. The successful reconstitution of long-term ex vivo hematopoiesis can perhaps be best simulated by mimicking in vivo conditions. In vivo, the marrow microenvironment plays a crucial role in the hematopoiesis. As mentioned earlier, the marrow microenvironment is composed of different stroma cells that secrete ECM components and various soluble growth factors distributed in three-dimensional space. Hematopoietic stem cells grow and mature in such a bone marrow microenvironment. Supporting this notion, Naughton et al. [253] hypothesized that growth of stem cells on a three-dimensional culture system will be a close replica of bone marrow microenvironment.

Following the concept of the three-dimensional culture system, murine bone marrow cells were cultured on porous collagen sphere in a packed-bed bioreactor [254]. Later, an air-lift packed-bed bioreactor was deployed for similar cell type [255]. Meissner et al. [256] studies on gelatin modified porous glass packed fixed-bed and fluidized-bed bioreactors for expansion of CB/PB derived progenitor stem cells in stroma contact culture. It has been concluded that for HSC, packed-bed bioreactor is most suitable [256]. Despite the feasibility of culturing progenitor cells on three-dimensional supports, this system was unsuccessful due to problems such as insufficient expansion of mononuclear cells and difficult to harvest cells from the scaffold matrix. Another shortcoming found in porous structure of collagen spheres was that the porous lattice consisting of large pores of 200–300 µm were closer to providing a two-dimensional local microenvironment for the cells, instead of three-dimensional space as presumed in the bone marrow.

In 1998, Mantalaris et al. [257] reported culture of human bone marrow cells in a similar system to that published earlier [254]. Mononuclear cells of healthy adult donors were cultured on collagen macroporous spheres in the presence of SCF, IL-3, GM-CSF, IGF-I, and Epo for four weeks. A few significant observations extracted from this report are: (a) microscopic examination of the thin section reveals that the marrow cells populated the pores of the microspheres in a three-dimensional fashion with cell-to-cell contact and the resulting interactions resembled the bone marrow tissue structure in vivo; (b) morphological and staining techniques ensured the presence of erythroblastic islands of matured red cells and monocytes; (c) no adipocytes and blanket cells were observed in this culture. In this culture system, cells of all lineages were present in significant levels, as compared to the two-dimensional Dexter's culture (Table 6). It is discernable from these results that the in vivo microenvironment is more prevalent in three-dimensional than in two-dimensional culture.

In the recent past, two different three-dimensional culture systems were reported for hematopoietic stem cells. The first was tantalum coated porous carbon matrix, known as Cellfoam [258], and the other was nonwoven polyethylene terephthalate (PET) fabric [259]. It was shown that low concentrations of early acting cytokines SCF and FltL result in higher yields of cells that were enriched with primitive progenitor and multipotent hematopoietic progenitor cells in a three-dimension Cellfoam bioreactor as compared to plastic culture flasks [258]. The conservation and amplification of primitive stem cells in

Table 6 Comparative performance of three- and two dimensional long-term bone marrow culture

Cell type	Fresh BM MNC (%)	Dexter's culture MNC (%)	Three-dimensional culture, MNC (%)
Erythroid	23.1	5.8	43.0
Lymphoid	16.5	1.0	8.0
Monocyte	1.7	2.9	5.0
Granulocyte	57.3	90.3	43.5

Experiments were conducted under identical conditions [257].

three-dimensional Cellfoam at low concentrations of early acting cytokines resembled that obtained in the presence of higher concentrations of cytokines in culture flasks [214]. Thus, the results in the Cellfoam system suggested that ligand saturation hypothesis [214, 216] is not critical in a three-dimensional culture system. Similarly, three-dimensional nonwoven PET matrix was reported to produce two- to threefold more CD34$^+$ cell when compared with the two-dimensional system in seven to nine weeks culture in the presence of Tpo and FltL [259]. It was shown that both stroma and hematopoietic cells were distributed spatially within the scaffold and facilitated cell-cell and cell-matrix interactions closure to the in vivo bone marrow microenvironment.

8
Conclusion

In this review, we have discussed ex vivo expansion of stem cells and outlined potential clinical applications of expanded cells. The technology for cell culture has been developed rapidly with advances in our knowledge of hematopoiesis. Mimicking of in vivo BM microenvironment in the ex vivo culture systems has been proved to be the best alternate for culture of hematopoietic stem cells. However, a few fundamental questions on stem cell biology and on engineering principle for rational design of a bioreactor need to be addressed together. Are stem cells truly expandable ex vivo? If so, for how many generations are asymmetric cell divisions possible? During the last ten years considerable progress has been made in understanding the molecular basis of hematopoiesis, and the roles of early acting cytokines in preserving and self-renewal of LTC-IC are still not clearly understood. Assuming that the ligand-receptor signaling threshold model is valid in vivo, then one must observe higher concentrations of cytokines in the BM microenvironment. If the same model is not valid in vivo, then it is necessary to rethink on the regulatory role of bone marrow microenvironment in hematopoiesis. Another important issue that should be readdressed is the engraftment potential of cytokine-induced ex vivo expanded cells; at present the knowledge is dubious. Though the results of clinical trials reflect potential benefits of the ex vivo expanded stem cells, the study is not adequate. For a logical conclusion, more clinical trials should be pursued with expanded BM and CB derived stem cells.

Current practice for expanding hematopoietic stem cells is primarily based on our previous knowledge of animal cell culture processes. Hematopoiesis is a very complex process involving different cell types, each cell type having few subsets that interact each other and also with the microenvironment. Therefore, it is important to remember that our previous knowledge in designing a bioreactor may not be particularly useful for the purpose. As mentioned in the main text, significant progress has been made on conceiving a bioreactor and its operation for expansion of stem cells, but still it is very much in its infancy. It is believed that if as close as possible BM microenvironmental conditions are maintained, the design and subsequent performance of the bioreactor will be similar to that of BM. Media feeding protocol, shear effect, and dissolved oxygen level are the three most important parameters for optimum performance of the stem cell bioreactor.

The presence of serum in the culture medium or of stroma in the expansion system may cause concern for the clinical use of expanded cells from allogenic sources. Therefore, attempts should be made to expand hematopoietic stem cells in the absence of these components, particularly stroma cells as shown by Xu et al. [260]. It will be interesting to design a bioreactor by grafting different ECM components, produced by BM stroma cells, on a three-dimension scaffold. Overall, the ultimate objective of ex vivo expansion of stem cells is to produce sufficient numbers of both short-term and long-term repopulating cells, which is expected to be achieved by reconstituting the BM microenvironment in vitro.

Acknowledgement The authors (AM and TM) are grateful to the Department of Biotechnology, Government of India for funding the National Institute of Immunology to carry out the work.

9
References

1. Kay HEM (1965) Lancet 11:418
2. Dexter TM, Allen TD, Lajtha LG, Schofield R, Lord BI (1973) J Cell Physiol 82:461
3. Moore MAS, Sheridan APC (1979) Blood Cells 5:297
4. Gartner S, Kaplan HS (1980) Proc Natl Acad Sci USA 77:4756
5. Bertolini F, Battaglia M, Pedrazzoli P, Antonio da prada G, Lanza A, Soligo D, Caneva L, Sarina B, Murphy S, Thomas T, Robustelli della cuma G (1997) Blood 89:2679
6. Brugger W, Heimfed S, Berenson RJ, Mertelsmann R, Kanz L (1995) New Engl J Med 333:283
7. Koller MR, Palsson BO (1993) Biotechnol Bioeng 42:909
8. Audet J, Zandstra PW, Eaves CJ, Piret JM (1998) Curr Opin Biotechnol 9:146
9. Conrad PD, Emerson SG (1998) J Leukoc Biol 64:147
10. LaIuppa JA, Papoutsakis ET, Miller WM (1997) Cancer Treat Res 77:159
11. McAdams TA, Miller WM, Papoutsakis ET (1996) Trends Biotechnol 14:341
12. Nielsen LK (1999) Annu Rev Biomed Eng 1:12
13. Haylock DN, To LB, Makino S, Dowse TL, Juttner CA, Simmons PJ (1995) Ex vivo expansion of human hematopoietic progenitors with cytokines. In: Levitt D, Mertelsmann R (eds) Hematopoietic stem cells. Marcel Dekker, New York, p 491
14. McNiece I, Briddell R (2001) Exp Hematol 29:3
15. William PL, Warwick R, Dyson M, Baunistner LH (1989) Gray's anatomy, 37th edn. Churchill Livingstone, Edinburgh

16. Wu A, Till J, Stiminovitch L, McCulloch E (1967) J Cell Physiol 69:177
17. Wu A, Till J, Stiminovitch L, McCulloch E (1968) J Exp Med 127:455
18. Wagers AJ, Christensen JL, Weissman, IL (2002) Gene Therapy 9:606
19. Morrison SJ, Weissman IL (1994) Immunity 1:661
20. Morrison SJ et al. (1997) Development 124:1929
21. Harrison DE, Zhong RK (1992) Proc Natl Acad Sci USA 89:10,134
22. Jones RJ, Celano P, Sharkis SJ (1989) Blood 73:397
23. Kondo M, Weissman IL, Akashi K (1997) Cell 91:661
24. Akashi K, Traver D, Miyamoto T, Weissman IL (2000) Nature 404:193
25. Lajtha LG (1979) Differentiation 14:23
26. Abboud CN, Lichtman AJ (1995) In: Beutler E, Lichtman AJ, Coller BS, Kipps TJ (eds) Hematology. McGraw-Hill, New York, p 25
27. Erslev AJ, Gabuzda TG (1985) In: Pathophysiology of blood, 3rd edn. Saunders, Philadelphia
28. Dexter TM (1989) Leukemia 3:469
29. Brockbank KGM, de Jong JP, Piersma AH, Voerman JSA (1986) Exp Hematol 14:386
30. Hunt P, Robertson D, Weiss D, Renick D, Lee F, Witte ON (1987) Cell 48:997
31. Bianco P, Riminucci M, Gronthos S, Robey PG (2001) Stem Cells 19:180
32. Garnett HM, Harigaya K, Cronkite EP (1982) Stem Cells 2:11
33. Funk PE, Stephan RP, Witte PC (1995) Blood 86:2661
34. Greenberger JS (1978) Nature 275:752
35. Tavassoli M, Maniatis A, Crosby WH (1974) Blood 43:33
36. Crocker PR, Gordon S (1987) In: Rich IN (ed) Molecular and cellular aspects of erythro-poietin and erythropoiesis. Springer, Berlin Heidelberg New York, p 259
37. Rich IN (1988) Anticancer Res 8:1015
38. Zuckerman KS, Wicha MS (1983) Blood 61:540
39. Bentley SA, Tralka TS (1983) Exp Hematol 11:129
40. Coulombel L, Vullet MH, Leroy C, Tehernia G (1988) Blood 71:329
41. Long MW, Dixit VM (1990) Blood 75:2311
42. Campbell AD, Long MW, Wicha MS (1987) Nature 329:744
43. Gordon MY, Riley GP, Clarke D (1988) Leukemia 2:804
44. Gallagher JT, Spooncer E, Dexter TM (1983) J Cell Sci 63:155
45. Roberts R, Gallagher JT, Spooncer E, Allen TD (1988) Nature 332:376
46. Armitage JO (1994) N Engl J Med 330:827
47. Gratwohl A, Passweg J, Baldomero H, Urbano-Ispizua A (2001) Bone Marrow Transplant 27,899
48. Socie G, Stone JV, Wingard JR et al. (1999) N Engl J Med 341:14
49. Thomas ED. (1999) N Engl J Med 341:50
50. McSweeney PA, Niederwieser D, Shizuru JA et al. (2001) Blood 97:3390
51. Appelbaum FR (2001) Nature 411:385
52. Rajkumar SV, Fonesca R, Dispenzieri A et al. (2000) Bone Marrow Transplant 26:979
53. Thomas DB (1993) Stem Cells 11(Suppl. 1):66
54. Korbling M, Anderlini P (2001) Blood 98:2900
55. Auditore-Hargreaves K, Heimfeld S, Berenson RJ (1994) Bioconjugate Chem 5:287
56. Berenson RJ, Bensinger WI, Hill RS et al. (1991) Blood 77:1717
57. Niederwieser D, Pepe M, Storb R et al. (1988) Br J Haematol 55:573
58. Korbling M, Burke P, Braine H, Elfenbein G, Santos G, Kaizer H (1981) Exp Hematol 9:684
59. Korbling M, Huh YO, Durett A et al. (1995) Blood 86:2842
60. Gianni AM, Siena S, Bregni M et al. (1989) Lancet 2:580
61. Siena S, Bregni M, Brando B et al. (1989) Blood 74:1905
62. Vesole DH, Jagannath S, Tricot G, Barlogie B (1995) Hematopoietic stem cell transplan-tation in multiple myeloma. In: Levitt D, Mertelsmann R (eds) Hematopoietic stem cells. Marcel Dekker, New York, p 571
63. McAdams TA, Winter JN, Miller MW, Papoutsakis ET (1996) Trends Biotechnol 14:388
64. Takahira H, Lyman SD, Broxmeyer HE (1996) Ann Hematol 72:131
65. Gluckman E, Rocha V, Garnier F, Ionescu I, Bittencourt H, Chevret S (2000) Blood 96:588a

66. Gluckman E, Rocha V, Boyer-Chammard A et al. (1997) N Engl J Med 337:373
67. Fernandez MN, Millan I, Gluckman E (1999) N Engl J Med 340:1287
68. Bambach B, Koc ON, Rizzieri DA et al. (2000) Blood 96:587a
69. Laughlin MJ (2001) Bone Marrow Transplant 27:1
70. Touraine JL, Roncarolo MG, Plotnicky H et al. (1994) Tolerance to alloantigens and recognition for "allo+X" induced in humans by fetal stem cells transplantation. In: Touraine JL, Traeger J et al. (eds) Transplantation and clinical immunology XXV. Kluwer, Dordrecht, The Netherlands, p 265
71. Touraine JL (1991) J Inher Metab Dis 14:619
72. Touraine JL, Raudrant D (1995) In utero stem-cell therapy, using fetal cells transplanted into fetal recipients. In: Levitt D, Mertelsmann R (eds) Hematopoietic stem cells. Marcel Dekker, New York, p 369
73. Stiff P, Parthasarathy M, Chen B, Moss T, Oldenberg D et al. (1998) Blood 92:530a
74. Koller MR, Emerson SG, Palsson BO (1993) Blood 82:378
75. Zandstra PW, Eaves CJ, Piret MJ (1994) Biotechnol 12:909
76. Zandstra PW, Conneally E, Petzer AL, Piret JM, Eaves CJ (1997) Proc Natl Acad Sci USA 94:4698
77. Lundell BI, Vredenburgh JJ, Tyer C, DeSombre K, Smith AK (1998) Bone Marrow Transpl 22:153
78. Szabolcs P, Moore MA, Young JW (1995) J Immunol 154:5851
79. Leary AG, Ikebuchi K, Hirai Y, Wong GG, Yang YC, Clark SC, Ogawa M(1988) Blood 71:1759
80. Spooncer E, Heyworth CM, Dunn A, Dexter TM (1986) Differentiation 31:111
81. Gualtieri RJ, Shadduck RK, Baker DG (1984) Blood 64:516
82. Dorshkind K (1990) Ann Rev Immuno l8:111
83. Quesenberry PJ, McNiece IK, McGrath HE, Temeles DS, Baber GB, Deacon DH (1989) NY Acad Sci 554:116
84. Eliason JF, Thorens B, Kindler V, Vassali P (1988) Exp Hematol 16:307
85. Fibbe WE, Van Damme J, Billiau A, Goselink HM, Voogt PJ, Van Eeden G, Ralph P, Altrock BW, Falkenburg JHF (1988) Blood 71:430
86. Charbord P, Tamayo E, Saeland S, Duvert V, Poulet J, Gown AM, Herve P (1991) Blood 78:1230
87. Slack JL, Nemunaitis J, Andrews DF III, Singer JW (1990) Blood 75:2319
88. Brandt JE, Bhalla K, Hoffman R (1994) Blood 83:1507
89. Verfaillie CM, Miller JS (1994) Blood 84:1442
90. Lemoli RM, Fortuna A, Fogli M, Motta MR, Rizzi S, Benini C, Tura S (1994) Exp Hematol 22:919
91. Li ML, Cardoso A, Sansilvestri P, Hartzfeld A, Kisselev S, Bartard P, Levesque JP, Hartzfeld J (1993) Nouv Rev Fr Hematol 35:81
92. Henscheler R, Brugger W, Luft T, Frey T, Mertelsmann R, Kanz L (1994) Blood 84:2898
93. Petzer AL, Zandstra PW, Piret JM, Eaves CJ (1996) J Exp Med 183:2551
94. Cicuttini FM, Welch KL, Boyd AW (1994) Exp Hematol 22:1244
95. Mayami H, Dragowska W, Lansdorp PM (1993) Blood 81:3252
96. Ruggieri L, Heimfeld S, Broxmeyer HE (1994) Blood Cells 20(2/3):436
97. Piacibello W, Sanavio F, Severino A et al. (1999) Blood 93:3736
98. Dravid G, Rao SGA (2002) Stem Cells 20:183
99. Hirayama F, Katayama N, Neben S, Donaldson D, Nickbarg EB, Clark SC, Ogawa M (1994) Blood 83:92
100. Ariyama Y, Misawa S, Sonoda Y (1995) Stem Cells Dayt 13:404
101. Yanai T, Sugimoto K, Takashita E, Aihara Y, Tsurumaki Y, Tsuji Y, Mori KJ (1995) Cell Struct Funct 20:117
102. Leary AG, Zeng HQ, Clark SC, Ogawa M (1992) Proc Natl Acad Sci USA 89:4013
103. McAdams TA, Sandstrom CE, Miller WM, Bender JG, Papoutsakis ET (1995) Cytotechnol 18:133
104. Mc Kenna HJ et al. (1995) Blood 86:3413

105. Piacibello W, Fubini L, Sanavio F, Brizzi MF, Severino A, Garetto L, Stacchini A, Pegoraro L, Aglietta M (1995) Blood 86:4105
106. Mackarehtschian K et al. (1995) Immunity 3:147
107. Hunte B, Hudak S, Campbell D, Xu Y, Rennick D (1996) J Immunol 156:489
108. Papayannopoulou T, Brice M, Blau CA (1993) Blood 81:299
109. Brandt JE, Bhalla K, Hoffman R (1994) Blood 83:1507
110. Kaushansky K et al. (1995) Proc Natl Acad Sci USA 92:3234
111. Broudy VC et al. (1995) Blood 85:1719
112. McDonald TP (1992) In: Murphy MJ (ed) Concise review in clinical experimental hematology. Alpha Medical Press, p 161
113. Luens KM, Travis MA, Chen BP, Hill BL, Scollay R, Murray LJ (1998) Blood 91:1206
114. Kusadasi N, Koevoet JL, van Soest PL, Ploemacher RE (2001) Leukemia 15:1347
115. Jazwiec B, Solanilla A, Grosset C, Mahon FX, Dupouy M, Pigeonnier-Lagarde V, Belloc F, Schweitzer K, Reiffers J, Ripoche J (1998) Leukemia 12:1210
116. Haynesworth SE, Baber MA, Caplan AI (1996) J Cell Physiol 166:585
117. Szilvassy SJ, Weller KP, Lin W, Sharma AK, Ho AS, Tsukamoto A, Hoffman R, Leiby KR, Gearing DP (1996) Blood 87:4618
118. Welzler M, Talpaz M, Lowe DG, Baiocchi G, Gutterman JU, Kurzrock R (1991) Exp Hematol 19:347
119. Smith AG, Heath JK, Donaldson DD et al. (1988) Nature 336:688
120. Lu L, Xiao M, Grigsby S, Wang WS, Wu B, Shen RN, Broxmeyer HE (1993) Exp Hematol 21:1442
121. Snoeck HW, Van Bockstaele DR, Nys G, Lenjou M, Lardon F, Haenen L, Rodrigus I, Peetermans ME, Berneman ZN (1994) J Exp Med 180:1177
122. Maciejewski JP, Selleri C, Sato T, Cho HJ, Keefer LK, Nathan CF, Young NS (1995) J Clin Invest 96:1085
123. Mayami H, Little MT, Dragowska W, Thornbury G, Lansdorp PM (1995) Exp Hematol 23:422
124. Xi X, Schlegel N, Caen JP, Minty A, Fournier S, Caput D, Ferrara P, Han ZS (1995) Br J Haematol 90:921
125. Sakamoto O, Hashiyama M, Minty A, Ando M, Suda T (1995) Blood 85:3487
126. Rogers JA, Berman JW (1994) J Immunol 153:4694
127. Jacobsen SE, Veiby OP, Myklebust J, Okkenhaug C, Lyman SD (1996) Blood 87:5016
128. Naughton BA, Tjota A, Sibander B, Naughton GK (1991) J Biochem Eng 113:171
129. Iscove NN, Shaw AR, Keller G (1989) J Immunol 142:2332
130. Nakahata T, Ogawa M (1982) Proc Natl Acad Sci USA 79:3843
131. McNiece IK, Williams NT, Johnson GR, Kriegler AB, Bradley TR, Hodgson GS (1987) Exp Hematol 15:972
132. Bodine DM, Karlsson S, Nienhuis AW (1989) Proc Natl Acad Sci USA 86:8897
133. Broxmeyer HE, Hangoc G, Cooper S, Riebeiro RC, Graves V, Yoder M, Wagner J, Vadhan-Raj S, Benninger L, Rubinstein P, Broun ER (1992) Proc Natl Acad Sci USA 89:4109
134. Koller MR, Bender JG, Papoutsakis ET, Miller WM (1992) Blood 80:403
135. Koller MR, Bender JG, Miller WM, Papoutsakis ET (1993) Biotechnol 11:358
136. Palsson BO, Paek S-H, Schwartz RM, Palsson M, Lee G-M, Silver S, Emerson SG (1993) Biotechnol 11:368
137. Sardonini CA, Wu Y-J (1993) Biotechnol Prog 9:131
138. Collins PC, Nielsen LK, Patel SD, Papoutsakis ET, Miller WM (1998) Biotechnol Prog 14:466
139. Koller MR, Miller WM, Papoutsakis ET (1998) Biotechnol Bioeng 59:534
140. Goltry KL, Lee JJ, Scott A, Armstrong RD (2001) ISHAGE 7th Annual Symposium, Quebec City, Abstract No. 39
141. Pecora AL, Preti R, Jennis A, Goldberg S, Stiff P et al. (1998) Blood 92:507a
142. Haylock DN, To LB, Dowse TL et al. (1992) Blood 80:1405
143. Williams SF, Lee WJ, Bender JG et al. (1996) Blood 87:1687

144. Noll T, Jelinek N, Schmidt S, Biselli M, Wandrey C (2002) Cultivation of hematopoietic stem and progenitor cells: biochemical engineering aspects. In: Schugerl K, Zeng A-P (eds.) Adv Biochem Eng/Biotechnol, vol 74. Springer, Berlin Heidelberg New York, p111

145. Verfaillie CM (1992) Blood 79:2821

146. Burroughs J, Gupta P, Blazar BR, Verfaillie CM (1994) Exp Hematol 22:1095

147. Verfaillie CM, Catanzarro PM, Li-W N (1994) J Exp Med 179:643

148. Breems DA, Blokland EAW, Siebel KE, Mayen AEM, Engles LJA, Ploemacher RE (1998) Blood 91:111

149. Brandt JE, Galy AHM, Luens KM et al. (1998) Exp Hematol 26:950

150. Koller MR, Palsson MA, Manchel I, Palsson BO (1995) Blood 86:1784

151. Ramsfjell V, Bryder D, Bjorgvinsdottir H et al. (1999) Blood 94:4093

152. Bender JG, Smith SL, Unverzagt KL (1995) Ex vivo expansion of neutrophil precursors from blood CD34+ cells: applications for the treatment of neutropenia. In: Levitt D, Mertelsmann R (eds) Hematopoietic stem cells. Marcel Dekker, New York, p 171

153. Caldwell J, Palsson BO, Locey B, Emerson SG (1991) J Cell Physiol 147:344

154. East CJ, Abboud CN, Borch RF (1992) Blood 80:1172

155. Lazzari L, Thrasher A, Lucchi S et al. (1999) Blood 94 (Suppl. 1):572a

156. Almeida-Porada G, Brown RL, MacKintosh FR, Zanjani ED (1997) Lab Invest 76:25

157. Dybedal I, Jacobsen SE (1995) Blood 86:949

158. Lill MC, Lynch M, Fraster JK, Chung GY, Schiller G et al. (1994) Stem Cells 12:626

159. Sandstrom CE, Collins PC, McAdams TA, Bender JG, Papoutsakis ET et al. (1996) J Hematother 5:461

160. Koller MR, Maher RJ, Manchel I, Oxender M, Smith AK (1998) J Hematother 7:413

161. Sutherland HJ, Eaves CJ, Lansdorp PM, Thacker JD, Hogge DE (1991) Blood 78:666

162. Martiat P, Ferrant A, Cogneau M, Bol A, Michel C, Rodhain J, Michaux JL, Sokal G (1987) Br J Haematol 66:307

163. Schwartz RM, Palsson BO, Emerson SG (1991) Proc Natl Acad Sci USA 88:6760

164. Lindop PJ, Rotblat J (1960) Nature 185:593

165. Wright EA, Bewley DK (1960) Radiat Res 13:649

166. Rich IN, Kubanek B (1982) Br J Haematol 52:579

167. Broxmeyer HE, Cooper S, Lu L, Miller ME, Langefeld CD, Ralph P (1990) Blood 76:323

168. Maeda H, Hotta T, Yamada H (1986) Exp Hematol 14:930

169. LaIuppa JA, Papoutsakis ET, Miller WM (1998) Exp Hematol 26:835

170. Rich N (1986) Exp Hematol 14:746

171. Williams DA (1993) Blood 81:3169

172. Stent GS (1998) Int J Dev Biol 42:237

173. Rosendaal M, Hodgson GS, Bradley TR (1976) Nature 264:68

174. Botnick LE, Hannon EC, Hellman S (1979) Blood Cells 5:195

175. Schofield R (1978) Blood Cells 4:7

176. Srour EF, Brandt JE, Leemhuis T, Ballas CB, Hoffman R (1992) J Immunol 148:815

177. Reems JA, Torok-Storb B (1995) Blood 85:1480

178. Watt FM, Hogan BLM (2000) Science 287:1427

179. Piacibello W, Sanavio F, Severino A et al. (1999) Blood 93:3736

180. Miller CL, Eaves CJ (1997) Proc Natl Acad Sci USA 94:13,648

181. Yagi M, Ritchie KA, Sitnicka E, Storey C, Roth GJ, Bartelmez S (1999) Proc Natl Acad Sci USA 96:8126

182. Zandstra PW, Conneally E, Petzer AL, Piret JM, Eaves CJ (1997) Proc Natl Acad Sci USA 94:4698

183. Ramsfjell V, Bryder D, Bjorgvinsdottir H et al. (1999) Blood 94:4093

184. Zandstra PW, Lauffenburger DA, Eaves CJ (2000) Blood 96:1215

185. Sauvageau G, Thorsleinsdottir U, Eaves CJ, Lawrence HJ, Largman C, Lansdorp PM (1995) Genes Dev 9:1753

186. Thorsteinsdottir U, Sauvageau G, Humphries RK (1999) Blood 94:2605

187. Karanu FN, Murdoch B, Gallacher L, Wu DM, Koremoto M, Sakano S, Bhatia M (2000) J Exp Med 192:1365

188. Varnum FB et al. (2000) Nat Med 6:1278
189. Bhardwaj G, Murdoch B, Wu DM, Baket DP et al. (2001) Nat Immunol 2:172
190. Asaumi N, Omoto E, Mahmut N, Katayama Y, Takeda K, Shinagawa K, Harada M (2001) J Biol Chem 276:37,069
191. Szilvassy SJ, Meyerrose TE, Ragland PL, Grimes B (2001) Exp Hematol 29:1494
192. Teixido J, Hemler ME, Greenberger JS, Anklesaria P (1992) J Clin Invest 90:358
193. Orschell-Traycoff CM, Hiatt K, Dagher RN, Rice S, Yoder MC, Srour EF (2000) Blood 96:1380
194. Dimitroff CJ, Lee JY, Rafi S, Fuhlbrigge RC, Sackstein R (2001) J Cell Biol 153:1277
195. Miyake K, Underhill CB, Lesley J, Kincade PW (1990) J Exp Med 172:69
196. Frenette PS, Subbarao S, Mazo IB, von Andrain UH, Wagner DD (1998) Proc Natl Acad Sci USA 95:14,423
197. Papayannopoulou T, Priestley GV, Nakamoto B, Zafiropoulos V, Scott LM (2001) Blood 98:2403
198. Williams DA, Rios M, Stephens C, Patel VP (1991) Nature 352:438
199. Rosenblatt M, Vuillet-Gaugler H, Leroy C, Coulombel L (1991) J Clin Invest 87:6
200. Peled A, Petit I, Kollet O et al. (1999) Science 283:845
201. Whetton AD, Graham GJ (1999) Trends Cell Biol 9:233
202. Schofield KP, Rushton G, Humphries MJ, Dexter TM, Gallagher JT (1997) Blood 90:1858
203. Prosper F, Stroneck D, McCarthy JB, Verfaillie CM (1998) J Clin Invest 101:2456
204. Levesque JP, Leavesley DI, Niutta S, Vadas M, Simons PJ (1995) J Exp Med 181:1805
205. Peters SO, Kittler EL, Ramshaw HS, Quesenberry PJ (1996) Blood 87:30
206. Szilvassy SJ, Bass MJ, Zant GV, Grimes B (1999) Blood 93:1557
207. Fruehauf S, Veldwijk MR, Kramer A, Haas R, Zeller WJ (1998) Stem Cells 16:271
208. Yamaguchi M, Ikebuchi K, Hirayama F, Sato N, Mogi Y, Ohkawara J, Yoshikawa Y, Sawada K, Koike T, Sekiguchi S (1998) Blood 92:842
209. Traycoff CM, Yoder MC, Kiatt K, Srour EF (1997) Exp Hematol 25:736
210. Gothot A, van der Loo JCM, Clapp DW, Srour EF (1998) Blood 92:2641
211. Becker PS, Nilsson SK, Li Z, Berrios VM, Dooner MS, Cooper CL, Hsieh C, Quesenberry PJ (1999) Exp Hematol 27:533
212. Berrios VM, Dooner GJ, Nowakowski G, Frimberger A, Valinski H, Quesenberry PJ, Becker PS (2001) Exp Hematol 29:1326
213. Long MW (1993) Exp Hematol 20:288
214. Zanjani ED, Flake AW, Almeida-Porada G, Tran N, Papayannopoulou T (1999) Blood 94:2515
215. Saeland S, Duvert V, Caux C (1992) Exp Hematol 20:24
216. Lapidot T (2001) Ann NY Acad Sci 938:83
217. Shih C-C, Hu MCT, Hu J, Medeiros J, Forman SJ (1999) Blood 94:162
218. Osawa M, Hanada KI, Hamada H, Nakauchi H (1996) Science 273:242
219. Bhatia M, Bonnet D, Murdoch B, Gan OI, Dick JE (1998) Nat Med 4:1038
220. Zhao Y, Lin Y, Zhan Y, Yang G, Loute J, Harrison DE, Anderson WF (2000) Blood 96:3016
221. Tsukamoto AS, Chen BP, Hoffman R, Uchida N (1995) Phenotypic and functional analysis of hematopoietic stem cells in mouse and human. In: Levitt D, Mertelsmann R (eds) Hematopoietic stem cells. Marcel Dekker, New York, p 85
222. Cardoso AA, Li ML, Batard P et al. (1993) Proc Natl Acad Sci USA 90:8707
223. Okada S, Nakauchi H, Nagayoshi K et al. (1991) Blood 78:1706
224. Ploemacher RE, van der Sluijs JP, Voerman JSA, Brons NHC (1989) Blood 74:2755
225. Sutherland HJ, Eaves CJ, Lansdorp PM et al. (1990) Proc Natl Acad Sci USA 87:3584
226. Gordon NY, Hibbin JA, Kearnet LU et al. (1985) Br J Haematol 60:129
227. Lord BI (1993) In: Testa NG, Molineux G (eds) Haemopoiesis – a practical approach. IRL Press, Oxford, United Kingdom, p 1
228. Magli MC, Iscove NN, Odartchenko N (1982) Nature 295:527
229. Metcalf D (1984) The hematopoietic colony stimulating factors. Elsevier, New York
230. Spangrude GJ, Heimfeld S, Weissman IL (1988) Science 241:58

231. Heyworh CM, Spooncer E (1993) In: Testa NG, Molineux G (eds) Haemopoiesis – a practical approach. IRL Press, Oxford, United Kingdom, p 37
232. Bradley TR, Hodgson GS, Bertoncello I (1989) High proliferative potential colony forming cells. In: Pollard JW, Walker JL (eds) Methods in molecular biology-animal cell culture. Human Press, Clifton, NJ, vol 5, p 289
233. Bradley TR, Hodgson GS (1979) Blood 54:1446
234. Brandt JE, Briddell RA, Srour EF et al. (1992) Blood 79:634
235. Frimberger AE, Stering AI, Quesenberry PJ (2001) Blood 96:1012
236. Zanjani ED, Pallavicini MG, Ascensao JL et al. (1992) J Clin Invest 89:1178
237. McCure JM, Namikawa R, Kaneshima H et al. (1988) Science 241:1632
238. Kerre TCCC, Smet GD, Smedt MD et al. (2001) J Immunol 167:3692
239. Guenechea G, Segovia JC, Albella B et al. (1999) Blood 93:1097
240. Fernandez MN, Regidor C, Cabrera R et al. (2001) Bone Marrow Transplant 28:355
241. Goltry KL, Lee JJ, Kurtzberg J (2000) Blood 96:774a
242. Andrews RG, Briddell RA, Gough M, McNiece IK (1997) Blood 90(Suppl. 1):92a
243. Schpall EJ, Quinones R, Hami L et al. (1998) Blood 92(Suppl. 1):646a
244. Gluckman E, Rocha V, Boyer-Chammard A et al. (1997) N Engl J Med 337:373
245. Kogler G, Nurberger W, Fisher J et al. (1999) Bone Marrow Transplant 24: 397
246. Kurtzberg J, Martin PL, Driscol T et al. (1999) Blood 94 (Suppl. 1):571a
247. Fernandez MN, Granena A, Millan I et al. (2000) Bone Marrow Transplant 25 (Suppl. 2): S60
248. Naparstek E, Hardan Y, Ben-Shahar M et al. (1992) Blood 80:1673
249. Reiffers J, Cailliot C, Dazey B et al. (1999) Lancet 354:1092
250. McNiece I, Jones R, Bearman SI et al. (2000) Blood 96:3001
251. Paquette R (2000) Clinical experience with ex vivo expanded peripheral blood progenitor cell. International Conference on Hematopoietic Stem cell Biology and Transplantation, Paris, France
252. Emerson SG (1996) Blood 87:3082
253. Naughton BA, Jacob L, Naughton GK (1990) Prog Clin Biol Res 333:435
254. Wang T-Y, Brennan JK, Wu JHD (1995) Exp Hematol 23:26
255. Highfill JG, Haley SD, Kompala DS (1996) Biotech Bioeng 50:514
256. Meissner P, Schroder B, Herfurth C, Biselli M (1999) Cytotechnology 30:227
257. Mantalaris A, Keng P, Bourne P, Chang AYC, Wu JHD (1998) Biotech Prog 14:126
258. Banu N, Rosenzweig M, Kim H, Bagley J, Pykett M (2001) Cytokine 13:349
259. Li Y, Ma T, Kniss DA, Yang S-T, Lasky LC (2001) J Hematol Stem Cell Res 10:355
260. Xu C, Inokuma MS, Denham J, Golds K et al. (2001) Nat Biotechnol 19:971

Received: October 2002

Adv Biochem Engin/Biotechnol (2004) 86: 255–278
DOI 10.1007/b12445

Adsorptive Separation of β-Lactam Antibiotics: Technological Perspectives

M. Dutta[1] · M. M. Borah[2] · N. N. Dutta[2]

[1] Department of Chemistry, Indian Institute of Technology, Guwahati 781 039, India
  E-mail: monalisaikia@hotmail.com
[2] Chemical Engineering Division, Regional Research Laboratory, Jorhat 785 006, India
  E-mail: chemengg@hq.csir.res.in

Abstract An overview on adsorptive separation of cephalosporin antibiotics has been presented. The fundamental aspects on adsorption mechanism, kinetic and column dynamics have been exhaustedly discussed. The importance of molecular modelling studies on deducing implications for design of adsorbents has been highlighted. Finally, some aspects of process design and scale-up of adsorption column have been addressed and research needs of pragmatic importance have been identified.

Keywords Adsorption · β-lactam antibiotics · Equilibrium · Kinetics · Adsorptive interaction

Abbreviations

C fluid phase concentration, mol/l

C_1 parameter of the Langmuir model, adsorption saturation capacity based on solid volume, mol/l

C_e equilibrium liquid phase concentration, mol/l

C_r concentration in liquid filled pores at radius r, mol/l

C_s fluid phase concentration near particle surface, mol/l

D_p dispersion coefficient, cm^2/s

D_s effective surface diffusity, cm^2/s

K_F parameter of Freundlich model

k_f film mass transfer coefficient, cm/s

K_L Langmuir constant, lmol^{-1}

n parameter of Freundlich model

N_r concentration of adsorbed solute at radius r of the particle, mol/l

q solid phase concentration, mol/g

q_e amount of solute adsorbed at equilibrium, mol/g

q_s solid phase concentration at particle surface, mol/l

r radial distance, cm

R   Universal gas constant

R   radius of particle, cm

t time, s

V_e interstitial fluid velocity, cm/s

z longitudinal distance in the column, cm

θ fractional coverage

ε bed void fraction [–]

ε_p porosity of the adsorbent particles

ϱ_p particle density, g/cm^2

1
Introduction

The β-lactam antibiotics are therapeutically important and advantageous for their broad antibacterial activity. The β-lactam family has developed into an array of chemical structure with six basic nuclei (Fig. 1). Out of the total world wide antibiotic production of around 5×10^7 kg/annum, the β-lactam group constitute an amount of the order of 3×10^7 kg/annum. Penicillins and cephalosporins are the main constituents of commercially important β-lactam as shown in Fig. 2. Since discovery of penicillins by Fleming [1] and cephalosporins by Newton and Abraham [2], the processes for production of natural and semisynthetic penicillins and cephalosporins have undergone various modifications. Most of the commercially important natural β-lactams are produced through biochemical processes. The semisynthetic penicillins and cephalosporins can be prepared by chemical or enzymatic processes, the latter having potential advantages in several respects.

The growing interest in various β-lactam antibiotics over the past decade has called upon improvement of their production methods via modification of either

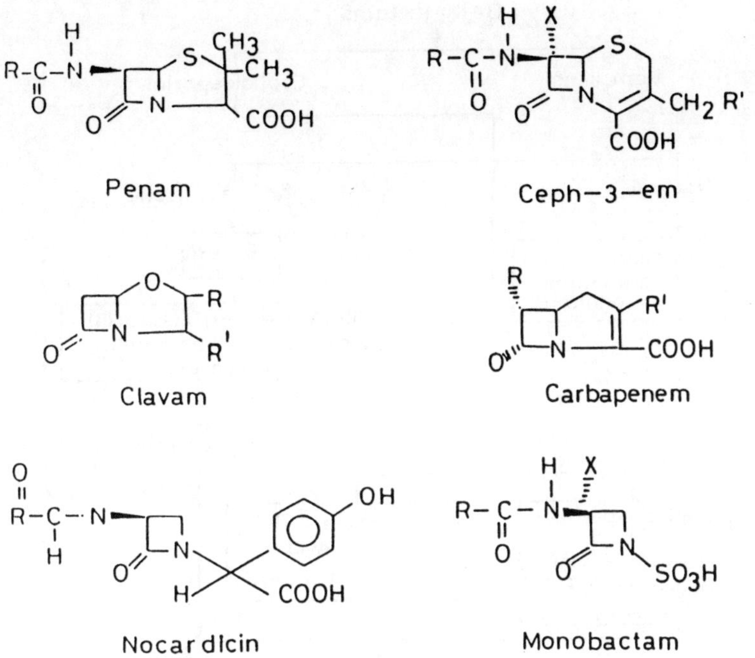

Fig. 1 Six distinct classes of naturally occurring β-lactam antibiotics

the basic process or microbial strain or downstream processing techniques. Recovery of β-lactams from the usually dilute product in the bioreactor is important for both analytical and preparative purposes. Product recovery may involve various methods of extraction and purification and plays an important role in the overall process economics [3, 4]. However, almost all the known processes for commercial scale extraction and purification are based on low-yield operation because of the unfavourable physical properties of the β-lactams, particularly the cephalosporins. While the less hydrophilic penicillins can be recovered by the well known solvent extraction technique, difficulties arise in the case of cephalosporins which are highly hydrophilic possessing zwitterionic properties.

Adsorption chromatography is often used for the isolation and purification of fermentation products, both lipophilic and hydrophilic [5]. In the literature, one can find countless examples of applications, but demonstrated mainly on a laboratory scale. However, the most elegant separation method is useless for biotechnological production processes, if it cannot be realised in practice on an industrial scale. The scale-up problem and the technical aspects of adsorption chromatography, applicable to low and high molecular weight products, need particular attention. Chromatographic media like octadecyl silane chemically bonded to silica particles, strong cation exchangers bonded to silica etc. which are known for analytical purposes can be used for preparative scale chromatography. C_8 or C_{18} sorbents with ion pairing agents such as tetrabutyl ammonium

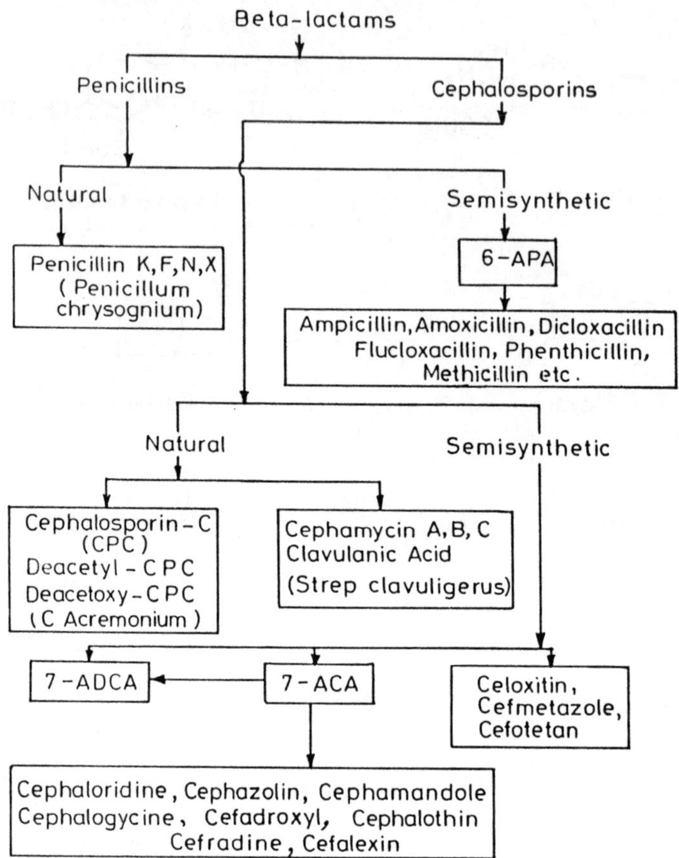

Fig. 2 Commercially important β-lactam antibiotics

hydroxide and cetyltrimethyl ammonium bromide have been used for resolution of penicillin broth [6]. Activated carbon and macroreticular porous resins were the early generation of adsorbents used for primary isolation of cephalosporin-C (CPC), cephalexin, clavulanic acid, nocardicin, etc. from the culture broth [7]. A number of neutral polymeric sorbents such as polyaromatics (Amberlite XAD-416, 1180, Diaion HP20), aliphatic esters (Amberlite XAD-7) and nitrated aromatics (nitrated Amberlite XAD-16) proved effective for penicillins and cephalosporins, but the aromatic sorbents provide the highest sorption capacity for both of the antibiotics [8]. The macroporous resin adsorbents are well suited for transforming salts of organic acids or bases into their respective free acids or bases. However, for the isolation of hydrophilic cephalosporins, only a few adsorbents (Table 1) are suitable due to capacity limitations and the large volume of liquid to be handled in commercial processes. This requires an exceptional physicochemical stability of the adsorbent to withstand frequent and often drastic regeneration conditions. Because of these requirements, inorganic adsorbents can rarely be used for the isolation of hydrophilic fermentation products. Organic gel type ion-exchange resins, mostly in salt form, are used on large industrial scales for the isolation of sugars and polyalchohols by adsorption chromatography [9]. The use of activated carbon is problematic for this kind of application because it can only be regenerated to a limited extent. The use of molecular sieves and cellulose is also limited practically to the chromatographic separation of components. However, some synthetic organic macroporous adsorbents fulfill the above requirements exactly, though their properties and general suitability are relatively unknown, especially on the industrial scale.

In 1965, Rohm and Haas Co. of USA commercialised the first synthetic adsorbents, the so called Amberlite XAD resins. They are characterised by a spectrum of surface polarities which range from non-polar at one extreme to highly polar at the other. Within the series of adsorbents, a variety surface areas, poro sities and pore size distribution is exhibited. Owing to these differences in surface properties, the Amberlite XAD adsorbents display a wide range of sorption behaviour. The sizeable amount of adsorption data so far accumulated has demonstrated that the XAD adsorbents are considered highly versatile and effective sorption media which have widespread practical applications in diverse fields.

Though adsorption chromatography has been practised for commercial scale separation of β-lactams, particularly the cephalosporin antibiotics, there appears to be limited understanding of the adsorption phenomenon and process design parameters. Only recently, some attempts have been made to provide a molecular basis of β-lactam adsorption using very limited adsorbents and solutes, however with limited success as evidenced from certain quantitative discrepancies between model predictions and experimental results. This apparently reveals that there is a lack of understanding of the molecular design of adsorbents for cephalosporin antibiotics as evident from only limited studies reported in the literature (Table 2). Furthermore, understanding of the basic phenomena of adsorption equilibrium and kinetics in relation to the sorbent surface chemistry appears to be quite meagre. It may be noted that this information is very essen-

Table 1 Adsorbents for isolation of hydrophilic bio-products

Type	Nature	Typical example
Activated carbon	Inorganic	Ambersorb (synthetic carbonaceous)
		Anthrasorb (from coal)
	Organic	Active carbon (from fossil fuel)
Molecular sieves	Inorganic	Zeolite
	Organic	Sephadex (G and LH), Biogel
Non-ionic	Inorganic	Silica gel, aluminum oxide, TiO
	Organic	Synthetic macroporous adsorbents, cellulose
Ion-exchange	Inorganic	Zeolite, ion-exchange crystals
	Organic	Strong acid or basic gel-type resin in salt form

Table 2 Adsorbents for beta-lactam and studies reported in literature

Sl. no.	Adsorbent type/ nature	Examples	Beta-lactams	References
1	Neutral polymeric sorbent	Amberlite XAD-4, Diaion HP20, Amberlite 1180	Cephalosporin-C	[10]
		Amberlite XAD-416 Amberlite 1180 Diaion HP20 Amberlite XAD-7 nitrated Amberlite XAD-16	Penicillin V tetracycline and cephalosporin -C	[8]
		Polystyrene resins crossed linked with divinylbenzene	Cephalosporin-C	[11]
		Styrene-divinyl benzene	Cephalosporins and penicillins	[12]
		brominated Diaion-HP20	Cephalosporins	[13]
		Amberlite XAD-4 Amberlite XAD-7	Cephalosporins	[14] [15]
2	Ion-exchange resin	Diaion, WA-30 HP-20, SK1B	Cephalosporin-C	[13]
3	Liquid ion-exchanger	Amberlite LA2-acetate	Cephalosporin-C	[16]
4	Polymeric reverse phase resin	Amberchrome	Cephalosporin-C	[17]
5	Organic	Activated carbon	Cephalosporins 6-amino penicillanic acid	[18] [19]

tial for designing and optimising processes for preparative scale separation of the antibiotics. Thermodynamic measurements as regards affinities and enthalpies of adsorption on all the known adsorbents are seldom reported in literature. These thermodynamic parameters may well be correlated with molecular descriptors such as the molecular connectivity/topological index [20, 21] etc.

In order to design better adsorbents, understanding of the interaction (through molecular modelling studies) between the surface of the adsorbent and the antibiotic molecule is essential. Molecular modelling has been used to predict adsorption of gases [22, 23], but very limited studies have been reported on modelling of adsorption from aqueous solutions [24]. Such a design of sorbents would complement current efforts to develop specialised sorbents through the molecular imprinting [25] technique. This article presents an overview of some important issues on liquid phase adsorption of β-lactam antibiotics with emphasis on equilibria, kinetics, adsorptive interaction and column dynamics.

2
Adsorption/Desorption Equilibria

Information concerning the adsorption/desorption equilibria is essential for designing and optimising processes for preparative scale separation of the antibiotics. There have been a lot of isotherm equations developed for either gas or liquid phase adsorption onto solids. For the sake of convenience, explicit and simple models are preferred for common engineering use. Although other types of isotherms have been used in describing adsorption phenomena in biological systems, the Langmuir isotherm is still the most commonly used expression in the study of both biospecific affinity [26–28] and ion exchange chromatographic process [29, 30].

2.1
Langmuir Isotherm

The most widely used two-parameter isotherm equation is the Langmuir model, which is represented as

$$\theta = q_e / C_1 = K_L\, C_e / \left(1 + K_L\, C_e\right) \tag{1}$$

Rearranging Eq. (1), the following form may be deduced

$$q_e = K_L\, C_1\, C_e / 1 + K_L\, C_e \tag{2}$$

or,

$$C_e\, q_e = \left(1 / K_L\, C_1\right) + \left(1 / C_1\right) C_e \tag{3}$$

A plot of (C_e/q_e) versus C_e would give K_L and C_1 from the slope and intercept. Equation (1) is essentially good only in higher concentration ranges [31]. An alternative form of the Langmuir isotherm was suggested for relatively lower concentrations [32, 33] as given by the following equation

$$1 / q_e = \left(1 / C_1\right) + \left(1 / K_L\, C_1\right)\left(1 / C_e\right) \tag{4}$$

and is known as the Langmuir equation II. The values of K_L and C_1 can be determined from the plot of $(1/q_e)$ versus $(1/C_e)$.

The so-called Langmuir analysis [34] can be made of the results over a given equilibrium liquid phase concentration, C_{ref} to characterise the type of isotherm as originally demonstrated by Vermeulan et al. [35] through use of the following relationship

$$\bar{r} = 1 / \left(1 + K_L\, C_{ref}\right) \tag{5}$$

The type of isotherm depends on the value of $\bar{r}$ as illustrated in Table 3.

Table 3 Characterization of the isotherm in Langmuir model

$\bar{r}$-value	Type of isotherm
$\bar{r} > 1$	Unfavourable
$\bar{r} = 1$	Linear
$0 < \bar{r} < 1$	Favourable
$\bar{r} = 0$	Irreversible

2.2
Freundlich Isotherm

The Freundlich isotherm is an empirical one used to describe the isotherm data and is given by

$$q_e = K_F\, C_e^n \tag{6}$$

A linear form of this equation can be obtained as follows

$$\ln q_e = \ln K_F + n \ln C_e \tag{7}$$

A plot of $\ln q_e$ versus $\ln C_e$ would give n and K_F from the slope and intercept, respectively. The Freundlich equation, which is suitable for highly heterogeneous surfaces, frequently gives a good representation of adsorption data over a restricted range of concentration and it does not reduce to Henry's law at concentrations approaching zero [36–38].

2.3
Redlich-Peterson Isotherm

Since in almost all systems, Langmuir and Freundlich equations give a relatively poor fit over the entire range of concentrations, the applicability of empirical equations containing three adjustable parameters has been examined from time to time. Each of these can reduce to Henry's law at extremely low concentrations. The Redlich-Peterson equation [39] is a well-known, three-parameter isotherm model

$$q_e = P_1\, C_e \,/\left(1 + P_2\, C_e^{P3}\right) \tag{8}$$

where P_1, P_2 and P_3 are the model parameters and the value of P_3 lies between 0 and 1. This equation gave a good representation of data for phenol adsorption from dilute aqueous solution on Amberlite XAD-8 [40] and XAD-4 and XAD-7 resins [33]. In a slightly different form, it was used also by Radke and Prausnitz [41]. Rewriting Eq. (8), the following form may be obtained

$$C_e\, q_e = \left(1/P_1\right) + \left(P_2\,/\,P_1\right) C_e^{P3} \tag{9}$$

For any system, the three parameters can be estimated by a least-squares fitting procedure by minimizing the deviation between calculated and experimental values. The three isotherms above have been tested for adsorption of β-lactams on activated carbon and polymeric resins at the author's laboratory [15, 18, 19]. The parameters were estimated by a nonlinear regression analysis using the well known Levenberg-Marquardt's method [42] using the package NUMERICAL RECIPES in C. It appears that the Langmuir model provides most satisfactory representation of the data at almost all the pH values studied. The Langmuir model was found to be quite satisfactory for adsorption of cephalosporin C and streptomycin on ion exchange resin over a range of practically relevant solute concentrations [8, 43].

Multilayer BET theory was used to analyse adsorption of cephalosporin C onto neutral polymeric resins like Diaion HP20, Amberlite XAD-4 and XAD-1180, whereas ion exchange resin exhibited Langmuir isotherm [10]. Casillas et al. [44] reported on Langmuir model for adsorption of cephalosporin C on Amberlite XAD-2 and XAD-4 whereas a modified XAD-2 i.e. XAD-2-CH_2-CH_2-Br resin exhibited a single linear isotherm at a pH 3 and the amount adsorbed at saturation was of the order of 0.96 mmol/g. Yang et al. [45] also used the Langmuir isotherm to model adsorption/desorption of cephalosporin C on XAD 2 and XAD 4 resins in a fixed bed adsorber.

2.4
Toth Isotherm

The Toth equation [45], originally proposed for gas adsorption, has been applied for adsorption of solute from dilute aqueous solutions [38]. It has three parameters i.e. q^∞, b_T, and M, in which M is a semi-empirical parameter characteristic of the adsorbent only, irrespective of the temperature and the nature of the adsorbate.

$$q_e = q^\infty C_e / \left(b_T + C_e^M\right)^{1/M} \tag{10}$$

Here q^∞ is an adjustable parameter indicating a maximum adsorption when $C_e \to \infty$. The procedures for determining the values of the constants, b_T and M involves the introduction of a dimensionless quantity, Ψ defined by [45]

$$\psi = \left(\mathrm{d}\ln C_e / \mathrm{d}\ln q_e\right) - 1 \tag{11}$$

which can be determined from the experimental results. The relation between ψ and C_e is

$$\psi = a C_e^b \tag{12}$$

with, a and b as constants. By setting $a = 1/b_T$ and $b = M$, integration of Eq. (11) gives the Toth Eq. (10) using the boundary condition, $q_e \to q^\infty$ when $C_e \to \infty$.

For the Langmuir equation, $b = 1$, and for the Freundlich equation, $b = 0$. If Henry's law is assumed to hold for all concentrations, 'a' would have to be zero. Furthermore, at very low concentrations, $b_T \gg C_e^M$ and the adsorption Eq. (10) follows Henry's law.

2.5
Isotherm of Jossen et al.

For a highly heterogeneous surface, the highest energy sites are filled first so that the enthalpy of adsorption declines rapidly with increased surface coverage. Assuming that an energy distribution function of adsorption sites, $f(E)$ obeys Eq. (13)

$$f(E) = A(E - E_0)^{\alpha} \tag{13}$$

the following relation between the isosteric enthalpy of adsorption, E, and the amount of adsorbate adsorbed per unit mass of adsorbent has been derived [28].

$$E - E_0 = D q_e^p \tag{14}$$

where, E_0 is the maximum enthalpy of adsorption (at zero coverage) and p is a constant related to the distribution of energy sites on the surface by the equation $\alpha = (1-p)/p$, $0 < p < 1$, $A = 1/pE_0^{1/p}$, and $D = E_0/N_s p$ (N_s is the number of total adsorption sites). Then the following isotherm equation is obtained

$$C_e = (q_e / H) \exp\left(K_j q_e^p\right) \tag{15}$$

where, H corresponds to Henry's law constant and is related to E_0 as follows

$$E_0 = -RT^2 \left(d \ln H / dT\right) \tag{16}$$

K_j with a function of temperature T only and related to 'D' as according to

$$D = -RT^2 \left(dK_j / dT\right) \tag{17}$$

These equations have two desirable characteristics. At low coverage, Eq. (15) reduces to Henry's law so that

$$\lim_{C_e \to 0} q_e = H C_e \tag{18}$$

and at highest coverage, an ultimate adsorption amount q can be obtained from Eq. (14), irrespective of temperature.

$$q^{\infty} = (E_0 / D)^{1/p} \tag{19}$$

Thus, p is a characteristic of the adsorbent only and independent of of both temperature and nature of the adsorbates. For example, $p = 0$ corresponds to a uniform surface for which all sites have the same energy. A good representation of equilibrium data using this equation was reported for aqueous phase adsorption of phenolic compounds onto activated carbon [38] and Amberlite XAD-4 and XAD-7 [33]. Rearrangement of Eq. (15) gives

$$\ln(C_e / q_e) = -\ln H + K_j q_e^p \tag{20}$$

If p is known, the values of H and K_j can be obtained either from the plot of $\ln(C_e/q_e)$ against q_e or using a least-square fitting procedure.

3
Adsorption/Desorption Kinetics

The kinetics of adsorption/desorption in liquid phase have been mostly modelled by incorporating the phenomenon of external mass transfer and solute diffusion in the adsorbent. The following models have, in general be applied for analysis of kinetics of aqueous phase adsorption.

3.1
Homogeneous Surface Diffusion Model

It has been shown by many investigators that the contribution of pore diffusion to the intraparticle transport during adsorption of solute from an aqueous solution on adsorbent is negligible as compared to that of surface diffusion [37, 46, 47]. Homogeneous surface diffusion (HSD) or solid diffusion model [47] can be used to describe the transport of solute from the liquid bulk to the interior of adsorbent particles. The model considers solute diffusion through the external film surrounding the particles, adsorption on the external film surrounding the particles and adsorption on the external surface of the particle. The particles are considered to be spherical and homogeneous with uniform pores. Intraparticle solute transport is assumed to take place only by diffusion along the pore walls in the adsorbed state (surface diffusion). The intrinsic adsorption kinetics is considered to be fast as compared to both the external film and the intraparticle diffusion. Therefore, the solid-phase and liquid-phase solute concentrations at the solid-liquid interface are considered to be in equilibrium. The mathematical formulation of the HSD model for adsorption in a batch system is given below

Adsorbent Particle

$$\frac{\delta q}{\delta t} = \frac{1}{r^2} \frac{\delta}{\delta r} \left[r^2 D_S (q) \frac{\delta q}{\delta r} \right] \tag{21}$$

Initial Conditions:

$$t = 0, \quad 0 \le r \le R, \quad q = 0 \tag{22}$$

Boundary Conditions:

$$t > 0, \; \frac{\delta q}{\delta t} \Big|_{r=0} = 0 \tag{23}$$

$$\varrho_S D_S \frac{\delta q}{\delta r} \Big|_{r=R} = K_f \left(C_b - C_S \right) \tag{24}$$

$$\bar{q} = \frac{3}{R^3} \int_0^R q \, r^2 \, dr \tag{25}$$

Bulk Solution:

$$V_f \frac{dC_b}{dt} = -\frac{3W}{\varrho_s R} K_f \left(C_b - C_s\right) \tag{26}$$

Initial conditions:

$$t = 0, \quad C_b = C_0 \tag{27}$$

Solid-Liquid Interface

$$q_{r=R} = \frac{\alpha_1 C_s^{\beta 1}}{1 + \alpha_2 C_s^{\beta 1}} \tag{28}$$

The orthogonal collocation method [48] may be employed to convert the set of the partial differential equations into a set of ordinary differential equations which could be subsequently solved with an integration based on Gear's implicit method using the IMSL package [49]. Initially, the solute surface diffusion coefficient (D_s) was assumed to be constant [50]. The values of kinetic parameters in the model (i.e., D_s and K_f) were evaluated by non-linear regression of data generated for each experiment separately using the Marquardt-Levenberg optimisation routine (IMSL) [51] to minimise the root mean square error (RMS) of the normalised residuals. Following this procedure, the predicted D_s values were found to increase as the initial solute concentration (C_0) in the solution increased and as the particle size (d_p) decreased [52]. Similar results on concentration dependence of D_s have been reported by several investigators [53, 54]. This is an indication that D_s is indeed not constant but increases with solute surface concentration.

Accordingly, the concentration-dependent surface diffusion coefficient ($D_s(q)$) has been represented by the following expression:

$$D_S(q) = D_0 \cdot e^k \left(q / q_{sat}\right) \tag{29}$$

where, D_0 and k are constants.

In order to minimize the total number of parameters to be determined by regression analysis, the following equation was employed to relate the value of external mass transfer coefficient (K_f) at any operating conditions (i.e. N, d_p) to a reference value ($K_{f,\,ref}$) at some reference operating conditions (N_{ref}, $d_{p,\,ref}$) [52]

$$K_f = K_{f,ref} \left(\frac{d_p}{d_{p,ref}}\right)^{(4\lambda-1)} \left(\frac{N}{N_{ref}}\right)^{3\lambda} \tag{30}$$

where, λ is a constant.

3.2
Linearised Model

The linearised model is a mere simplification of the one that [55] considers three steps: (1) external mass transfer, (2) intraparticle diffusion, and (3) adsorption at an interior site. It is assumed that step 3 is rapid with respect to the first two steps. Particularly for a well agitated slurry adsorber where mixing in the liquid phase is rapid. Hence the concentration C of adsorbate and concentration 'm' of adsorbent in the liquid are nearly uniform throughout the vessel. Then the change in concentration C_t at a time (t), is related to the fluid-particle mass transfer coefficient by the equations:

$$\frac{dC_t}{dt} = -K_f\, S_s \left(C_t - C_s \right)$$
(31)

with the initial condition given by

$$C_t = C_0 \left(t = 0 \right)$$
(32)

The differential mass balance of solute within the particles assuming a constant total effective intraparticle diffusivity, D_t is

$$D_t \left(\frac{\delta C_r^2}{\delta r^2} + \frac{2}{r}\frac{\delta C_r}{\delta r} \right) - \varrho_P \frac{\delta n_r}{\delta t} = \frac{\delta C_r}{\delta t}$$
(33)

The boundary and initial conditions for solution of the above equation are

$$D_r \left(\frac{\delta C_r}{\delta r} \right) = K_f \left(C_t - C \right)$$
(34)

$$\left(\frac{\delta C_r}{\delta r} \right) = 0 \quad (r = 0)$$
(35)

$$C_r = 0 \quad (t = 0)$$
(36)

Since equilibrium is assumed for step 3, n_r and C_r are related by the isotherm as

$$\frac{\delta n_r}{\delta t} = \frac{\delta}{\delta C_r} \left(\frac{K\, C_r}{1 + K_L\, C_r} \right) \frac{\delta C_r}{\delta t}$$
(37)

Equations (31) to (37) can be solved numerically to relate the observed C_t versus t curves to rate coefficients, K_f and D_t.

When intraparticle diffusion resistance is negligible and the isotherm is linear, an analytical solution for C_t versus t is possible. Also since $C_r = C_s$ and n_r is uniform throughout the particle, Equations (33) and (34) may be replaced by

$$m\, \frac{dn}{dt} = K_t\, S_s \left(C_t - C \right)$$
(38)

and Eq. (37) becomes

$$\frac{dn}{dt} = K \left(\frac{dC_s}{dt} \right)$$
(39)

The outer surface area of the particle, S_s may be obtained by assuming spherical particles of diameter d_p using the following expression

$$S_s = \frac{6W}{d_p \, \varrho_p \left(1 - \varepsilon_p\right)} \tag{40}$$

Equations (31), (38) and (39) with the initial conditions (Eq. 32) and $n = 0$ at $t = 0$ can be solved analytically to give:

$$\frac{C_t}{C_0} = \frac{1}{1 + mK} + \frac{mK}{1 + mK} \exp\left(-\frac{1 + mK}{mK} K_f \, S_s \, t\right) \tag{41}$$

From the contact time results obtained, the mass transfer coefficients between bulk liquid and outer surface of particles, K_f can be obtained from the initial slope of the C_t/C_0 versus t curve. However, the uncertainty in evaluating the slope of a curve at $t = 0$ can be avoided by using Eq. (41). This expression shows that a plot of: $\ln \left(C_t/C_0 - 1/1 + mk\right)$ versus t will give a straight line at $t = 0$ since the influence of intraparticle diffusion does not yet affect the results and the isotherm becomes linear as time approaches zero. This model has exhibited a reasonable fit of experimental data on adsorption of cephalosporins on activated carbon and polymeric resins [15].

3.3
Two-Resistance Model

A two-resistance model based on external mass transfer and particle surface diffusion is frequently used to generate a theoretical concentration-time decay curve for adsorption on small porous particles [56, 57]. The model equation is based on the hypothesis that the rate of change of concentration of the solid phase is equal to the rate of mass transfer of the solute from the fluid phase through the film and is represented by

$$\frac{dq_t}{dt} = \frac{K_f}{V_p \, \varrho_p \left(1 - \varepsilon_p\right)} \left(C_t - C_{s,t}\right) \tag{42}$$

The average solid-phase concentration of the particle is obtained by averaging the point concentration over the volume of the particle

$$q_t = \frac{3}{R^3} \int_0^R qr^2 \, dr \tag{43}$$

The Langmuir isotherm can be used to represent equilibrium adsorption at the surface of the particle. For a spherical particle, the variation of the solid-phase concentration in the radial direction is given by:

$$\frac{\delta q}{\delta t} = D \frac{\delta^2 q}{\delta r^2} + \frac{2}{r} \frac{\delta q}{\delta r} \tag{44}$$

The material balance for the solute is given by:

$$V\frac{\delta C}{\delta t} = -W\frac{\delta q}{\delta t}$$

(45)

The initial conditions are:

$$q(r,0) = 0 \quad C(0) = C_0$$

(46)

and the boundary conditions are:

$$q(R,t) = q_s(t)\frac{\delta q}{\delta t}(0,t) = 0$$

(47)

For solutions, Eq. (44) can be changed to dimensionless variables using the transformations $\tau = Dt/R^2$ and $x = r/R$. Changing the variables with $U = xq$ and rearranging gives:

$$\frac{\delta U}{\delta \tau} = \frac{\delta^2 U}{\delta x^2}$$

(48)

$$U(0,\tau) = 0 = U(x,0) \quad \text{and} \quad U(x,\tau)$$

(49)

$$\int_0^1 \frac{\delta U}{\delta \tau} x\,dx = \frac{K_f\,R(C-C_s)}{1000\varepsilon_p(1-\varrho_p)}$$

(50)

$$q = 3\int_0^1 U\,x\,dx$$

(51)

$$V\frac{dC}{dt} = -W\frac{dq}{dt}$$

(52)

Equations (50) to (52) with Eq. (45) can be solved by the finite difference technique illustrated by Mathew and Weber [56]. The value of the external mass transfer coefficient (when all the mass transfer resistance is restricted to the external boundary layer and the surface concentration is assumed to be negligible) is given by

$$K_f = -\frac{V_L}{S_s}\left[\frac{d(C/C_0)}{dt}\right]_{t=0}$$

(53)

The concentration-time decay curve can be analysed by a single resistance model of the form:

$$\frac{C}{C_0} = \exp\frac{6W\,K_f\,t}{\varrho_p\,d_p\,V_L(1-\varepsilon_p)}$$

(54)

The models have been tested for adsorption of cephalosporins on activated carbon/polymeric resins at our own laboratory [15, 18, 19]. The external mass transfer effect on adsorption kinetics was analysed by using the McKay Eq. (41) valid for negligible intraparticle diffusion resistance and for linear isotherms. The McKay plots of $\ln [C_t/C_0 - 1/(1 + mk)]$ versus t was found to be linear from which it is apparent that external mass transfer plays prominent role in the adsorption process. However, deviation of the plots at the initial stage of adsorption could be attributed to a different rate controlling step. The role of intraparticle diffusion on the adsorption kinetics was examined from a graphical relationship between 'q' (amount adsorbed) and the square root of the time and the linearity of the plots indicate the effect of intraparticle diffusion on the adsorption process. However, as the plots do not appear to pass through the origin, the pore diffusion alone may not be the rate controlling step. It is therefore preferable to simulate the adsorption kinetic data using more logistic models such as the homogeneous surface diffusion model.

4
Analysis/Interpretation of Adsorptive Interaction

It appears that only a few adsorbents are suitable for the isolation of cephalosporins perhaps because of capacity limitations. Despite the success in producing sorbents with well-defined physical properties, there has been relatively few studies on the molecular details of the adsorption process. Because of this limited understanding of the solute solvent binding interaction, adsorption operations have been developed empirically and this empirical knowledge is insufficient for addressing practical issues of improving binding strength to facilitate on-off chromatographic operation, enhancing binding selectivities to permit separation of the desired product from other chemically similar compounds, and reducing nonspecific adsorption by components of the complex fermentation broth. The theoretical treatment of molecular interaction is usually based on Molecular Orbital (MO) theory. It has been developed to describe several molecular phenomenon involving molecular interactions by calculating interaction energies for the corresponding geometric structures of the interacting molecules. From these results, implications may be deduced for the design of suitable adsorbents.

4.1
Molecular Orbital Calculation

In order to calculate electronic states of adsorbents by the MO method, it is necessary to provide a structural (molecular) model of the adsorbent surface. Tamon et al. [58] have used a cluster model for the adsorbent surface. Kaneko et al. [59] have evaluated the size of graphitic crystallite in carbonaceous materials such as an activated carbon (AC), a non-porous carbon black (NPC) and activated carbon fibers (ACF) from diamagnetic susceptibility measurements. They have reported that the sizes of the micrographitic unit are 39 carbon numbers for pitch-based ACF, 24 for phenol-based ACF, 37 for AC and 31 for NPC.

In our recent work [14], we used a polycyclic aromatic compound for the surface model of activated carbon. This model was also successfully used for computation of interaction energy for adsorption of phenol [59]. A styrene divinylbenzene polymer was used as the cluster model of synthetic polymeric adsorbent, Amberlite XAD-4 and an aliphatic carboxylic acid polymer for synthetic adsorbent Amberlite XAD-7.

Because the cluster model of the adsorbent surface is large, it is difficult to execute the ab initio MO calculation of the adsorbent surface from the viewpoint of computational time [58]. A semiempirical MO method is used at a marginal loss of calculation accuracy. From several semiempirical MO methods such as AM1, CNDO, MINDO, MNDO, PM3, etc., the electronic states of adsorbent, adsorbate and solvent were calculated using MINDO/3 because the ionisation potential and the electron affinity calculated by the method reproduce the experimental data reasonably well [14, 58]. In the MO calculation, the structures of the antibiotics have been determined using the geometry optimisation method by MINDO/3.

4.2
Frontier Orbital Theory

In order to estimate the accurate values of interaction energy between adsorbate and adsorbent, the frontier orbital theory proposed by Fukui et al. [60] for two body interaction has been applied to simplify the treatment. The concept of HOMO and LUMO mixing according to frontier orbital theory is shown in Fig. 3 in which molecule 'a' is electron-donating and molecule 'b' is electron-attracting. We consider charge transfer from atom 'r' of molecule 'a' to atom 's' of molecule 'b'. The energy level of HOMO of molecule 'a' changes to the more stable level by orbital mixing. On the other hand, the energy of LUMO of molecule 'b' attains a more stable level.

The energy difference, ΔE shown in Fig. 3 is called the perturbation energy. The second order perturbation expression for the energy that accompanies the interaction can be derived, and the perturbation energy ΔE caused by the HOMO-LUMO interaction can be calculated by the method reported in the literature [61, 62]. When molecule 'a' is electron-attracting, ΔE between atom 'r' of molecule 'a' and atom 's' of molecule 'b' is given by

$$\Delta E = 2(C_r^* C_s^* \Delta\beta)^2 \ |E_a^* - E_b^*| \tag{55}$$

where E_a^* and C_r^* are the HOMO energy and the orbital coefficient of atom 'r' of molecule 'a', respectively. Whereas E_b^* and C_s^* are the LUMO energy and the orbital coefficient of atom 's' of molecule 'b', respectively.

By executing the MO calculation of molecules 'a' and 'b' using the well-known package MOPAC 6.0 [63] based on MINDO/3, HOMO and LUMO energies and orbital coefficients for all constitutional atoms of the molecules were obtained. The square of the coefficient indicates the existence probability of an electron and the $1s$ orbital for H atom, $2p_x$, $2p_y$ and $2p_z$ orbitals for C, O, S and N atoms were taken into account to estimate ΔE. $\Delta\beta$ is the interorbital interaction integral between the interacting orbitals of atoms, which is proportional to the overlap integral. The values between interacting atoms at specific distance have been reported [64]. The interaction distance depends on the kinds of

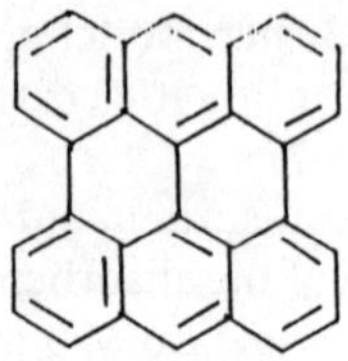

Activated carbon

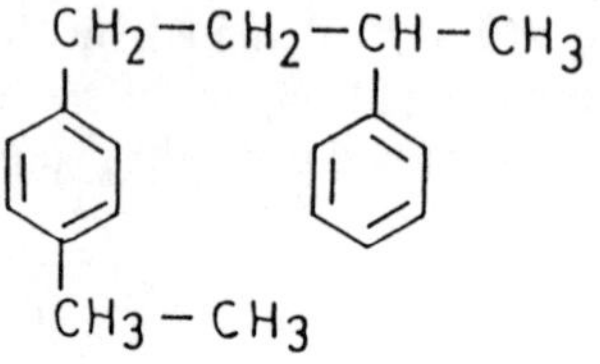

Amberlite XAD - 4

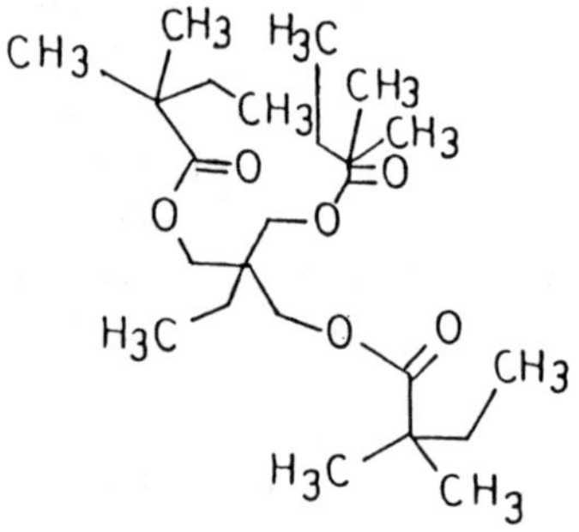

Amberlite XAD-7

Fig. 3 Concept of molecular orbital mixing

adsorbates. In order to determine the accurate distance, ab initio study is needed. The characteristic energy ΔE_{ad} i.e. adsorptive interaction between adsorbate and adsorbent calculated by using Eq. (55) is scaled with characteristic energy of a reference adsorbate. The ratio of the characteristic energy of the adsorbate in question to that of the reference adsorbate $\Delta E_{ad}/\Delta E_{ad(ref)}$ means the relative strength of adsorptive interaction with a particular adsorbent in aqueous solution. The plots of the values of adsorption affinity (q/C_e) versus $\Delta E_{ad}/\Delta E_{ad(ref)}$ for adsorption of cephalosporins on activated carbon, amberlite XAD-4 and XAD-7 [14] indicate a unique relationship between adsorption affinity and relative strength of adsorptive interaction for all the ß-lactam molecules studied. This suggests that the equilibrium of adsorption is a strong function of the strength of the solute-sorbent binding interaction. The experimentally observed adsorption enthalpy (ΔH_0) values, estimated from the Van't Hoff's relation are plotted against adsorptive interaction energies (ΔE_{ad}) which shows a reasonably good linear relationship. This implies that the enthalpies of adsorption are in conformity with the adsorptive interaction energy.

5
Dynamics of Column Adsorption

Generally preparative scale separation and purification by adsorption in liquid phase followed by desorption are carried out in packed columns. The sequence of two column operation is of practical relevance for solute recovery by simultaneous adsorption cum desorption and regeneration of the adsorbent. Therefore, the overall dynamics of the column, rather than the adsorption kinetics for a single particle system alone control the design and optimisation of adsorptive separation process. The importance and relevance of column dynamics study have been well documented in standard text [65].

The basic element on design of a fixed-bed adsorber is the reliable prediction of a predetermined degree of exhaustion or effluent "breakthrough" for a given set of influent conditions. For analysis of the dynamics of fixed-bed adsorption column, it is convenient to consider the response of an initially sorbate free column to either a step change in sorbate concentration at the column inlet (step input) or to the injection of a small pulse of sorbate at the column inlet (pulse input). The response to a step input is commonly called the breakthrough curve while the pulse response is often referred to as the chromatographic response. The choice is however, determined by practical convenience rather than by more fundamental considerations. In general, the breakthrough curve is a common practice representing the effluent concentration versus time profile of the solute in the column.

5.1
Mathematical Models

For analysing isothermal adsorption cum desorption, mathematical models capable of describing the mass balances for each adsorbate present in the column are necessary. The models which are essentially partial differential equations account for adsorption equilibria and kinetics along with the flow characteristics. The column dynamics can be analysed by solution of the equation using appropriate initial and boundary conditions.

For the simple case of adsorption of a single dilute solute of concentration C_0 from a solvent in a column of length L, which is initially solute free and packed with spherical adsorbent particles of radius, r, the following fixed bed models are commonly used.

Solid phase mass balance

$$\frac{\delta q}{\delta t} = D_s \frac{\delta^2 q}{\delta r^2} + \frac{2}{r} \frac{\delta q}{\delta r} \tag{56}$$

Liquid phase mass balance

$$- D_{\mathrm{p}} \frac{\delta^2 C}{\delta z^2} + V_{\mathrm{e}} \frac{\delta C}{\delta z} + \left(\varrho_{\mathrm{p}} \frac{1-\varepsilon}{\varepsilon} \right) \frac{\delta q}{\delta t} + \frac{\delta C}{\delta t} = 0 \tag{57}$$

The above equations may be coupled with either Langmuir or Freundlich isotherm equations for solution with the following

Initial and boundary conditions

$$C = C_0 + \frac{D_{\mathrm{p}}}{V_{\mathrm{e}}} \frac{\delta C}{\delta z} \quad \text{at } z = 0 \tag{58}$$

$$\frac{\delta C}{\delta z} = 0 \quad \text{at } z = L \tag{59}$$

$$\frac{\delta q}{\delta r} = 0 \quad \text{at } r = 0 \tag{60}$$

$$\frac{\delta q}{\delta r} = \frac{k_{\mathrm{f}}}{D_{\mathrm{s}} \varrho_{\mathrm{p}}} (C - C_{\mathrm{s}}) \quad \text{at } r = \underline{R} \tag{61}$$

$$q = 0 \quad \text{at } t = 0 \tag{62}$$

$$C = 0 \quad \text{at } t = 0 \tag{63}$$

The above equations are based on the assumption of mass transfer and surface diffusion controlled adsorption mechanism. The simulation of the fixed bed adsorption and desorption providing an analysis of the column dynamics can be carried out by solving the above set of equations by the orthogonal collocation technique using finite element analysis [48]. In order to test the validity of the model, the model parameters i.e. k_{f}, D_{s} and D_{p} are usually derived from micro column breakthrough curves as well as stirred tank transient adsorption data. The value of the mass transfer coefficient, k_{f} can also be estimated from standard correlations i.e. $Sh = 2.0 + 1.1 \, S_{\mathrm{e}}^{1/3} \, R_{\mathrm{e}}^{0.6}$ where $Sh = 2R_{\mathrm{p}} \, k_{\mathrm{f}}/D_{\mathrm{m}}$, $S_{\mathrm{e}} = \mu/\varrho_{\mathrm{f}} \, D_{\mathrm{m}}$ and $R_{\mathrm{e}} = 2R_{\mathrm{p}} \, \varrho_{\mathrm{f}} \, V/\mu$ reported in the literature [66]. Studies have shown that the simulation results are often more sensitive to the equilibrium isotherm than to the mass transfer resistance. For instance, the analytical solution of the model equation is possible for the cases of linear isotherm or constant pattern zone formation for the LDF model.

Column dynamics for adsorption of dilute two or three component mixtures have also been studied for the case of pore and surface diffusion resistances. Diffusivities of the solute could be measured from batch kinetic test or by column data on breakthrough curve. The dynamics of adsorption cum desorption in laboratory or industrial column may be influenced by liquid maldistribution, channeling and non-uniform particle size distribution which also affect the scale-up and design of adsorption column. There is practically no report in the literature on column dynamics for cephalosporin adsorption excepting the study made by Whitehall [67] who analysed the breakthrough curve for adsorption of cephalosporin-C onto XAD-2 polymeric resin in a micro fixed bed. Axial dispersion plug flow models with the so called Danckwert's boundary condition were used under pore diffusion controlled conditions with the non-linear Langmuir isotherm. Breakthrough was observed in 300 minutes of column operation after which pH swing desorption was observed. The experimental data for adsorption exhibited a reasonable fit with simulation results, but a poor fit was observed for the desorption step. The reason was attributed to oversimplification of instantaneous equilibrium and differences in pore diffusivity value during the desorption step for simulation of which pore diffusivity value for adsorption was used. It is thus apparent that there is ample scope and prospects for carrying out column dynamics studies for cephalosporin adsorption using more elaborate mathematical models capable of describing the mechanistic aspects precisely.

6
Some Scale-Up and Process Design Considerations

Scale-up of the adsorption column is not a simple multiplicative procedure owing to involvement of many process variables which change while translating a laboratory scale to pilot plant. The recovery of cephalosporin by adsorption chromatography is rather more complicated particularly when a fermentation medium is the feed stock. In this case, a combination of method is generally required for a technoeconomically feasible recovery process. Under the operating conditions, the microorganism must remain active and viable. In a process, optimisation of the overall yield of isolation is essential because even a 1% increase in yield of CPC for a plant of 100 tonne/annum at a product cost typically of $100/kg will result in an extra annual benefit of $100,000 which is quite attractive. Scale up and process design studies are however, the prerequisite for optimisation of process leading to commercialisation.

Several adsorption chromatographic separation processes are amenable for easy scale up and design. The technology for large scale chromatography using reverse phase (at 100–150 g loading capacity) has been developed for recovery of semisynthetic antibiotics [68]. Since, the preparative chromatography is performed at much higher sample column ratios than those used in analytical column, it is important to select a chromatographic system under load condition, similar to those used on the process scale rather than those used in the analytical system. The column load (g sample/g support) primarily determines the process throughput and product purity. By properly optimising the column load in the laboratory experiments, the results can be easily translated to large scale processes [69].

The commercialisation of macroporous resins such as XAD-2, 4, 16 etc. as adsorbents opens up wide possibilities for translating analytical procedures into pilot plant demonstration through a knowledge of the elution and regeneration properties, selectivites and hydraulic behaviour of the sorbents. Though several companies have commercialised preparative scale adsorption chromatographic equipments for R&D purpose, large scale systems are still in their early stage of commercialisation. This is particularly true in case of downstream processing of the cephalosporin group of antibiotics which necessitates a combination of viable techniques for cell separation, desalting, product recovery etc. because of complicated nature of the fermentation broth. Combined processes involving MF, UF, RO membranes and adsorption chromatography using WA-30, HP-20, XAD-2000, SK-18 resins have been shown to be attractive in preparative scale isolation of CPC [13,70], however large scale commercial application of such processes is yet to be established based on appropriate scale-up and process design criteria developed through a systematic experimental investigation.

Proper scale up and design of adsorption processes rely on adsorption isotherm as well as kinetic data. Literature report on these aspect is rather scanty and there is little information from adsorbent manufacturer who usually quote uptake capacity at specific solute concentration. As evident from literature, neutral polymeric resins such as XAD-2, 4, 16 etc. will gain prospect for recovery of cephalosporins by adsorption chromatographic processes. In an integrated process for recovery from fermentation broth, ion exchange chromatography as well cell mass separation by a suitable technique will be involved. Currently practised limited commercial processes appear to rely on empirical approach of the licensers who have the advantages of holding proprietary technological rights.

7
Conclusion and Recommendation

Knowledge on technological developments of cephalosporin antibiotic recovery by adsorption chromatography appears to be quite meagre and there is no recommendation on tailor-made adsorbents useful for specific duty. However, neutral polymeric and ion exchange resins merit attention for further development and exploitation in technological perspectives. Adsorptive separation alone cannot be a technologically feasible proposition for downstream processing of fermentation products of cephalosporins, rather a combination of adsorption chromatography with membrane separation process appears to be the most feasible future generation technology.

8
References

1. Fleming A (1929) Br J Exp Pathol 10: 226
2. Newton GF, Abraham EP (1955) Nature 175
3. Ghosh AC, Mathur RK, Dutta NN (1997) In: Scheper T (ed) Adv Biochem Eng/Biochem Springer, Berlin Heidelberg New York, p 111
4. Ghosh AC, Bora MM, Dutta NN (1996) Bioseparation, 6: 91
5. Voser W (1982) J Chem Technol Biotechnol 32: 109
6. Margosis M (1989) In: Giddings JC, Gruscha E, Brown RP (eds) Advances in Chromatography, vol 28. Marcel Dekker, New York, p 33
7. Belter PA (1985) In: Murray J (ed) Comprehensive Biotechnology, vol 2. Pergamon Press, p 473
8. Chaubal MV, Payne GF, Reynolds CH, Albert RL (1995) Biotech Bioeng 47:215
9. Martinola F, Sieger G (1973) Verfahrenstechnik 13:32
10. Hicketier M, Buchholz K (1990) Appl Microbiol Biotechnol 32:680
11. Sacco D, Dellacherie E (1984) Anal Chem 56:1521

12. Salto F, Prieto JG (1981) J Pharm Sci 70: 994
13. Lee HJ, Sohn YS, Ahn DH, Kim HS, Hyun HH (1992) Kor J Appl Microbiol Biotechnol 20: 178
14. Dutta M, Dutta NN, Bhattacharyya KG (2000) J Chem Eng Japan 32:303–307
15. Dutta M, Dutta NN, Bhattacharyya KG (1999) Sep and Purif Technol 16: 213
16. Andrisana R, Guerra G, Mascellani G (1976) J Appl Chem Biotechnol 26: 459
17. Firoujtale E, Maikner JJ, Diessler CK, Cartier GP (1994) J Chromatography 658 (2): 361
18. Dutta M, Baruah R, Dutta NN, Ghosh AC (1997) Colloids Surf A 96: 219
19. Dutta M, Baruah R, Dutta NN (1997) Sep Purf Technol 12: 99
20. Kier LB, Hall LH (1976) Molecular Connectivity in Chemistry and Drug Research, Academic Press,2
21. Kier LB, Hall LH (1979) J Pharmacol Sci 168:120
22. Karavis F, Myers AL (1991) Mol Simul 8:23
23. Roth AM, Lenhoff AM (1993) Langmuir 9: 962
24. Dr Lu, Park K (1990) J Biomateri Sci Polymer Educn 1:243
25. Mosbach K (1994) Trends Biochem Sci 19:9
26. Anspach FB, Johnton A, Wirth HJ, Unger KK, Hearn MTW (1989) J Chromatogr 476:205
27. Arve BH, Liapis AI (1987) AIChEJ 33:179
28. Chase HA (1984) J Chromatogr 297:179
29. Cowan GH, Gosling IS, Sweetenham WP (1989) J Chromatogr 484: 187
30. Gosling LS, Cook D, Fry MDM (1989) Chem Eng Res Des 67:232
31. Ruey-Shin J, Feng-Chin W, Ru-Ling T (1996) J Chem Eng Data 41:487–492
32. McKay G (1982) J Chem Technol Biotechnol 32:759–772
33. Itaya A, Kato N, Yamamato J, Okamato KI (1984) J Chem Eng Japan 17:389
34. McConvey IF, McKay G (1985) Chem Eng Process 19:267
35. Vermeulan Th, Hall KR, Eggleton LC, Acrivos A (1966) Ind Eng Chem Fundam 5:212
36. Cal MP, Larson SM, Rood MJ (1994) Environ Prog 13:26
37. Fritz W, Schlunder EU (1981) Chem Eng Sci 36: 721
38. Jossen L, Prausnitz JM, Fritz W, Schlunder EU, Myer AL (1978) Chem Eng Sci 33: 1097
39. Redlich O, and Peterson DL (1959) J Phys Chem 63: 1024
40. Farrier DS, Hines AL, Wang SW (1979) J Colloid Interface Sci 69: 233
41. Radke CJ, Prausnitz JM (1972) Ind Eng Chem Fundam 11: 445
42. Himmelblau DM (1970) Process Analysis by Statistical Methods. John Wiley and Sons, Inc., Newyork
43. Prasad R, Gupta AK, Bajpai RK (1980) J Chem Technol Biotechnol 30: 324–331
44. Casillas JL, Martinez M, Addo-Yabo F, Aracil J (1993) Chem Engng. J 52: 1371
45. Toth J (1971) Acta Chim Acad Sci Hung 69:311
46. Neretnieks I (1976) Chem Eng Sci 31:1029
47. Fritz W, Merk W, Schlunder EU (1981) Chem Eng Sci 36:731
48. Villadsen J, Michelsen ML (1978) Solution of Differential Equation Models by Polynomial approximation. Prentice-Hall, Englewood Caffs, New Jersey
49. IMSL Math/Library (1989b) FORTRAN Subriutunes for Mathematical Application, Houston, Texas
50. Chatzopoulos D, Varma A (1994) Environmental Process 13:21
51. IMSL Stat/Library (1989a) FORTRAN Subroutine for Mathematical Applications, Houston, Texas
52. Chatzopoulos D, Varma A, Irvine RL (1993) AIChEJ 39: 2027
53. Al-Duri B, McKay (1991) Chem Eng Sci 46:193
54. Sudo Y, Misic DM, Suzuki M (1978) Chem Eng Sci 33:1278
55. Furusawa T, Smith JM (1973) Ind Eng Chem Fundam 12:197
56. Mathew AP, Weber WJ (1976) AIChE Symp Ser 73:91–98
57. Ghosh AC, Satyanarayan K, Srivastava RC, Dutta NN (1995) Colloids Surfaces A 96: 219
58. Tamon H, Aburai MA, Okazaki M (1995) J Chem Eng Japan 28: 823
59. Kaneko K, Tamaguchi K, Ishii C, Ozeki S, Hagiwara S, Suzuki T (1991) Chem Phys Lett 176: 75

60. Fukui K, Yonezawa T, Shingu H (1952) J Chem Phys 20:722
61. Salem L (1968) J Am Chem Soc 543
62. Klopman G (1968) J Am Chem Soc 90:223
63. Stewart JPP (1993) MOPAC V6.0, QCEP # 455
64. Houk KN, Sims J, Watts CE, Luskus LJ (1973) J Am Chem Soc 95: 7301
65. Douglas M Ruthvan (1984) Principles of adsorption and Adsorption Processes. Wiley, New York
66. Levius DM, Glastonbury JR (1972) Chem Eng Sci 25: 537
67. Whitehall IN (1995) Bsc thesis, The Dept of Food Science and Tech, The University of Reading, UK
68. Shih CK, Suavely CM, Molnor TE, Myer SL, Caldwel WB, Paul EI (1983) Chem Eng Prog 10: 53
69. Cautwell AM, Calderane R, Sienko M (1994) 316: 133
70. Kalyaupur M, Seka S, Siwak M (1985) Isolation of CPC from fermentation broth using liquid membrane system and high performance liquid chromatography. In: Developments in Industrial Microbiology, vol 26, pp 455–470, Society of Industrial Microbiology Washington DC

Received: April 2003

Author Index Volumes 51–86

Author Index Volumes 1–50 see Volume 50

Subject Index

Acetaldehyde 139
Acetic acid 90
Acetoin 128
α-Acetolactate 94, 128
α-Acetolactate decarboxylase 124, 128, 149
Acetone 177
Acid guanidinium thiocyanate-phenol-
 chloroform method 202
Acid phosphatase 165
Aconitase 90
Acremonium chrysogenum 3, 5, 9, 11
Acridine orange 32
Acrolein-acrylic acid copolymer 176
Acrylamide 131
N-Acryloxysuccinimide 167, 175
ACTCS (2-(D-4-amino-4-carboxybutyl)
 thiazol-4-carbonic acid) 39, 40
Actin filament regulatory protein 204
Actinomycetes 116
Activated charcoal 116
ACV-synthetase (ACVS) 4, 5, 13–16, 34, 37,
 42
–, K_M-value 31
Acyltransferase reaction 107
Adenine nucleotides 67
Adenosine triphosphate-linked protein,
 periplasmatic 100
Adenyl deaminase 105
Adenylosuccinate synthetase 67
Adhesion molecules 236
Adsorption chromatography 257, 259, 276
Agarose gel electrophoresis 55, 75, 201
Agrobacterium tumifaciens 70
AIDS 75
Aijinomoto 103, 106
Air compressors and agitators 96
Air-lift bioreactor 231
Alanine 90, 93, 102, 119
Alanine racemase 67
Albedo 124
Alcohol dehydrogenase, yeast 128
Aleurone cells 110
Alginate/alginic acid 118, 163, 172, 176, 181

Allicin 121
Allin lyase (allinase) 122, 148
Amano 106
Amberlite XAD resins 259
5-Amino-4-imidaxole carboxamide ribose
 106
Aminocyclase columns 103
Aminopeptidase 149
Aminothiazoline carboxylate 149
2-Amino-Δ2-thiazoline-4-carboxylate
 (L-ATC) 103
α-Amino-ε-caprolactam racemase 103
Ammonia synthesis 154
Ammonium fumarate 143
Amoxycillin 102
Ampicillin 53, 64, 66
Amplification 197, 201
Amycolatopsis 132
Amylase 13, 25, 35, 38, 43, 94, 110, 123, 135,
 139–141, 153, 165, 177, 179
Amyloglucosidase 123, 153
Anchor ligation 194
Animal feeds 94
Antidote 68
Antioxidants 89, 133
Anti-RNase 200
Arabinogalactans 108
α-Arabinose 125
Arabinoxylans 108, 120
Arthrobacter ramosus 115
Arthrospores 32, 43
Ascorbic acid (vitamin C) 86, 90, 95, 98,
 135, 140
Ashi Chemical Co. 130
Aspartame 92, 96, 98, 103, 137, 138, 153
Aspartase 96, 101, 134
Aspartate ammonia lyase 96
Aspartate semialdehyde dehydrogenase 67
Aspartic acid 93, 96, 102
Aspergillus niger 90
– – xylanase 182
Aspergillus oryzae 89, 181
Aspergillus sojai 89

Printing: Druckhaus Berlin-Mitte GmbH
Binding: Buchbinderei Stein & Lehmann, Berlin